Adaptive Sampling

Adaptive Sampling

STEVEN K. THOMPSON

The Pennsylvania State University

GEORGE A. F. SEBER

The University of Auckland

A Wiley-Interscience Publication

JOHN WILEY & SONS, INC.

New York • Chichester • Brisbane • Toronto • Singapore

Library of Congress Cataloging-in-Publication Data:
Thompson, Steven K., (date)
    Adaptive sampling / Steven K. Thompson, George A. F. Seber.
      p. cm. — (Wiley series in probability and statistics.
  Probability and statistics)
    Includes bibliographical references (p. -).
    ISBN 0-471-55871-0 (alk. paper)
    1. Adaptive sampling (Statistics)   I. Seber, G. A. F. (George
Arthur Frederick), (date) . II. Title. III. Series.
QA276.6.T577   1996
519.5'2—dc20   95-42060
                                           CIP

Contents

Preface

Adaptive sampling refers to designs in which the procedure for selecting units to include in the sample may depend on values of the variable of interest observed during the survey. For example, in a survey to assess the abundance of a rare animal species, neighboring sites may be added to the sample whenever the species is encountered during the survey. In an epidemiological survey of a contagious or genetically linked disease, sampling intensity may be increased whenever prevalence of the disease is encountered. Adaptive designs such as those preceding are in marked contrast to conventional sampling designs, in which the entire sample of units to observe may be selected prior to the survey. In planning a survey, consideration of adaptive possibilities considerably widens the class of potential procedures from which to choose.

This monograph explores the theory and methods of adaptive sampling. It was written to make recent research results more accessible, to make adaptive design and estimation methods available to practitioners in a variety of fields, to help investigators devise new designs for a variety of situations, and to facilitate further research and development of promising new sampling design and inference methods. It is intended for researchers in the methods and theory of statistical sampling and for researchers in a variety of the natural and social sciences and other fields faced with inherently difficult sampling situations—including in particular rare, clustered, unpredictable, elusive, spatially and temporally uneven, and hard-to-detect populations.

The book is organized into 10 chapters. Chapter 1 gives a broad introduction to sampling design and estimation possibilities, describes some of the situations motivating adaptive methods, and provides some preliminaries that will be useful in subsequent chapters. Chapter 2 examines sampling theory from the classical fixed-population point of view, with particular attention to results pertaining to adaptive designs. For adaptive sampling, the theory requires more generality than is standard in sampling theory texts. Chapter 3 considers sampling theory with a stochastic population, so that the sampling situation involves both the design and the population model. The emphasis again is on results useful for adaptive sampling.

Chapter 4 describes adaptive cluster sampling, in which an initial sample is selected with a conventional design and then additional units are added in the vicinity of any sufficiently "interesting" observed values (such as high abundance of a rare species, high levels of pollution, or high reported incidence of drug use). In Chapter 5, factors affecting the efficiency of adaptive cluster sampling in comparison with conventional sampling designs are evaluated to help determine which will be better in a given situation. Practical considerations of time limitations and unforeseen exigencies in surveys are also dealt with in that chapter. Since an appropriate condition for extra sampling is sometimes difficult to specify prior to a survey, Chapter 6 describes adaptive cluster sampling based on order statistics. For example, in a survey of an environmental contaminant, additional observations could be made in the neighborhoods of the initial sites with the ten largest values. Chapter 7 describes a wide range of adaptive allocation strategies, in which the allocation of sampling effort in stratified sampling is determined during the survey based on observed values. A wide range of adaptive allocation strategies are described.

In Chapter 8, adaptive sampling is applied to situations in which the variable of interest is multivariate—for example, a survey of rare and endangered birds may be used to estimate the abundance of several species simultaneously, and the adaptive condition for adding sites may depend on the whole combination of species observed. The multivariate extension allows also for the use of a "rapid assessment" variable, such as a quick visual count, for determining when to add neighboring sites to the sample. A common source of nonsampling error in many surveys of animal populations as well as of elusive human populations is imperfect detectability. Adjusting estimates for imperfect detectability with adaptive designs is discussed in Chapter 9.

Chapter 10 examines the issue of optimal sampling strategies from a number of approaches. Various optimal strategies are described and some results are given on when the optimal strategy is a conventional one and when it is adaptive. Though the theoretically optimal strategy can be difficult to implement in practice, some of the optimal designs described are suggestive of practical procedures for real situations.

We have endeavored to keep each of the chapters as self-sufficient as possible. For readers interested primarily in methodologies, Chapters 1 and 4 through 9 are of interest; the more theoretical Chapters 2, 3, and 10 may be skipped at first reading. Readers whose first interest is the theory and the possibilities it suggests may wish to focus on Chapters 1, 2, 3, and 10; such readers may skip the more methodological Chapters 4 through 9 at first reading. We see the book as an integrated whole, however, since the development of practical sampling methods has motivated extensions to the theory, while theoretical results on design and model-based sampling have opened the way to some of the new design and estimation methods developed. The reader is assumed to have some acquaintance with sample survey theory and basic ideas such as simple random sampling and stratified sampling. Throughout the book, practical methods and theoretical results are illustrated with simple worked examples.

The importance of adaptive sampling strategies comes from a number of sources. (1) Adaptive sampling designs are the natural extension of conventional designs, and the collection of conventional and adaptive designs together comprise a fuller set of possibilities from which to choose a design for a given situation. (2) Many real-world sampling situations—for example in environmental pollution studies, surveys of rare animal and plant species, studies of contagious diseases, drug use epidemiology, marketing surveys, and assessments of mineral and fossil fuel reserves—motivate the development of adaptive sampling procedures. (3) Some theoretical results point to adaptive procedures as being optimal in many situations, in the sense of giving the most precise estimates with a given amount of sampling effort. (4) In addition to giving more precise estimates in some situations, adaptive designs can increase the yield of the sample, so that more animals of the rare species are observed or more individuals with the disease are monitored. (5) Data from observational studies where no design was deliberately applied may have inherent adaptive features. For example, a study of ground water pollution may have to rely on existing wells, but in selecting locations for their wells landowners have taken into account the water quality measurements from the wells of their neighbors. Analysis of the data from such studies can be assisted by taking the adaptive selection procedures into account.

Although ideas of adaptive sampling have been floated in the past, some of the most practical methods along with the supporting theory have been developed quite recently. Recent research by the authors and others has developed a number of new designs and inference procedures as well as theoretical results suggesting further possibilities. Much of the literature is therefore recent, and this monograph is an attempt to bring this material together in a unified way. The new design and inference possibilities made available, as well as many practical questions regarding which design will be most effective in a given situation, leave much to be done in the field and many unsolved problems to be investigated. We view this monograph as a beginning only and hope that it will stimulate further research and encourage researchers to further explore the possibilities of adaptive methods.

We would like to thank several authors for permission to quote from unpublished material, namely Patricia Munholland, John Borkowski, and Jenny Brown. We also thank John Wiley & Sons, Inc., and various journals for permission to reproduce published material. Acknowledgements are made in the text. Support for research leading to this monograph has been provided for one author (SKT) by the National Science Foundation and the Environmental Protection Agency.

University Park, Pennsylvania						STEVEN K. THOMPSON
Auckland, New Zealand							GEORGE A. F. SEBER

CHAPTER 1

Introduction and Preliminaries

1.1 ADAPTIVE SAMPLING

The basic problem of sampling is to estimate some characteristic of a population by observing only part of the population. For example, to estimate the total number of a rare species of trees in a forest, the forest could be partitioned into units ("plots"). A sample of the units is selected for observation. For each unit in the sample the variable of interest, the number of trees of the species, is recorded. An estimate of the total number of trees of the species is then made based on the number in the sample, expanding as appropriate to reflect the size of the study region.

The procedure by which the sample is selected is called the sampling design. In adaptive sampling, the procedure for selecting the sample may depend on values of the variable of interest observed during the survey. Thus, the sampling plan has the flexibility to change during the course of the survey in response to observed patterns in the population. Situations that have motivated researchers and field workers to use, or consider using, adaptive sampling designs include the following.

1.2 SOME SURVEY SITUATIONS

Surveys of commercial species of shrimp in the Gulf of Alaska involve trawl sampling, in which a net is towed behind the research vessel at selected locations and estimates of shrimp abundance for the whole study region are made by expanding the amounts caught, taking into account the area swept by the net. The survey covers a vast area and the sampling method is very time consuming. Biologists aboard the research vessel found that typically shrimp were concentrated in just a few areas, whereas tows in other areas would produce hardly any shrimp. Yet the regions of concentration changed from survey to survey due to the schooling movements of shrimp, and the concentration areas were not entirely predictable in advance. The field biologists expressed the desire for a plan in which they could cut down on effort in the low-abundance areas. For this reasons an adaptive procedure was proposed and used on some surveys (Thompson and Ramsey [1983], Thompson et al. [1992]) in which, whenever the catch was 50 pounds or less, the following tow would be for one-half

1

a nautical mile, whereas when the catch was over 50 pounds, the following tow would be one nautical mile in length. Similar designs were investigated for surveys of rare and endangered species of Hawaiian birds (Thompson and Ramsey [1983]) (see Chapter 7).

Surveys of whales in which whales or their spouts are sited by observers on research vessels are carried out in either "passing" or "closing" mode (Hiby and Hammond [1989]). In passing mode, the research vessel sticks to the preselected transect as whales are observed and counted. In closing mode, when a pod of whales is sighted the vessel departs from the preselected transect to close in on the pod. Because of the clustered nature of pods of whales, more pods of whales are observed in closing than in passing mode. However, because typically the vessel does not return to the point at which it departed from the transect, some pods that would have been seen from the preselected transect are missed (Kishino and Kasamatsu [1987]), and many design and estimation problems for such surveys remain to be solved (cf. Schweder et al. [1993]).

Spatial patterns of hardwood tree species in northeastern North America were examined by Roesch [1993], who found that whereas overall forest density appeared fairly random in distribution, the individual species of interest were apt to be rare and highly clustered in their distributions and therefore difficult to sample. For estimating the prevalence of insect infestation in some of these species, Roesch therefore investigated an adaptive sampling procedure in which once a tree of the species was found, an area of specified radius around it would be searched for additional individuals of the species (see Chapter 4).

Moose surveys in interior Alaska are carried out with aircraft flying over selected spatial units with observers counting moose detected. Estimates of moose abundance not only take into account the sizes of the units in the sample and the size of the study region but also adjust for imperfect detectability of moose, with adjustments based on intensive searches in a subsample of the region as well as observations from the air of known numbers of marked moose. As is standard in survey practice, larger sample sizes are used in habitat strata having higher variability of moose abundance. Since this variability is not known prior to the survey, it is estimated at the end of each day during the survey, and the allocation of the next day's effort is based on the estimated relative variabilities (Gasaway et al. [1986]; see Chapter 7).

Waterfowl wintering in a federal refuge have a patchy distribution due partly to the patchiness of the habitat caused by uneven flooding, so that a sampling design with units selected with probabilities proportional to available habitat can be used for aerial surveys of abundance. Additional spatial unevenness due to the birds' social behavior, unobserved habitat variables, and other factors is less predictable. Smith et al. [1995] therefore investigated sampling plans combining initial unequal probability sampling with subsequent addition of neighboring units to the sample whenever sufficiently high abundance was observed (see Chapter 4).

Environmental restoration of areas contaminated with hazardous wastes is a very expensive procedure; failure to remedy this problem on the other hand carries serious social, health, and economic costs. The decision to remedy the situation is based on the level of contamination, which must be estimated from measurements on a sample of sites. Because the chance of making an error in either direction is highest for areas near the threshold contamination level, Englund and Herari [1995] investigated an adaptive design which resulted in more of the sample sites being located near the threshold contour (see Chapter 10).

Household surveys for estimating rare characteristics, such as drug use, infection with a rare disease, or specialty consumer purchases, also present difficult sampling and estimation problems. A marketing survey to estimate purchases of disposable diapers in a region of New Zealand ran into difficulty when it was discovered that only 2% of households in the region had children under 2 years of age. The distribution of such households was not known but was known to be clustered spatially. Danaher and King [1994] therefore investigated the use of an adaptive cluster sampling design in which a systematic sample of households was augmented with neighboring households whenever a household in the sample had the rare characteristic (see Chapter 4).

Many commercially important fish species exhibit aggregated patterns of abundance that are difficult to predict prior to a survey and that make conventional survey methods inefficient. Because of these recurring patterns, adaptive allocation designs, with increased sampling effort devoted to strata in which high abundance or variability is observed in the initial part of the survey, were investigated by Francis [1984] for mackerel and orange roughy surveys, and by Jolly and Hampton [1990] for anchovy surveys (see Chapter 7).

Estimation of the contribution of hatchery salmon to an ocean salmon fishery is crucial both for determining the economic effectiveness of the hatchery program and to protect wild stocks of fish from ecological displacement. Because the proportion may change during the course of the fishing season, Geiger [1994] investigated the use of an adaptive Bayes design for estimating the proportion from fish otoliths (see Chapter 7).

In an article summarizing statistical challenges in the environmental sciences, Cormack [1988] described two examples in which spatial considerations raised the question of whether more efficient sampling schemes would be possible. The first involved the caterpillar of the pine beauty moth that killed large areas of planted pine forest in the Scottish Highlands. The decision of whether to spray is based on an estimate of the number of pupae in the survey. The distribution of pupae is highly clustered, with no visual indicator of which areas have high density under the soil. The second example involved forest damage due to bark stripping by red deer. The extent of damage is highly variable, even after stratification on known factors. In each example it is cheap to observe the damage at neighboring points. But conventional cluster sampling leads to inefficient estimates because of the positive local correlations. Cormack therefore raised several questions about the possible use of adaptive procedures to deal with the spatial clustering issues.

In each of the preceding situations investigators are dealing with characteristics of populations that for one reason or another are inherently difficult for sampling and estimation. These studies deal with issues of great importance in terms of economics, health, environmental quality, or scientific understanding. In each case the investigators are motivated to search for sampling and estimation methods that go beyond the conventional set of techniques, while at the same time using the conventional methods to the greatest advantage possible. Some of these examples are discussed in greater detail later. In this monograph we endeavor to take a look at what is possible and what might be practical in sampling and inference methods, with particular attention to adaptive strategies.

1.3 TYPES OF SAMPLING DESIGNS

A sampling design is a procedure for selecting the sample of units or sites to observe for estimating a characteristic of the population. In searching for the most effective design for sampling a population, it is useful to divide the set of all possible sampling designs into three types: conventional, adaptive, and nonstandard.

In a conventional design, the procedure for selecting the sample does not depend on any observations of the variable of interest, so that the entire sample may be selected prior to the survey. The purposive selection of a set of units thought to be "representative" of the population as a whole is one example of a conventional design. Other examples include all of the familiar probability designs—simple random sampling, stratified random sampling, sampling with probability proportional to size, systematic sampling, cluster sampling, multistage sampling—and some recent and less well-known ones including network sampling and various "balanced" sampling designs. Auxiliary information known about the population as a whole can be used at both the design and the estimation stages to improve effectiveness. Examples of auxiliary variables include spatial size of units, vegetation cover in an animal survey, elevation or depth of a sample site, past survey data on a human population, and visual estimates of the variable of interest.

In adaptive sampling, the procedure for selecting the sample may depend on values of the variable of interest observed in the sample. Examples include simple random sampling with a stopping rule based on observed values, "inverse" sampling, adaptive cluster sampling, adaptive stratification during a survey, adaptive allocation of sampling effort in stratified sampling, and optimal Bayes designs. Auxiliary variables known for the population as a whole may be used at both the design and the inference stages in adaptive sampling. Auxiliary variables known only for units in the sample, on the other hand, are most usefully regarded as components of a multivariate variable of interest, as in Chapter 8.

Designs in which the selection of the sample is influenced by values of the variable of interest for units outside the sample or by unknown parameter values will be referred to as nonstandard. An example described in Chapter 3 involves an environmental agency receiving a sample of automobiles from a manufacturer for fuel efficiency testing; the manufacturer has inspected a larger number of the vehicles and selected the most efficient ones for the sample. With nonstandard designs the inference problem is complicated by the linkage of unknowns to the design, as noted in Chapter 3. Although it is questionable whether survey practitioners would choose to deliberately implement such a design, nonstandard selection procedures may arise in practice more commonly than is generally realized, and in such cases the selection

procedure needs to be taken into account in any inference about population characteristics. The class of adaptive—together with conventional—designs appears to be
the natural class within which to look for practical sampling designs to deliberately
implement.

1.4 APPROACHES TO INFERENCE IN SAMPLING

Once the sample is obtained, the sample data can be used for inference about characteristics of the population. The inference can take the form of estimates, predictions,
confidence and prediction intervals, and tests of hypotheses. The methods and criteria brought to bear in sampling for inference about the population can be broadly
classified into three approaches: a design-based approach, a frequentist model-based
approach, and a likelihood model-based approach. In addition, the approaches can
be combined and modified in various ways.

In the design-based approach, little or nothing is assumed about the population
itself. Probability enters the situation only in terms of the design-induced probability
of selecting one sample as opposed to another. Inference procedures are sought
that have good properties over hypothetically repeated selections of the sample. For
example, a design-unbiased estimator of the population total has an expected value,
over all possible samples that might be selected under the design, equal to the actual
value of the population total.

In the frequentist model-based approach, the values of the variable of interest in the
population, such as the number of plants in each plot of a study region or the income
in each household in the population, are viewed as a realization of a set of random
variables. A stochastic model ("superpopulation model") is assumed describing the
distribution of possible realizations of the population values. Inference procedures
are sought that have good properties over hypothetically repeated realizations of the
population values.

In the likelihood approach, inference is based on the likelihood function of the
unknowns given the sample data. Likelihood-based methods include maximum likelihood estimation and prediction, and Bayesian methods. Bayesian methods rely on
a known prior distribution for the population or its parameters, while maximum
likelihood methods do not assume a prior distribution.

Each of the three inference approaches has broad implications in terms of each of
the three types of designs. These connections are detailed in Chapters 2 and 3. The
broad conclusions can be summarized as follows: (1) With the design-based approach,
the selection probabilities determined by the design are central to inference. Whether
the design is conventional or adaptive, a design-unbiased strategy will be unbiased
no matter what the population itself is like. (2) With the frequentist model-based
approach, inference can be based solely on the population model without taking
the design into account—*provided the design used was conventional*. Thus, a model-
unbiased estimator will be unbiased under the assumed population model irrespective
of whether the sample was selected purposively or by simple random sampling. If,
on the other hand, the sample was selected by an adaptive design, the same estimator

will not necessarily be unbiased. Model-unbiased strategies with adaptive designs, such as the model-unbiased method of adaptive allocation described in Chapter 7, take account both of the model assumed and the design used. For adaptive designs, the frequentist properties of an inference procedure will depend on both model and design. (3) With the likelihood-based approach, once the sample is selected the inference can be based solely on the model without taking the design into account— *provided the design used was conventional or adaptive*. Thus, a maximum likelihood estimator or predictor is maximum likelihood under the assumed model whether the design was conventional or adaptive, and even if the field workers diverged from the prescribed design for reasons such as shortage of time. If the data were obtained from a nonstandard design, on the other hand, the design is tied to the unknown variables or parameters in the likelihood, and the design is therefore involved in likelihood-based inference.

All three of the above approaches to inference are represented in this monograph. Much attention is given to design-unbiased adaptive strategies because of the practical advantage that the unbiasedness of such procedures does not rely on any assumptions about the population itself. For many of the populations of interest, including natural animal and plant populations and elusive human populations, statistical models of the populations themselves are seldom realistic enough at present to give confidence in using them as a sole basis for inference. Further, there are often conflicts of interests involved in the use of survey estimates. Commercial fishermen may prefer a high estimate of the population total so that the regulatory agencies will allow higher catches. Environmental groups may feel a low estimate of abundance for an endangered species is preferable in supporting protective measures. Law enforcement agencies and rehabilitation centers may receive increased funding if estimates of illegal drug use are high. In such cases, design unbiasedness or consistency provides a measure of public fairness. On the other hand, maximum likelihood predictors often seem to provide in some respect the natural estimates for situations such as stratified sampling, coinciding with the classical estimators when the allocation design is conventional or adaptive.

Recent methods modifying or combining the three basic approaches include selecting "balanced" samples to provide robustness against departures from an assumed model, imposing design unbiasedness while seeking low mean-square error under an assumed model, and incorporating design-based inclusion probabilities into model-based estimators. These modified and combined approaches are occasionally used in the succeeding chapters and are briefly summarized in Chapter 10.

In the sampling literature, considerable disagreement has been expressed regarding the appropriate approach to inference in sampling. The likelihood approach is the only one that adheres to the "likelihood principle," which says inference should be based on the likelihood function, with the same inference resulting from two data values having proportional likelihood functions—that is, the proportionality constant not depending on any unknowns. The design-based and frequentist model-based approaches, on the other hand, appeal to a "repeated sampling" principle, with the hypothetical repetitions being over the set of possible samples in the one case and over realizations of the population values in the other. In the heat of the argument, it is not always noticed that the three approaches, while leading to different conclusions on exactly what estimate or interval should be computed from the sample data, lead to the same conclusion on something more basic: the importance of choosing a good sampling design to start with.

1.5   OPTIMAL SAMPLING STRATEGIES

A sampling strategy consists of a design for selecting the sample together with the inference procedure that will be used. Suppose one wishes to have a strategy that will produce an unbiased estimator of the population total, the estimator having the lowest possible variance or mean-square error. Does such a strategy exist, and if so, what is it? The results on optimal sampling strategies are the subject of Chapter 10 and are briefly sketched here.

With the fixed population approach, where no model is assumed about the population, with the population values being considered fixed constants and probability entering only through the design, no optimal strategy exists. Rather, a wealth of practical results exists on the types of populations for which one specific type of design and estimator are preferable to another. For conventional designs, results of this type are found in the standard sampling texts. Available results of this type for adaptive strategies and comparable conventional ones are contained in this monograph.

At the other end of the scale, where a model of the population is known exactly—that is, the joint probability distribution of the population values is known—the optimal strategy is known in general to be adaptive. Only in special cases will the optimal strategy be a conventional one. This result also applies when the population model depends on parameters for which a prior distribution is assumed. The nature of optimal designs and the conditions under which they are conventional are detailed in Chapter 10.

Between the two extremes of no model and an exact statistical model assumed for the population are the cases where partial knowledge is assumed about the probability distribution for the population. For example, the distribution may be assumed to be of known parametric form, but with unknown parameters. Or, without assuming a specific parametric form, the covariance matrix of the population values may be assumed proportional to a known matrix. For such cases, the only optimality results available are restricted to the class of conventional designs. Frequentist model-based results with such assumptions lead to the familiar and widely used regression predictors of survey sampling and the "kriging" predictors of spatial statistics. From the frequentist model-based view, the optimal design is to purposively select the sample that gives the minimum mean-square prediction error under the model. Rather less well known are the optimal conventional strategies with design unbiasedness imposed on the linear predictors, so that the selection cannot be purposive but the prediction will be unbiased even if the assumed model is not correct. These results are summarized in Chapters 3 and 10 but apply unfortunately only to the class of conventional designs. The question remains whether a better strategy could be found by considering the larger class containing the adaptive designs.

Although the optimal strategies described in Chapter 10 may be interesting from a theoretical viewpoint, suggesting for example that the optimal strategy will often be an adaptive one, the optimal procedures can be difficult to implement and computationally complex to analyze. In addition, determination of the optimal strategy may require unrealistic amounts of prior knowledge about the population. The greatest potential of the optimal strategies may be in suggesting procedures which, though not optimal, are efficient yet practical and simple. The adaptive strategies described in Chapters 4 through 9, on the other hand, while not optimal, are relatively simple to implement and analyze, require a minimum of assumptions about the population, and can be shown in certain situations to provide improvements upon comparable conventional designs.

1.6 SOME MORE DETAILED EXAMPLES

In this section some examples of adaptive sampling are described in more detail than in Section 1.2; these examples are referred to from time to time in the text of subsequent chapters.

In recent years many of the sample survey methods have been applied to natural populations in an attempt to estimate total numbers or population density (cf. Seber [1982, 1986, 1992] and Thompson [1992]). The population study area is divided up into spatial units (plots) of generally the same size, and the numbers of animals or organisms are counted on a selection (such as a random sample) of the units. However, in applying the traditional sampling theory for finite populations, problems have been encountered with some populations. For example, suppose that the animals are spread over a very wide area so that the average population density is low but the population is highly clustered as is the case with some fish populations. There are then large regions between the clusters where there are no animals. If we apply the traditional sampling methods such as choosing a random sample of units without replacement, we find that many of the units will have zero counts while most of the units with nonzero counts are concentrated in a few locations. This can lead to population estimates with large variances and wide confidence intervals. To overcome this inefficiency of estimation, some modifications to the sampling procedure can be made. For instance, if the population is highly clustered and the unit currently being sampled has a zero count, then most of the neighboring units would also have zero counts. There would be little point in sampling further in the neighborhood of that unit. However, if a unit contains animals, then further units in the neighborhood may be worth sampling. One might say, if the fishing is good keep fishing, otherwise move on!

The preceding type of sampling scheme, in which each step of the sampling process depends on what is observed from the previous steps, is adaptive: the sampling is adapted to the data. A wide variety of adaptive schemes is possible. For example, the population area could be divided up into subareas (strata) and a number of units are then sampled from each stratum. Further sampling is carried out in those strata where nonempty units are found. This adaptive procedure and the one above have two features in common. Firstly, the total number of units sampled overall is random and, secondly, a stopping rule is needed to decide when to stop sampling in a neighborhood (or in a given stratum) and move on to the next stage of the sampling process. The preceding loose description of adaptive sampling also includes sequential sampling, where sampling is continued until some condition is fulfilled. If we use traditional estimators such as the sample mean for adaptive samples, these will be biased and new estimators are needed. This monograph is an attempt to introduce such estimators and open the way for further research. An introduction to this work is given by Seber and Thompson [1994]. We now give a number of examples of adaptive sampling schemes.

Example 1.1   Rare Species.   Suppose that a certain species of animal is fairly rare but when such an animal is encountered there is often at least one other close by. To estimate population density, the following sampling scheme proposed by Thompson [1990] is used.

An initial simple random sample (without replacement) of square plots from a grid of plots are searched. If an animal is found on one of these plots, then all the plots in its neighborhood are also sampled. For this experiment we define a neighborhood as the plot itself together with the four plots with adjacent boundaries (thus forming a cross). If further animals are found on an adjacent plot, its neighborhood is also searched. This process of augmentation continues until no further animals are encountered. We finally end up with a group of plots each containing animals, but surrounded by a boundary of empty plots as in Figure 1.1. Each plot containing animals is therefore part of a cluster of plots so that when that particular plot is chosen in the initial sampling scheme, the rest of the cluster is then sampled as well. This is a type of

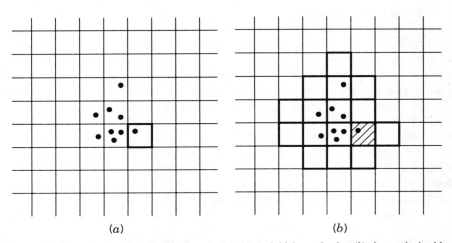

(a) (b)

Figure 1.1. The procedure described in Example 1.1. (a) An initial sample plot, (b) cluster obtained by adding adaptively.

cluster sampling in which the clusters can be identical or overlap on their boundaries. We shall refer to it as adaptive cluster sampling and the theory of this technique is described in Chapter 4.

If we are investigating more than one species at the same time, then we are in a multivariate situation where we have a vector of counts instead of a single count for each plot. This multivariate extension is described in Chapter 8 where it is applied not only to numbers of animals but also to characteristics measured on animals.

In the above survey for a single species we are assuming that all animals are clearly visible. However, if the animal is a bird in dense bush, then there is a possibility that it may not be detected even though it is present. This further complication of incomplete detectability is discussed in Chapter 9.   □

Example 1.2   Rare Disease.   As noted by Thompson [1990], an adaptive scheme can be used to investigate a rare contagious disease. A simple random sample of people are selected and tested for the disease. If a person tests positively, then all the friends and contacts of that person are also tested. If one of the contacts tests positively, then all that person's contacts are tested, and so on. In contrast to Example 1.1, the units are now people rather than plots. Also it is more difficult to define what we mean by a "neighborhood", that is the group of people close in a relevant social sense to the person being sampled.   □

Example 1.3   Pollution Study.   When carrying out environmental pollution studies, the following situation is commonly encountered. In most places sampled the pollution is light or negligible, but a few scattered pockets of high pollution are encountered. Two questions of interest are: What is the average level of pollution for the whole area, and where are the pockets or "hot spots" likely to be? A traditional approach would be to select a random or systematic sample of sites (plots) and measure the pollutant at each of the selected sites. The average of these measurements provides an unbiased estimate of the population average, and the individual observations might then provide a contour map so that the pollution peaks could be located. However, in environmental sampling, there can be problems with this procedure. For example, in the extreme case where the pollution is negligible over most of the area, a majority of the measurements will record a zero level of pollution, and there is a good chance that most of the pockets will be missed. In addition, the contour map will have little accuracy in the areas of high concentration, these being of particular interest.

Clearly the above problem lends itself to an adaptive sampling strategy. When we strike appreciable pollution on a site, we would like to look in the neighborhood of the site for hot spots. However, we may not want to do this for every such site, and a compromise would be to investigate the neighborhood of just those sites with the higher pollution levels. This brings in the idea of using order statistics as in the following scheme due to Thompson [1995]. A smaller random sample is taken and, provided we don't strike all zeros, we can then go back to the sites corresponding to, say, the three highest readings, and sample all sites in their neighborhoods. If the reading on any of these additional sites exceeds the fourth largest reading in the initial sample, then we add its neighboring sites as well. Each of the initial sites then gives rise to a cluster of sites with each cluster tending to grow toward the locations of highest pollution. With such an adaptive scheme the conventional estimators of the population mean or total will be biased, and new estimators are needed. The theory for this strategy is discussed in Chapter 6. □

Example 1.4   Schools of Shrimp.   Shrimp resources in the Gulf of Alaska are assessed with trawl surveys conducted by the Alaska Department of Fish and Game. The net is towed for a measured distance along the ocean floor and shrimp density is estimated from the amount of shrimp caught, taking into account the width of the net and the distance towed. It is known that shrimp have distinct schooling tendencies but it is impossible to predict in advance where the shrimp will concentrate within a survey area. Also it is common experience on surveys for many consecutive tows to have zero or near-zero catches of shrimp and then for the catches to increase more or less steadily as a concentration is approached. An adaptive strategy therefore seems appropriate here. One such strategy considered by Thompson et al. [1992] is as follows.

The survey area, such as a bay, is divided into a grid of strata, with each stratum approximately 1 mile square. Within a stratum a tow is located at random. If more than 50 lbs of shrimp per mile are caught, a full mile tow is made in the next stratum. If 50 lbs or less per mile are caught, then a short tow of 1/2 mile is made at the next stratum. The process begins with a 1-mile tow.

This strategy differs from that described in the previous examples in that the concept of "neighborhood" is not used. Instead of adapting the sampling locally we allocate the samples to the strata in an adaptive manner. This methodology, which we call adaptive allocation, is discussed in Chapter 7. □

Example 1.5   Aggregations of Orange Roughy.   Off one short stretch of the East Coast of the South Island in New Zealand and reaching out to the Chatham Islands, the sea bed rises to form a comparatively shallow plateau called the Chatham Rise. Deep sea fish called the orange roughy come to this plateau to spawn forming very large aggregates of fish, especially round "pinnacles" rising from the sea floor. The fish are trawled at depths ranging from about 700 m to 1200 m and, when an aggregate is struck, very large catches are obtained. To determine seasonal quotas for the fishing industry, the Ministry of Agriculture and Fisheries needed to obtain an estimate of the total size of the spawning population. To overcome the problems associated with such a high degree of aggregation, Francis [1984] introduced the following adaptive scheme.

As the depth contour lines run almost parallel to the lines of latitude, the area is divided into strata by depth and longitude. The sampling is then carried out as a two-phase operation. The total number of tows is roughly fixed in advance and is determined by time constraints. In the first phase more than 75% of the tows are allocated at random in the strata in a roughly optimal fashion, the number of tows in a given stratum being proportional to the product of the stratum area and some intuitive weighting factor subjectively derived from past data. Using the information from these tows, an estimate of the total weight of fish (biomass) and its variance estimate are calculated for each stratum. The tows in phase 2 are then chosen in the following fashion. The first tow of phase 2 is chosen at random in the stratum that would give the greatest reduction in its variance estimate. Suppose this is stratum j. Then the variance estimate of stratum j is recalculated by adding 1 to its number of phase-1 tows. The process is repeated for the second phase-2 tow. It is allocated at random to the stratum that will give the greatest variance reduction. Thus, all the phase-2 tows are allocated one at a time before any phase-2 sailing is carried out. A sailing strategy is then worked out to cover the strata with phase-2 tows. In practice the number of such tows actually made will depend on such things as weather conditions and distance to be traveled. The mathematical details are described in Section 7.2.1.

□

In conclusion, we see that adaptive methods adapt the sampling effort to the population patterns encountered on the survey. Furthermore, more individuals or higher values are sampled so that there is more information available on any other associated variable of interest. For example, in animal populations we may be interested in estimating the population total. However, each time an animal is sampled (caught), we may also wish to record information such as weight, age, and so on. Clearly adaptive sampling can be applied to any patchy population, whether it be animals, plants, humans, minerals, or fossil fuels.

1.7 INDICATOR FUNCTIONS

In sample survey theory it is often of interest to know whether a unit is included in the sample or not. This information can be described by means of an indicator function I which takes the value 1 if the unit is included and zero if it is not. Because such indicators can be applied to any event, they are a very useful tool in deriving means and variances of estimators. We therefore give a formal definition of an indicator function and list some of its properties.

Let Ω be a sample space and $\mathcal{F}$ a family (σ-field) of events. Then, for each event A in $\mathcal{F}$, we can define the indicator function $I_A: \Omega \to \mathbb{R}$ of A to be

$$I_A(\omega) = \begin{cases} 1, & \omega \in A, \\ 0, & \omega \notin A. \end{cases}$$

This function takes the value 1 if event A occurs and zero otherwise. The following properties hold:

1. $I_\Omega = 1$.
2. $I_\phi = 0$, where ϕ is the empty set.
3. $I_A^2 = I_A$.
4. $I_{\bar{A}} = 1 - I_A$, where $\bar{A}$ is the complement of A.
5. $I_{A \cap B} = I_A I_B$, for all A, B in $\mathcal{F}$.
6. $I_{A \cup B} = I_A + I_B - I_A I_B$, for all A, B in $\mathcal{F}$.

These results are proved in a straightforward fashion by showing that both sides of an equation are equal to 1 or 0 under the same conditions. For example, $I_{A \cap B} = 1$ only when both A and B occur, that is, when $I_A = 1$ and $I_B = 1$, or equivalently $I_A I_B = 1$. We now link up these indicator functions with probability.

Let $P: \mathcal{F} \to \mathbb{R}$ be a probability measure defined on $\mathcal{F}$. Since I_A is a function on the sample space Ω, it is a random variable with expected value

$$\mathrm{E}[I_A] = \int I_A(\omega) dP(\omega)$$

$$= \int_A dP(\omega)$$

$$= P(A). \tag{1.1}$$

Alternatively, $I_A(\omega) = 1$ if and only if $\omega \in A$, that is, if and only if A occurs, so that

$$P\big(I_A(\omega) = 1\big) = P(A), \qquad P\big(I_A(\omega) = 0\big) = P(\bar{A})$$

and

$$\mathrm{E}\big[I_A(\omega)\big] = 1 \cdot P(A) + 0 \cdot P(\bar{A}).$$

Furthermore, from property (3) above,

$$\mathrm{var}[I_A] = \mathrm{E}[I_A^2] - \big(\mathrm{E}[I_A]\big)^2$$

$$= \mathrm{E}[I_A] - \big[P(A)\big]^2$$

$$= P(A)\big[1 - P(A)\big]. \tag{1.2}$$

One other result that we shall need follows from property (5), namely,

$$\mathrm{cov}[I_A, I_B] = \mathrm{E}[I_A I_B] - \mathrm{E}[I_A]\mathrm{E}[I_B]$$

$$= \mathrm{E}[I_{A \cap B}] - P(A)P(B)$$

$$= P(A \cap B) - P(A)P(B)$$

$$= \pi_{AB} - \pi_A \pi_B, \quad \text{say.} \tag{1.3}$$

Often A is the event that unit i is in the sample. In this case, it is convenient to alter the notation and write I_i instead of I_A, as in the following example.

Example 1.6   Mean and Variance of a Sum.   A typical application of indicator variables is the following. Suppose we wish to find the mean and variance of

$$Z = \sum_{i=1}^{m} z_i I_i ,$$

where the I_i are indicator variables such that $I_i = 1$ with probability π_i and $I_iI_j = 1$ with probability π_{ij} ($i \neq j$). Then

$$E[Z] = \sum_{i=1}^{m} z_i E[I_i] = \sum_{i=1}^{m} z_i \pi_i \tag{1.4}$$

and, using (1.2) and (1.3),

$$\begin{aligned} \text{var}[Z] &= \sum_{i=1}^{m} \text{var}[z_iI_i] + \sum_{i=1}^{m}\sum_{j \neq i} \text{cov}[z_iI_i, z_jI_j] \\ &= \sum_{i=1}^{m} z_i^2 \pi_i(1 - \pi_i) + \sum_{i=1}^{m}\sum_{j \neq i} z_iz_j(\pi_{ij} - \pi_i\pi_j) \\ &= \sum_{i=1}^{m}\sum_{j=1}^{m} z_iz_j(\pi_{ij} - \pi_i\pi_j), \end{aligned} \tag{1.5}$$

with the convention that $\pi_{ii} = \pi_i$. Since $E[I_i] = E[I_i^2] = \pi_i$ and $E[I_iI_j] = \pi_{ij}$, an unbiased estimate of var[Z] is given by

$$\begin{aligned} \widehat{\text{var}}[Z] &= \sum_{i=1}^{m} z_i^2(1 - \pi_i)I_i + \sum_{i=1}^{m}\sum_{j \neq i} z_iz_j \frac{(\pi_{ij} - \pi_i\pi_j)}{\pi_{ij}} I_iI_j \\ &= \sum_{i=1}^{m}\sum_{j=1}^{m} z_iz_jI_iI_j \left(\frac{\pi_{ij} - \pi_i\pi_j}{\pi_{ij}} \right). \end{aligned} \tag{1.6}$$

This can be obtained from (1.5) by simply multiplying each term by $I_iI_j/E[I_iI_j]$. Expressed this way, (1.6) is obviously unbiased. $\qquad\square$

1.8 DESIGN AND MODEL-UNBIASED ESTIMATORS

1.8.1 Design-Unbiased Estimators

At this stage we do not introduce all the notation but just enough to consider one or two ideas about design. We assume that we have a population of N units—representing, for example, objects, people, households, or plots—which are labeled $1, 2, \ldots, N$. Associated with the ith member is a characteristic or measurement y_i, for example, the number of animals on the ith plot. Usually we are interested in estimating such parameters as the mean per unit

$$\mu = \frac{1}{N} \sum_{i=1}^{N} y_i,$$

or the total $\tau = N\mu$, for example, the total number of animals on all the plots. Given that a is the area of each plot, then $D = \mu/a$ is the population density per unit area. If our population consists of N people and we define y_i to be 1 if the ith person has a particular disease or characteristic and 0 otherwise, then τ is the total number of people in the population with the characteristic and μ is the proportion.

Suppose we take a random sample of size n with or without replacement from the population. The individuals are selected one at a time in such a way that each individual has the same probability of being selected as all the others currently in the population. If each individual is returned to the population after it is selected, then we call the design random sampling with replacement (*rswr*). If each selected individual is not returned so that sampling is without replacement and all remaining individuals have equal probability of being selected at the next draw, we refer to the design as simple random sampling (*srs*), or random sampling without replacement.

The results of a random sample of size n can be listed as the unordered set of sample unit labels and associated y-values,

$$\{(i_1, y_{i_1}), (i_2, y_{i_2}), \ldots, (i_n, y_{i_n})\}.$$

A more rigorous development is given in Chapter 2. If simple random sampling is used, then all the labels $i_1, i_2, \ldots, i_n$ are distinct and represent an unordered set $s = \{i_1, i_2, \ldots, i_n\}$ of n of the labels $(1, 2, \ldots, N)$. If sampling is with replacement, then some of the labels may be the same. Double subscripts are clumsy, so we shall use alternative notations for the labels such as $i \in s$ or $i = 1, 2, \ldots, n$, where the second notation is taken to mean that any n labels are chosen and not just the first n.

A natural estimate of the population mean μ is the sample mean

$$\bar{y} = \frac{1}{n} \sum_{i \in s} y_i$$

$$= \frac{1}{n} \sum_{i=1}^{n} y_i$$

$$= \frac{1}{n} \sum_{i=1}^{N} y_i I_i, \tag{1.7}$$

where I_i is the indicator function taking the value one if the ith label is selected in the sample and 0 otherwise. If the random sampling is without replacement (*srs*), then all samples of size n are equally likely, so that any particular sample has a probability of $1/\binom{N}{n}$ of being selected. The number of samples containing label i is $\binom{N-1}{n-1}$, so that the probability that $I_i = 1$ is

$$P(I_i = 1) = \binom{N-1}{n-1} \Big/ \binom{N}{n}$$

$$= \frac{n}{N}. \tag{1.8}$$

Hence, as in Equation (1.4) with $\pi_i = n/N$, we have

$$\mathrm{E}[\bar{y}] = \frac{1}{n} \sum_{i=1}^{N} y_i P(I_i = 1)$$

$$= \frac{1}{N} \sum_{i=1}^{N} y_i$$

$$= \mu, \tag{1.9}$$

and $\bar{y}$ is design unbiased for μ. At present we are not interested in finding the variance var$[\bar{y}]$, though this can be obtained readily from (1.5) once we know $\pi_{jk} = \mathrm{E}[I_j I_k]$.

If sampling is with replacement, then some of the labels in the sample could be the same. Suppose that only ν of the labels are distinct, that is, s contains ν distinct elements, and that label i occurs n_i times in the sample $\left(n_i = 0, 1, 2, \ldots, n; \sum_{i=1}^{N} n_i = n\right)$. Here ν is called the *effective sample size*, and it is a random variable. Then

$$\bar{y} = \frac{1}{n} \sum_{i=1}^{n} y_i$$

$$= \frac{1}{n} \sum_{i=1}^{N} y_i n_i.$$

Since each individual has the same probability $1/N$ of being selected each time (as it is returned to the population) and successive selections are independent, we effectively have n binomial trials. Hence n_i has a binomial distribution Bin$(n, 1/N)$ with mean n/N so that

$$\mathrm{E}[\bar{y}] = \frac{1}{n} \sum_{i=1}^{N} y_i \mathrm{E}[n_i] = \mu,$$

and $\bar{y}$ is unbiased for both sampling with or without replacement. We say that such an estimate is design unbiased as the unbiasedness depends on the design used (i.e., random sampling with or without replacement) and not on any assumptions about the y_i values in the population.

In the case of sampling with replacement, another design-unbiased estimator is also available, namely the average of the y's associated with the distinct labels, that is,

$$\bar{y}_\nu = \frac{1}{\nu} \sum_{i=1}^{\nu} y_i$$

$$= \frac{1}{\nu} \sum_{i=1}^{N} y_i I_i.$$

Since ν is random we shall prove unbiasedness by considering first

$$\mathrm{E}[\bar{y}_\nu \mid \nu] = \frac{1}{\nu} \sum_{i=1}^{n} y_i P(I_i = 1 \mid \nu). \qquad (1.10)$$

To evaluate the above probabilities we look at the model as an occupancy model in which exactly ν cells are occupied when n identical balls are dropped at random into N cells. Thus,

$$
\begin{aligned}
P(I_i = 1 \mid \nu) &= P(i \in s \mid \nu) \\
&= \frac{P(i \in s \text{ and } \nu \text{ cells occupied})}{P(\nu \text{ cells occupied})} \\
&= \frac{P(i\text{th cell occupied and } \nu - 1 \text{ other cells occupied})}{P(\nu)}. \qquad (1.11)
\end{aligned}
$$

The model is also equivalent to choosing ν cells to have one ball each (in $\binom{N}{\nu}$ ways) and then dropping the remaining $n - \nu$ balls at random into those ν cells (in $a_{n,\nu}$ ways, say). Suppose that the total number of ways of dropping n balls into N cells is b_n. Then, from (1.11),

$$
\begin{aligned}
P(I_i = 1 \mid \nu) &= \frac{\binom{N-1}{\nu-1} a_{n,\nu}/b_n}{\binom{N}{\nu} a_{n,\nu}/b_n} \\
&= \frac{\binom{N-1}{\nu-1}}{\binom{N}{\nu}} \\
&= \frac{\nu}{N}. \qquad (1.12)
\end{aligned}
$$

Hence from (1.10) we have

$$\mathrm{E}[\bar{y}_\nu \mid \nu] = \frac{1}{\nu} \sum_{i=1}^{N} y_i \frac{\nu}{N} = \mu$$

and

$$\mathrm{E}[\bar{y}_\nu] = \mathrm{E}_\nu \mathrm{E}[\bar{y}_\nu \mid \nu] = \mu,$$

so that $\bar{y}_\nu$ is unbiased. Which of $\bar{y}$ and $\bar{y}_\nu$ is preferred in *rswr*? It turns out, as we shall see later, that

$$\mathrm{var}[\bar{y}_\nu] \le \mathrm{var}[\bar{y}], \qquad (1.13)$$

so that $\bar{y}_{\nu}$ is preferred. This raises a further question about design-unbiased estimators. Is there an unbiased estimator of μ that has the smallest variance of all unbiased estimators? This question is discussed in Chapter 2.

1.8.2 Model-Unbiased Estimators

In ecological studies another concept is useful, namely model-unbiased estimation. Let y_i be the number of animals on the ith plot. Suppose that n plots of area a are now chosen in some purposive (nonrandom) fashion from the population of N plots. However, we now make the assumption that the τ animals in the population are all randomly distributed throughout the population area. We can imagine a process whereby each animal is dropped onto the population area in such a way that the animal has the same chance of falling on any location. The probability that it falls on any particular plot will be the area of the plot divided by the total population area, namely $a/(Na)$, or $1/N$. Because the animals are all "randomly" distributed they can be regarded as being independent of one another. Hence the number y_i falling onto the ith plot will be the outcome of τ binomial trials with probability $1/N$ of "success." We therefore have $y_i \sim \text{Bin}(\tau, 1/N)$ with mean τ/N, and we have

$$\mathrm{E}[\bar{y}] = \frac{1}{n}\sum_{i=1}^{n}\mathrm{E}[y_i] = \frac{\tau}{N} = \mu. \tag{1.14}$$

To find the variance of $\bar{y}$ we need the joint distribution of the y_i, which can be shown to be multinomial. Alternatively, $\sum_{i=1}^{n} y_i$ animals are found on a total sample area of na, and the probability of a single animal being on the sample area is na/Na or n/N. Hence $n\bar{y} \sim \text{Bin}(\tau, n/N)$ and

$$\text{var}[\bar{y}] = \frac{1}{n^2}\text{var}[n\bar{y}]$$

$$= \frac{1}{n^2}\tau\frac{n}{N}\left(1 - \frac{n}{N}\right). \tag{1.15}$$

From (1.14) we see that, once again, $\bar{y}$ is an unbiased estimator of μ. However, the unbiasedness now follows from the binomial (or multinomial) model used. We say that $\bar{y}$ is model unbiased for μ, because the unbiasedness depends on assumptions about the population.

Another model used for randomly distributed animals is the Poisson distribution. If τ and N are both large, then we can approximate the binomial distribution of y_i by the Poisson distribution with mean τ/N, and the y_i are approximately independent. The sum of independent Poisson variables is also Poisson, so that $n\bar{y}$ is approximately Poisson$(n\tau/N)$ and, once again,

$$\mathrm{E}[\bar{y}] = \frac{\tau}{N} = \mu,$$

with

$$\text{var}[\bar{y}] = \frac{1}{n^2}\text{var}[n\bar{y}] = \frac{1}{n^2}\frac{n\tau}{N} = \frac{\tau}{nN}. \tag{1.16}$$

The main disadvantage of model-unbiased estimators is that they assume that some underlying model is true. In addition, we shall see that with adaptive designs, model-unbiased strategies depend on the design as well as on the model.

1.9   UNEQUAL PROBABILITY SAMPLING

In some sampling schemes, including many of the adaptive sampling designs, each unit will have a different probability, called the *inclusion probability*, of being included in the sample. Suppose that unit i has an inclusion probability of π_i. Then as in (1.9) and (1.4) we have

$$\text{E}[\bar{y}] = \frac{1}{n}\sum_{i=1}^{N} y_i P(I_i = 1)$$

$$= \frac{1}{n}\sum_{i=1}^{N} y_i \pi_i, \tag{1.17}$$

which will generally be different from μ so that $\bar{y}$ will be biased. To handle unequal inclusion probabilities in standard types of sampling schemes, a number of estimators have been introduced. Two of these are discussed next, because they give some idea as to how we might construct unbiased estimators in the adaptive situation.

1.9.1   Horvitz–Thompson Estimator

Let π_i $(= P(I_i = 1) = \text{E}[I_i])$ be the probability under the design used where unit i is included in the sample, and assume π_i is strictly greater than zero for every unit in the population. Then the Horvitz–Thompson estimator (Horvitz and Thompson [1952]) of μ is

$$\hat{\mu}_{HT} = \frac{1}{N}\sum_{i=1}^{\nu}\left(\frac{y_i}{\pi_i}\right), \tag{1.18}$$

where $y_1, y_2, \ldots, y_\nu$ represent the y-values from the ν distinct labels in the sample, and sampling may be with or without replacement. If the π_i's are unequal, then the estimator is label dependent, as it depends on specific knowledge of which labels are sampled. Now

$$\hat{\mu}_{HT} = \frac{1}{N}\sum_{i=1}^{N}\frac{y_i I_i}{\pi_i}, \tag{1.19}$$

so that

$$E[\hat{\mu}_{HT}] = \frac{1}{N} \sum_{i=1}^{N} \frac{y_i}{\pi_i} E[I_i]$$

$$= \frac{1}{N} \sum_{i=1}^{N} y_i$$

$$= \mu,$$

and $\hat{\mu}_{HT}$ is unbiased. Using (1.5) with $\pi_{ij} = P(I_i I_j = 1), m = N, z_i = y_i/(\pi_i N)$, and $\pi_{ii} = \pi_i^2$, we have

$$\text{var}[\hat{\mu}_{HT}] = \frac{1}{N^2} \sum_{i=1}^{N} \sum_{j=1}^{N} \left(\frac{\pi_{ij} - \pi_i \pi_j}{\pi_i \pi_j} \right) y_i y_j. \tag{1.20}$$

By (1.6) an unbiased estimate of this is

$$\widehat{\text{var}}[\hat{\mu}_{HT}] = \frac{1}{N^2} \sum_{i=1}^{N} \sum_{j=1}^{N} \left(\frac{\pi_{ij} - \pi_i \pi_j}{\pi_i \pi_j \pi_{ij}} \right) y_i y_j I_i I_j$$

$$= \frac{1}{N^2} \sum_{i=1}^{\nu} \sum_{j=1}^{\nu} \left(\frac{\pi_{ij} - \pi_i \pi_j}{\pi_i \pi_j \pi_{ij}} \right) y_i y_j, \tag{1.21}$$

provided $\pi_{ij} > 0$ for all i and $j, i \neq j$. We note that although the variance estimator (1.21) is unbiased, it can be negative. An alternative estimator, given by Yates and Grundy [1953] and Sen [1953], is invariably nonnegative. However, this estimator is only unbiased for fixed sample sizes (cf. Särndal et al. [1992, Section 2.8]), which is not the case with many adaptive designs. A class of approximately unbiased variance estimators that are always nonnegative has been studied by Brewer and Hanif [1983, p. 68] (cf. Thompson [1992, p. 50]).

We note that $\hat{\mu}_{HT}$ will perform well, that is, have a low variance, if π_i is approximately proportional to y_i. In this case $\hat{\mu}_{HT}$ will be approximately constant and will therefore have little variability. However, if π_i and y_i are unrelated, the estimate could be unreliable. For example, if one unit with a very small selection probability π_i and a large value of y_i happened to be selected, then y_i/π_i could be very large and therefore dominate $\hat{\mu}_{HT}$. For this reason the following modification has been suggested (Hájek [1971]):

$$\mu_{HT}^* = \sum_{i=1}^{\nu} \frac{y_i}{\pi_i} \Big/ \sum_{i=1}^{\nu} \frac{1}{\pi_i}. \tag{1.22}$$

This estimate will be robust to the above problem, though it has a small amount of bias. It can be motivated by assuming, for the moment, that N is unknown (which may be the case in some situations such as forestry where the total number of trees is usually unknown). Then an unbiased estimate of N is

$$\hat{N} = \sum_{i=1}^{\nu} \frac{1}{\pi_i}$$

$$= \sum_{i=1}^{N} \frac{I_i}{\pi_i}.$$

Using this estimate of N then leads to (1.22). Other advantages of μ_{HT}^* over $\hat{\mu}_{HT}$ are described by Särndal et al. [1992, pp. 182–184], who use the theory of ratio estimation to obtain the following approximate expression for the variance:

$$\text{var}[\mu_{HT}^*] \approx \frac{1}{N^2} \sum_{i=1}^{N} \sum_{j=1}^{N} \left(\frac{\pi_{ij} - \pi_i \pi_j}{\pi_i \pi_j} \right) (y_i - \mu)(y_j - \mu). \qquad (1.23)$$

They also give the variance estimator

$$\widehat{\text{var}}[\mu_{HT}^*] = \frac{1}{N^2} \sum_{i=1}^{\nu} \sum_{j=1}^{\nu} \left(\frac{\pi_{ij} - \pi_i \pi_j}{\pi_i \pi_j \pi_{ij}} \right) (y_i - \mu_{HT}^*)(y_j - \mu_{HT}^*). \qquad (1.24)$$

Finally we note for future reference that $\hat{\mu}_{HT}$ can be expressed in the form

$$\hat{\mu}_{HT} = \frac{1}{N} \sum_{i=1}^{N} y_i \frac{I_i}{E[I_i]}. \qquad (1.25)$$

Example 1.7 Simple Random Sampling. In the case of simple random sampling we have $\nu = n$ and, from (1.8), $\pi_i = n/N$. Using a similar argument that led to (1.8), the probability π_{ij} that the ith and jth units are both included in the sample is given by

$$\pi_{ij} = \frac{\binom{N-2}{n-2}}{\binom{N}{n}} = \frac{n(n-1)}{N(N-1)}. \qquad (1.26)$$

Then, from (1.18), $\hat{\mu}_{HT} = \bar{y}$ and, substituting in (1.20),

$$\text{var}[\bar{y}] = \frac{1}{N^2} \left\{ \sum_{i=1}^{N} \left(\frac{N-n}{n} \right) y_i^2 + \sum_{i=1}^{N} \sum_{j \neq i} \left(\frac{n(n-1)N^2}{N(N-1)n^2} - 1 \right) y_i y_j \right\}$$

$$= \frac{1}{N^2} \left\{ a \sum_{i=1}^{N} y_i^2 + b \sum_{i=1}^{N} \sum_{j \neq i} y_i y_j \right\}$$

$$= \frac{1}{N^2} \left\{ a \sum_{i=1}^{N} y_i^2 + b \left(\sum_{i=1}^{N} y_i \right)^2 - b \sum_{i=1}^{N} y_i^2 \right\}$$

$$= \frac{1}{N^2} \left\{ (a - b) \sum_{i=1}^{N} y_i^2 + bN^2 \mu^2 \right\},$$

where

$$a - b = \frac{N - n}{n} - \frac{(n - 1)N}{(N - 1)n} + 1 = \frac{N(N - n)}{n(N - 1)}$$

and

$$b = \frac{(n - 1)N}{(N - 1)n} - 1 = -\frac{N - n}{(N - 1)n}.$$

Hence

$$\text{var}[\bar{y}] = \frac{N - n}{nN(N - 1)} \left\{ \sum_{i=1}^{N} y_i^2 - N\mu^2 \right\}$$

$$= \frac{\sigma^2}{n} \left(1 - \frac{n}{N} \right), \tag{1.27}$$

where

$$\sigma^2 = \frac{1}{N - 1} \sum_{i=1}^{N} (y_i - \mu)^2.$$

In a similar fashion it follows from (1.21) with $\nu = n$ that

$$\widehat{\text{var}}[\bar{y}] = \frac{s^2}{n} \left(1 - \frac{n}{N} \right), \tag{1.28}$$

where

$$s^2 = \frac{1}{n - 1} \sum_{i=1}^{n} (y_i - \bar{y})^2. \qquad \Box$$

Example 1.8 Sampling with Replacement.   Suppose that sampling is now with replacement and that p_i, the probability of selecting the ith unit on a draw, is known. Then the probability that unit i is excluded from the sample is $(1 - p_i)^n$, and the inclusion probability is given by

$$\pi_i = 1 - (1 - p_i)^n. \tag{1.29}$$

Also, the probability of not including either unit i or j in a draw is $1 - p_i - p_j$ (as $\sum_{r=1}^{N} p_r = 1$) so that

$$P\big([I_i = 1] \cup [I_j = 1]\big) = 1 - P\big([I_i \neq 1] \cap [I_j \neq 1]\big) \qquad (1.30)$$
$$= 1 - (1 - p_i - p_j)^n.$$

Hence

$$\begin{aligned}
\pi_{ij} &= P(I_i I_j = 1) \\
&= P\big([I_i = 1] \cap [I_j = 1]\big) \\
&= P(I_i = 1) + P(I_j = 1) - P\big([I_i = 1] \cup [I_j = 1]\big) \qquad (1.31) \\
&= \pi_i + \pi_j - \big[1 - (1 - p_i - p_j)^n\big] \\
&= (1 - p_i)^n + (1 - p_j)^n - \big[1 - (1 - p_i - p_j)^n\big]. \qquad (1.32)
\end{aligned}$$

These values of π_i and π_{ij} can be used in equations (1.20) and (1.21).   □

1.9.2   Hansen–Hurwitz Estimator

If sampling is with replacement, then another estimator has been suggested, called the Hansen–Hurwitz estimator (Hansen and Hurwitz [1943]):

$$\hat{\mu}_{HH} = \frac{1}{nN} \sum_{i=1}^{n} \frac{y_i}{p_i}, \qquad (1.33)$$

where p_i is the probability of selecting the ith unit in a draw. If the p_i's are distinct, then, like $\hat{\mu}_{HT}$, we have another label-dependent estimator. Now

$$\hat{\mu}_{HH} = \frac{1}{nN} \sum_{i=1}^{N} \frac{y_i n_i}{p_i}, \qquad (1.34)$$

where n_i, the number of times unit i is selected, is $\mathrm{Bin}(n, p_i)$. Hence

$$\hat{\mu}_{HH} = \frac{1}{N} \sum_{i=1}^{N} y_i \frac{n_i}{E[n_i]} \qquad (1.35)$$

and

$$E[\hat{\mu}_{HH}] = \frac{1}{N} \sum_{i=1}^{N} y_i = \mu,$$

so that $\hat{\mu}_{HH}$ is unbiased. Now the n_i have a multinomial distribution as there are n trials and N categories with p_i the probability of being in the ith category ($i = 1, 2, \ldots, N$). Hence $E[n_i] = np_i$, $\text{var}[n_i] = np_i(1 - p_i)$ and $\text{cov}[n_i n_j] = -np_i p_j$ ($i \neq j$). We can now find from (1.34) and $\sum_i p_i = 1$ that

$$
\begin{aligned}
\text{var}[\hat{\mu}_{HH}] &= \frac{1}{n^2 N^2} \left\{ \sum_{i=1}^{N} \frac{y_i^2}{p_i^2} \text{var}[n_i] + \sum_{i=1}^{N} \sum_{j \neq i} \frac{y_i y_j}{p_i p_j} \text{cov}[n_i, n_j] \right\} \\
&= \frac{1}{nN^2} \left\{ \sum_{i=1}^{N} \frac{y_i^2}{p_i}(1 - p_i) - \sum_{i=1}^{N} \sum_{j \neq i} y_i y_j \right\} \\
&= \frac{1}{nN^2} \left\{ \sum_{i=1}^{N} \frac{y_i^2}{p_i} - \left(\sum_{i=1}^{N} y_i \right)^2 \right\} \\
&= \frac{1}{n} \sum_{i=1}^{N} p_i \left(\frac{y_i}{Np_i} - \mu \right)^2.
\end{aligned}
\tag{1.36}
$$

Unfortunately this approach does not lend itself readily to finding an unbiased estimate of $\text{var}[\hat{\mu}_{HH}]$. Another, more general, approach is to write $\hat{\mu}_{HH}$ as a sample mean of independently and identically distributed random variables as follows.

Let T be a random variable taking the value y_i/Np_i with probability p_i ($i = 1, 2, \ldots, N$). Then

$$
\mu_T = E[T] = \sum_{i=1}^{N} \frac{y_i}{Np_i} p_i = \mu
$$

and

$$
\sigma_T^2 = \text{var}[T] = \sum_{i=1}^{N} \left(\frac{y_i}{Np_i} - \mu \right)^2 p_i.
$$

Since we have sampling with replacement, the n selections are independent and we can write

$$
\hat{\mu}_{HH} = \frac{1}{n} \sum_{i=1}^{n} t_i = \bar{t},
$$

where the t_i are a random sample from the distribution of T. Hence

$$
E[\hat{\mu}_{HH}] = E[\bar{t}] = \mu_T = \mu
$$

and

$$\text{var}[\hat{\mu}_{HH}] = \frac{\sigma_T^2}{n} = \frac{1}{n} \sum_{i=1}^{N} \left(\frac{y_i}{Np_i} - \mu \right)^2 p_i. \tag{1.37}$$

An unbiased estimate of $\text{var}[\hat{\mu}_{HH}]$ is then

$$\widehat{\text{var}}[\hat{\mu}_{HH}] = \frac{s_T^2}{n}$$

$$= \frac{1}{n(n-1)} \sum_{i=1}^{n} (t_i - \bar{t})^2$$

$$= \frac{1}{n(n-1)} \sum_{i=1}^{n} \left(\frac{y_i}{Np_i} - \hat{\mu}_{HH} \right)^2. \tag{1.38}$$

If we have the usual random sample with replacement, that is, $p_i = 1/N$, then $\hat{\mu}_{HH} = \bar{y}$ and, by (1.37),

$$\text{var}[\bar{y}] = \frac{1}{nN} \sum_{i=1}^{N} (y_i - \mu)^2.$$

From (1.38) we have

$$\widehat{\text{var}}[\bar{y}] = \frac{1}{n(n-1)} \sum_{i=1}^{n} (y_i - \bar{y})^2. \tag{1.39}$$

In this case the estimator is label independent.

1.10 DISCRETE UNIFORM DISTRIBUTION

In the next chapter, we develop some statistical theory for survey designs. As the theory has some features peculiar to sample surveys the reader may find the following example a helpful introduction to some of the ideas there.

Let X be a discrete random variable with probability function

$$P_\theta(x) = \begin{cases} \frac{1}{\theta}, & x = 1, 2, \ldots, \theta; \\ 0, & \text{otherwise,} \end{cases}$$

where θ is an unknown positive integer. Then

$$E[X] = \frac{1}{\theta} \sum_{x=1}^{\theta} x = \frac{\theta + 1}{2}. \tag{1.40}$$

If $X_1, X_2, \ldots, X_n$ represent a random sample from the above distribution, then the likelihood function for θ, that is, the joint probability function of the X_i's, is

$$L(\theta; x_1, x_2, \ldots, x_n) = \prod_{i=1}^{n} P_\theta(X_i = x_i)$$

$$= \begin{cases} \dfrac{1}{\theta^n}, & \theta \geq x_{\max}; \\ 0, & \theta < x_{\max} \end{cases}$$

where $x_{\max} = \max(x_1, x_2, \ldots x_n)$. In sample survey theory discussed in Chapter 2, we shall find that the range of θ for the likelihood function also depends on the data. Furthermore, in the designs of interest to us, the nonzero probability is independent of θ, for example, $1/\theta^n$ is replaced by a constant.

When it comes to inferences about θ, we note that once we know $x_{\max}$, the other x_i's tell us nothing more about θ and we say that $X_{\max}$ is sufficient for θ. If $Y = X_{\max}$, we can find the distribution of Y using distribution functions or directly as follows. If y is a positive integer ($y \leq \theta$), then

$$P_\theta(Y = y) = \sum_{k=1}^{n} P_\theta(k \, X_i\text{'s equal to } y, n - k \, X_i\text{'s less than } y)$$

$$= \sum_{k=1}^{n} \binom{n}{k} \left(\frac{1}{\theta}\right)^k \left(\frac{y-1}{\theta}\right)^{n-k}$$

$$= \left(\frac{1}{\theta} + \frac{y-1}{\theta}\right)^n - \left(\frac{y-1}{\theta}\right)^n$$

$$= \left(\frac{y}{\theta}\right)^n - \left(\frac{y-1}{\theta}\right)^n. \tag{1.41}$$

If we consider the conditional probability

$$P_\theta(X_1 = x_1, X_2 = x_2, \ldots, X_n = x_n | Y = y),$$

then this probability will be nonzero for all $x_1, x_2, \ldots, x_n$ taking values from 1 to y, with at least one x_i equal to y. Since these combinations are all equally likely, we see that the conditional probability does not depend on θ. (In fact it is equal to $1/[y^n - (y-1)^n]$.) This property is used to define what we mean by Y being sufficient for θ.

Unfortunately Y is not an unbiased estimate of θ. For example, when $n = 2$, $E[Y] = (\theta + 1)(4\theta - 1)/6\theta$. However, we find that we can use Y to "improve" upon unbiased estimators. For example, from (1.40), a trivial unbiased estimator of θ is

$2X_1 - 1$. Then, using similar arguments that led to (1.41), we find that

$$P_\theta(X_1 = x \text{ and } Y = y) = \begin{cases} \frac{1}{\theta}\left[\left(\frac{y}{\theta}\right)^{n-1} - \left(\frac{y-1}{\theta}\right)^{n-1}\right], & x < y; \\ \frac{1}{\theta}\left(\frac{y}{\theta}\right)^{n-1}, & x = y. \end{cases}$$

Hence

$$E[X_1|Y = y] = \sum_{x=1}^{y} xP_\theta(X_1 = x|Y = y)$$

$$= \sum_{x=1}^{y} x\frac{P_\theta(X_1 = x \text{ and } Y = y)}{P_\theta(Y = y)}$$

$$= \frac{1}{2}\left\{\frac{y^{n+1} + y^n - (y-1)^{n+1} - (y-1)^n}{y^n - (y-1)^n}\right\}$$

$$= g(y), \quad \text{say.}$$

From $EE[X_1|Y = y] = E[X_1]$ we see that

$$h(y) = 2g(y) - 1 = \frac{y^{n+1} - (y-1)^{n+1}}{y^n - (y-1)^n}$$

is an unbiased estimate of θ. Since $E[\bar{X}|Y = y] = E[X_1|Y = y]$, we find that we arrive at $h(y)$ again if we start with $\bar{X}$ instead of X_1. We shall see in the next chapter that if our original unbiased estimator is not a function of the sufficient statistic (here $\bar{X}$ is not a function of Y), then $h(y)$ will have a smaller variance. This technique of finding a sufficient statistic and then deriving the conditional expected value of an unbiased estimator to obtain an improved unbiased estimator is used extensively throughout this book.

CHAPTER 2

Fixed-Population Sampling Theory

2.1 THE SAMPLING SITUATION

For the finite population sampling situation, there is a population of N units $(u_1, \ldots, u_N)$ indexed by their labels $(1, \ldots, N)$. With unit i is associated a variable of interest y_i, for $i = 1, \ldots, N$. The vector of population y-values may be written $\mathbf{y} = (y_1, \ldots, y_N)'$.

In the fixed population view, the population vector $\mathbf{y}$ is considered a fixed set of unknown constants, and we may write $\boldsymbol{\theta} = \mathbf{y}$ to represent the values as a vector of unknown parameters. In the "model-based" or "superpopulation" view, the vector $\mathbf{y}$ is considered a realization of a random vector $\mathbf{Y}$ having a joint distribution F, which may in turn depend on an unknown parameter $\boldsymbol{\phi}$ ranging in a parameter space Φ. Assuming the distribution F has a density f, the population model is given by $f(\mathbf{y}; \boldsymbol{\phi})$ with $\boldsymbol{\phi} \in \Phi$. In a Bayesian viewpoint $\boldsymbol{\phi}$ has a known prior distribution. This chapter will consider the fixed population approach while approaches incorporating stochastic population models will be considered in Chapter 3.

An ordered sample of size n is a sequence $s_O = (i_1, \ldots, i_n)$ of n of the labels (some of which may be the same, as in sampling with replacement). The set of all possible samples is denoted by $\mathcal{S}$. The set of possible values of $\mathbf{y}$, the vector of population y-values, is denoted by $\mathcal{Y}$.

The objective is to select a sample, observe the y-values for the units in the sample, and estimate some function $z(\mathbf{y})$ of the population y-values. Typical examples of population quantities to be estimated include the population total $z(\mathbf{y}) = \sum_{i=1}^{N} y_i = \tau$, the population mean $z(\mathbf{y}) = (1/N) \sum_{i=1}^{N} y_i = \mu$, and the finite-population variance $z(\mathbf{y}) = \sum_{i=1}^{N} (y_i - \mu)^2 / (N - 1) = \sigma^2$.

The *data* d_O consist of the y-values for the units in the sample, together with the associated unit labels, namely the ordered pairs

$$d_O = \big((i_1, y_{i_1}), (i_2, y_{i_2}), \ldots, (i_n, y_{i_n})\big) = ((i, y_i) : i \in s_O).$$

The notation may be shortened to $d_O = (s_O, \mathbf{y}_O)$, where $\mathbf{y}_O$ is the set of ordered sample y-values; that is, $\mathbf{y}_O = (y_i : i \in s_O)$. We will also be interested in s, consisting of the reduced set of $\nu (= \nu(s))$ *distinct* labels in s_O but uniquely ordered from the smallest to the largest label. If $\mathbf{y}_s$ is the ordered set of corresponding y-values (in the same order as s) then, given s_O, we only need $\mathbf{y}_s$ to provide all the information about d_O. To see this, we get s from s_O and then link up the y-values in $\mathbf{y}_s$ with the correct labels. Finally we can match the y-values with the labels in s_O to get $\mathbf{y}_O$. In some cases it will be convenient to redefine d_O to be equal to $(s_O, \mathbf{y}_s)$ as both representations contain exactly the same information. We do this in Chapter 3.

We note that the values of one or more auxiliary variables x_i may be known for every unit in the population. The matrix $\mathbf{X}$ of these values is then a known function of the unit labels, so that in what follows the auxiliary information is implicit in the unit labels. Auxiliary information may be used at both the design and inference stages.

The *sampling design* is the procedure by which the sample is selected. The sampling design may be specified as a conditional probability $p(s_O \mid \mathbf{y})$ of selecting the sample s_O for each possible sample $s_O \in \mathcal{S}$. The selection probability may depend on the configuration $\mathbf{y}$ of y-values in the population. The design probabilities satisfy $p(s_O \mid \mathbf{y}) \geq 0$ for all $s_O \in \mathcal{S}$ and $\sum_{s_O \in \mathcal{S}} p(s_O \mid \mathbf{y}) = 1$ for all $\mathbf{y} \in \mathcal{Y}$.

Designs for which $p(s_O \mid \mathbf{y}) = p(s_O)$ will be referred to as *conventional* designs. With such designs, the selection probabilities do not depend on any values of the variable of interest or on any parameter values. The selection probabilities may, however, depend on any auxiliary variables $\mathbf{x}$ known for the population; this dependence will be left implicit rather than writing $p(s_O \mid \mathbf{x})$. With a conventional design the entire sample of labels may be selected prior to the survey. Examples of conventional selection procedures include standard purposive selection, simple random sampling, stratified random sampling, unequal probability designs such as probability-proportional-to-x sampling, systematic, cluster, and multistage designs, certain double sampling designs, network sampling, and many other designs.

In this book, special interest focuses on designs for which $p(s_O \mid \mathbf{y}) = p(s_O \mid \mathbf{y}_s)$; that is, designs in which the procedure for selecting units may depend on values of the variable of interest, but only through units included in the sample. This class includes random sampling with a sequential stopping rule, adaptive cluster sampling designs, adaptive allocation in stratified designs, optimal Bayes designs, and many other possibilities. As a whole we refer to this class as *adaptive designs*. In addition to designs which deliberately make use of sample values, the class includes sampling situations in which the selection procedure is unintentionally or unavoidably sample-response dependent. It is often useful to view the class of conventional designs as a special case of the adaptive designs, in which $p(s_O \mid \mathbf{y}_s) = p(s_O)$. The class of conventional and adaptive procedures together will be termed the *standard* designs.

Designs of the form $p(s_O \mid \mathbf{y})$, in which selection probabilities are influenced by values of units not included in the sample or by unknown parameter values will be termed *nonstandard*. The desirability of deliberately implementing nonstandard procedures is open to question. However, nonstandard selection procedures may be inherently involved in the way data are obtained. Examples can include "representative" samples selected purposively by experts: A botanist looks over the vegetation in a study region and selects what he/she perceives as typical plots for inclusion in the sample; the choice is made based partly on observations of the variable of interest and auxiliary variables in units not included in the sample. Sometimes the intention is to select the most interesting rather than the most representative units, after observing the variable of interest in a larger set of units. For example, in selecting sites at which to set up pollution monitoring devices to determine if problems exist in a region, field workers look throughout the study region and select sites at which high pollution is apparent.

Mapping of ground water pollution levels is often based on linear unbiased prediction (kriging) methods using measured levels at existing wells. Suppose some wells have already been abandoned because of extremely low water quality and that no record exists of those sites. Then the selection procedure is nonstandard and the unbiasedness of the mapping methods may be compromised. Whenever the sample selected is based on observations of a larger set of units, as when the editor of a newspaper selects letters for publication, the design is nonstandard. An example in which a manufacturer selects a sample of products for testing by a regulatory agency by first observing a larger number of the products and selecting the best is described in Chapter 3. Certain "size-biased" sampling situations examined in Chapter 3, such as surveys of groups of animals in which the detection probability increases with group size, are nonstandard in that the selection probability for a given sample can depend on y-values outside the sample as well as within it. (With conventional unequal probability sampling designs, the probability of selecting a given sample depends only on the inclusion probabilities involving units in the sample.) Nonstandard selection processes also occur when the people designing the survey use auxiliary information about the population which is not available to the people analyzing the survey (Scott [1977]).

Conceivably, nonstandard design situations could exist in which the selection probability $p(s_O \mid \mathbf{y}; \boldsymbol{\phi})$ depends also on an unknown parameter $\boldsymbol{\phi}$, as when prior information about the population model is known to the designer of the survey but not the analyzer (Berger and Wolpert [1988]). When nonstandard selection influences are known to arise in a survey, whether deliberately or inadvertently, then inference based on the data needs to take the nonstandard selection procedure into account, as will be seen in the sections of this chapter on likelihood and expectations.

Terminology for the different types of designs has been used somewhat inconsistently in the literature. The term "sequential" has been used variously to describe algorithms for selecting conventional samples, for only those designs in which the selection is based unit by unit on the values of the units already in the sample, as well as for any design depending on sample y-values. Zacks [1969] uses the term "single phase" to describe designs not depending on y-values, and "sequential" to describe designs depending on y-values in the sample. Cassel et al. [1977] refer to conventional designs as "noninformative" and designs which depend in any way on y-values as "informative," so that "noninformative" designs would include only conventional procedures. Berger and Wolpert [1988], in discussing stopping rules for sequential designs, reserve the term "informative" for designs depending on y-values outside the sample or on unknown parameter values, so that the "noninformative" designs would include adaptive as well as conventional procedures. Basu [1969] and others use the term "informative" in a different sense, describing one sample as more "informative" than another. Rubin [1987] uses the term "confounded" to describe designs depending on y-values in any way and "ignorable" to characterize designs not depending on y-values or parameters outside the sample, since from a Bayesian or likelihood point of view such designs can be ignored at the inference stage. With this terminology, conventional and adaptive designs would be ignorable, and adaptive and nonstandard designs would be confounded.

2.2 NOTATION

In the fixed-population view the parameter of interest is $\boldsymbol{\theta} = (y_1, y_2, \ldots, y_N)'$, and functions of it such as the mean μ and total $\tau (= N\mu)$ are typical objects of inference. We assume that $\boldsymbol{\theta} \in \Theta$, a subset of $\mathbb{R}^N$. We defined the ordered sample s_O to be the sequence of labels $(i_1, i_2, \ldots, i_n)$, and we introduced the reduced sequence of labels s. The order of s_O is usually provided by the order in which the sample is selected, while the reduced sample s is arbitrarily arranged in ascending label order for uniqueness. In what follows it is helpful to consider the unordered *reduced set* $s_R = \{i_1, i_2, \ldots, i_\nu\}$ of the ν distinct labels in the sample. Here s_R contains less information than s_O, which is particularly appropriate when repeated labels are possible. However, s and s_R contain the same information in that they have the same components, unordered in s_R and arbitrarily ordered in s. The finite-population, repeated sampling view received early articulation from Neymann [1934], Godambe [1955], and other authors and is reviewed extensively in recent works including Cassel et al. [1977], Chaudhuri and Stenger [1992], and Hedayat and Sinha [1991].

A sampling design is a probability function $p_\theta(s_O)$ [$= p(s_O \mid \mathbf{y})$ in Section 2.1] which assigns a probability for every possible sample s_O. Where appropriate, s_O can be replaced by s_R (or equivalently by s). In a conventional design the process for selecting the units does not depend on the associated y-values; that is, $p_\theta(s_O)$ does not depend on $\boldsymbol{\theta}$ and we write $p(s_O)$. For example, in simple random sampling $p(s_R) = 1/\binom{N}{n}$ as all $\binom{N}{n}$ unordered samples of n distinct labels are equally likely. However, in adaptive sampling we noted in the previous section that the probability of obtaining a particular sample can depend on $\boldsymbol{\theta}$ or, more strictly, on the y-values associated with the labels in s.

Summarizing the notation for this section, the data usually consist of both the y-values and their associated unit labels. The data in original order are denoted $d_O = ((i, y_i) : i \in s_O)$. The reduced data ordered from smallest to largest label in the sample are $d = (s, \mathbf{y}_s)$. As noted in Section 2.1, other definitions for d_O are $d_O = (s_O, \mathbf{y}_s)$ and $d_O = (s_O, \mathbf{y}_O)$, the latter being used in Chapters 3 and 10. All three definitions are equivalent in that they contain exactly the same information. The reduced data expressed as an unordered set are $d_R = \{(i, y_i) : i \in s_R\}$. We shall use d and d_R somewhat interchangeably. However, we shall use d when we want to think of the reduced data as a vector rather than a set. The reduction from d_O to d_R or d is achieved by ignoring information about order and the multiplicities of the units in the sample. We denote the reduction by the function r_d, so that $d_R = r_d(d_O)$. For notational simplicity we shall always use $r(.)$, with an appropriate subscript, to denote any such reduction, for example, $\mathbf{y}_R = r_y(\mathbf{y}_O)$ and $s_R = r_s(s_O)$. We note that $\mathbf{y}_s$, on its own, provides essentially the same information as $\mathbf{y}_R$. However, $\mathbf{y}_s$ cannot be constructed from the data without information about the labels.

In this chapter, random variables will usually be denoted by capital letters so that D_O takes the value of d_O, for example. Thus, we can write $P_\theta(D_O = d_O)$, where P denotes "probability."

A parameter vector $\boldsymbol{\theta}$ is said to be *consistent* with a data value $d = (s, \mathbf{y}_s)$ if the ith component of $\mathbf{y}_s$ equals the s_ith component, say, of $\boldsymbol{\theta}$ for all units i in the sample. That is, $\boldsymbol{\theta}$ and d are consistent if the y and θ values coincide for all sample units. Consistency depends only on the values of the distinct units in the sample, not on the order or multiplicity of selection. Therefore, if $\boldsymbol{\theta}$ is consistent with the reduced data d (or equivalently with the reduced data d_R), it will also be consistent with the original data d_O. We associate with every d a subset Θ_d of Θ whose vectors are consistent with d. Since consistency depends only on the values of the distinct units in the sample, consistency with d_O is equivalent to consistency with the reduced data d, and

$$\Theta_{d_O} = \Theta_d = \Theta_{d_R}. \tag{2.1}$$

Also, if d_O and d_O^* are different data values with $\Theta_{d_O} = \Theta_{d_O^*}$, then d_O and d_O^* have the same reduction; that is, $r_d(d_O) = r_d(d_O^*) = d_R$.

Example 2.1 Notation. Suppose that $n = 4$ units are selected by a with-replacement design from a population of $N = 50$ units, and the data in the order selected are

$$d_O = \big((37, 4.2), (45, 9.3), (37, 4.2), (11, 7.3)\big).$$

The first unit selected was unit number 37, the value of the variable of interest associated with that unit was 4.2, and so on. The sample of units in order selected is

$$s_O = (37, 45, 37, 11).$$

Since unit number 37 was selected twice, the reduced sample s, which is arranged in ascending unit order, is

$$s = (11, 37, 45).$$

The y-values of the reduced sample, arranged in the order corresponding to the units in s, are

$$\mathbf{y}_s = (7.3, 4.2, 9.3).$$

In some developments it will be appropriate to view s and $\mathbf{y}_s$ as column rather than row vectors, but in most instances the distinction will not be crucial for clarity.

The unordered reduced sample s_R, on the other hand, is the set consisting of the three elements 37, 45, and 11. The unordered reduced data d_R are the set consisting of the three pairs $(37, 4.2)$, $(45, 9.3)$, and $(11, 7.3)$. Thus, d_R and d contain the same information, but in d the elements are ordered uniquely, which will be convenient in some of the developments that follow.   □

Example 2.2   ***Notation.***   Suppose the population has $N = 4$ units, from which a sample of $n = 2$ units are chosen without replacement, and the data are $d = (s, \mathbf{y}_s)$, where $s = (1, 3)$ and $\mathbf{y}_s = (3.1, 2.6)$. Let the parameter space Θ be $\mathbb{R}^4$. Then the set of parameter vectors consistent with the data is

$$\Theta_d = \{\boldsymbol{\theta} : \theta_1 = 3.1, \ -\infty \le \theta_2 \le \infty, \ \theta_3 = 2.6, \ -\infty \le \theta_4 \le \infty\}.$$

Here θ_1 and θ_3 are fixed by the sample values, but θ_2 and θ_4 can vary.   □

2.3 LIKELIHOOD FUNCTION

Consider any adaptive or conventional design $p(s_O \,|\, \mathbf{y}_s)$. The selection probability does not depend on any values of the variable of interest for units outside the sample. The probability may depend on values of the variable of interest within the sample and may depend on order of selection. With such a design the probability of obtaining a specific data value $d_O = (s_O, \mathbf{y}_s)$ is $P_\theta[D_O = (s_O, \mathbf{y}_s)]$. The probability is zero for any value of $\boldsymbol{\theta}$ not consistent with d_O. If $\boldsymbol{\theta}$ is consistent with d_O, i.e. $\boldsymbol{\theta} \in \Theta_d$, then d_O is obtained if and only if the ordered sample s_O is selected. Hence, in this case, the probability of obtaining d_O is simply the probability $p(s_O \,|\, \mathbf{y}_s)$ under the design of obtaining the sample s_O. Since this probability does not depend on unobserved components of $\boldsymbol{\theta}$, it is constant as a function of $\boldsymbol{\theta}$ for all $\boldsymbol{\theta}$ in Θ_{d_O} $[= \Theta_{d_R} = \Theta_d$, by (2.1)]. We can therefore write

$$P_\theta(D_O = d_O) = \begin{cases} p(s_O \,|\, \mathbf{y}_s), & \boldsymbol{\theta} \in \Theta_{d_R}, \\ 0, & \boldsymbol{\theta} \notin \Theta_{d_R}, \end{cases}$$

$$= p(s_O \,|\, \mathbf{y}_s) I_{\Theta_{d_R}}(\boldsymbol{\theta}), \qquad \boldsymbol{\theta} \in \Theta, \tag{2.2}$$

where d_R (or equivalently $d = (s, \mathbf{y}_s)$) is the reduction of d_O, and $I_A(\boldsymbol{\theta})$ is an indicator function taking the value 1 whenever $\boldsymbol{\theta} \in A$ and 0 otherwise.

Viewing the above probability distribution as the likelihood function of the parameter $\boldsymbol{\theta}$, it is convenient to re-express $I_{\Theta_{d_R}}(\boldsymbol{\theta})$ as $I(d_R, \boldsymbol{\theta})$ to emphasize that the latter is a function of d_R and $\boldsymbol{\theta}$, as follows:

$$L(\boldsymbol{\theta}; d_O) = P_\theta(D_O = d_O)$$

$$= p(s_O \,|\, \mathbf{y}_s) I(d_R, \boldsymbol{\theta}). \tag{2.3}$$

Given the value of d_O, $p(s_O \,|\, \mathbf{y}_s)$ is fixed and the likelihood takes only two values. Therefore the likelihood is maximized when $\boldsymbol{\theta} \in \Theta_{d_R}$ so that *any* value of $\boldsymbol{\theta}$ consistent with the data will be a maximum likelihood estimator. We note that we can replace d_R by d in the above equations.

The flatness of the likelihood function in the fixed-population survey sampling situation was pointed out by Godambe [1966] and Basu [1969]. Note that the flatness applies to any conventional or adaptive design, but not necessarily to a nonstandard design, for which the likelihood function is $L(\boldsymbol{\theta}; d_O) = p(s_O \mid \mathbf{y}_s, \mathbf{y}_{\bar{s}})I(d_R, \boldsymbol{\theta})$. Here $\bar{s}$ is the set of labels, ordered from the smallest to the largest, corresponding to the unobserved labels.

2.4 SUFFICIENT STATISTICS

A helpful concept both for data reduction and for developing estimators in sampling is that of sufficiency. In general terms, a statistic $W = g(D_O)$ that is a function of the data is said to be sufficient for $\boldsymbol{\theta}$ if the conditional distribution of D_O given W does not depend on $\boldsymbol{\theta}$, assuming that the conditional distribution is well defined. In practice this means that once we know W, then D_O provides no additional information when it comes to estimation. A formal definition is:

Definition $W = g(D_O)$ is sufficient for $\boldsymbol{\theta}$ if $P_\theta(D_O = d_O \mid W = w)$ is independent of $\boldsymbol{\theta}$ for all $\boldsymbol{\theta}$ such that $P_\theta(W = w) > 0$. □

Technically, the statistic W is sufficient for the family P_θ, $\boldsymbol{\theta} \in \Theta$.

We now show that the reduction $D_R = r_d(D_O)$ is sufficient for $\boldsymbol{\theta}$ for adaptive as well as conventional designs.

Theorem 2.1. For any conventional or adaptive design, namely, one satisfying (2.2), D_R is sufficient for $\boldsymbol{\theta}$.

Proof: Let Ω_{d_R} be the set of all $\boldsymbol{\theta}$ for which $P_\theta(D_R = d_R) > 0$. For $\boldsymbol{\theta} \in \Omega_{dR}$, the definition of conditional probability gives

$$P_\theta(D_O = d_O \mid D_R = d_R) = \frac{P_\theta(D_O = d_O \text{ and } D_R = d_R)}{P_\theta(D_R = d_R)}. \tag{2.4}$$

We note that in (2.4), D_R is the reduction of D_O, but d_R is not necessarily the reduction of d_O. Now $P_\theta(D_R = d_R) > 0$ implies both that $p(s_R) > 0$ and that $\boldsymbol{\theta}$ is consistent with d_R. Thus $\boldsymbol{\theta} \in \Omega_{d_R}$ implies $\boldsymbol{\theta} \in \Theta_{d_R}$, so that $\Omega_{d_R} \subset \Theta_{d_R}$.

Let $A = \{s'_O : r_d(s'_O) = s_R\}$, the set of all s'_O which reduce to s_R. Then, for $\boldsymbol{\theta} \in \Omega_{d_R}$,

$$P_\theta(D_R = d_R) = \sum_{s'_O \in A} p(s'_O \mid \mathbf{y}_s), \tag{2.5}$$

since $\mathbf{y}_s$, which is determined by s_R, is the same for all the s'_O. Thus, for $\theta \in \Omega_{d_R}, P_\theta(D_R = d_R)$ does not depend on θ.

Now for $\boldsymbol{\theta} \in \Omega_{d_R}$ (so that $\boldsymbol{\theta} \in \Theta_{d_R}$),

$$P_\theta(D_O = d_O \text{ and } D_R = d_R) = P_\theta(D_O = d_O)$$
$$= p(s_O \mid \mathbf{y}_s). \qquad (2.6)$$

when d_R is the reduction of d_O (so that $\Theta_{d_R} = \Theta_{d_O}$) and $P_\theta(D_O = d_O \text{ and } D_R = d_R) = 0$ when d_R is not the reduction of d_O.

Therefore, when $\boldsymbol{\theta} \in \Omega_{d_R}$

$$P_\theta(D_O = d_O \mid D_R = d_R) = \begin{cases} \dfrac{p(s_O \mid \mathbf{y}_s)}{\sum_{s_O' \in A} p(s_O' \mid \mathbf{y}_s)}, & d_R = r_d(d_O), \\ 0, & d_R \neq r_d(d_O), \end{cases}$$

which does not depend on $\boldsymbol{\theta}$; that is, is constant as a function of $\boldsymbol{\theta}$ for all $\boldsymbol{\theta} \in \Omega_{d_R}$.

$\square$

The preceding proof is based directly on the definition of sufficiency. Alternatively, from (2.3), we have

$$P_\theta(D_O = d_O) = p(s_O \mid \mathbf{y}_s) I(d_R, \boldsymbol{\theta}),$$

which represents a factorization into two functions, one involving only the data and the other involving the sufficient statistic and the parameter. Sufficiency then follows directly by the Factorization Theorem (Lehmann [1983]).

We have proved in the preceding theorem that the reduced data d_R (or d) are sufficient for any conventional or adaptive design. The result in this generality was presented in Basu [1969]. Discussions of the sufficiency of d_R for conventional designs are found in Cassel et al. [1977, pp. 35–39] and Chaudhuri and Stenger [1992, pp. 17–18]. As an example of a conventional design, consider simple random sampling, with order of selection retained in the original data. Then $p(s_O) = 1/[N(N-1)\dots(N-n+1)]$, which is constant when $\boldsymbol{\theta} \in \Theta_{d_O}(= \Theta_{d_R})$, and the criterion (2.2) for the above theorem to hold is satisfied. Also $p(s_R) = p(s_O) \times n! = 1/\binom{N}{n}$, and any statistic based on the order of the sample rather than on the unordered elements in s_R will not be sufficient for $\boldsymbol{\theta}$.

As an example of an adaptive design, suppose we use simple random sampling and continue sampling until a y-value exceeds some prechosen level, say c. Then n is random and the probability of getting a particular sample will depend on the y-values in the sample and therefore on the value of $\boldsymbol{\theta}$. However, the condition (2.2) states that $P_\theta(D_O = d_O)$ can depend on $\boldsymbol{\theta}$ but only through the definition of Θ_{d_R}. For example, suppose that just the first unit in the population has a y-value exceeding c, that is, $y_1 > c$. Then $p_\theta(s_O) = 1/N$ for $s_O = (1)$; $p_\theta(s_O) = 1/[N(N-1)\dots(N-k)]$ for

$s_O = (i_1, i_2, \ldots, i_k, 1)$, where $i_j \neq 1$ for $j = 1, 2, \ldots, k$, and $k = 1, 2, \ldots, N - 1$; and $p_\theta(s_O) = 0$ otherwise. If θ was such that $y_2 > c$ instead of y_1, then $p_\theta(s_O) = 0$ for $s_O = (1)$ as θ is no longer compatible with the sample consisting of just the first unit. Hence for each d_O (and corresponding s_O) there is a Θ_{d_R} such that $p_\theta(s_O)$ does not depend on θ for $\theta \in \Theta_{d_R}$ and $p_\theta(s_O) = 0$ for $\theta \notin \Theta_{d_R}$.

For some statistical experiments there may exist more than one sufficient estimator. Since sufficiency is a good property for estimators, we would like to have a method that selects the "best" sufficient estimator. A theorem which helps us take the first step in this direction follows.

Theorem 2.2. If W is sufficient for θ and W is a function of Z, say, $W = W(Z)$, then Z is also sufficient.

Proof: Suppose that $P_\theta(Z = z) > 0$; then

$$P_\theta\big(W = h(z)\big) \geq P_\theta(Z = z) > 0,$$

since more than one value of z may lead to the same value of $h(z)$. Hence $P(D_O = d_O \mid W = h(z))$ is well defined and is independent of θ, by the definition of sufficiency. Now, since z is a function of the data, z is wholly determined by d_O so that

$$
\begin{aligned}
P_\theta(D_O = d_O \mid Z = z) &= \frac{P_\theta(D_O = d_O \text{ and } Z = z)}{P_\theta(Z = z)} \\
&= \frac{P_\theta(D_O = d_O)}{P_\theta(Z = z)} \\
&= \frac{P_\theta[D_O = d_O \mid W = h(z)] P_\theta[W = h(z)]}{P_\theta[Z = z \mid W = h(z)] P_\theta[W = h(z)]} \\
&= \frac{P[D_O = d_O \mid W = h(z)]}{P_\theta[Z = z \mid W = h(z)]}.
\end{aligned}
$$

Also

$$P_\theta[Z = z \mid W = h(z)] = \sum_{d_O \in B} P[D_O = d_O \mid W = h(z)], \tag{2.7}$$

where $B = \{d_O : Z(d_O) = z\}$ and (2.7) does not depend on θ. Hence $P_\theta(D_O = d_O \mid Z = z)$ does not depend on θ for all θ such that $P_\theta(Z = z) > 0$, and Z is sufficient for θ. □

We note that if h is a one-to-one function, then $W = w [= h(z)]$ if and only if $Z = z$. Thus, Z is, in a sense, equivalent to W in providing information about θ. If h is not a one-to-one function, then W can be regarded as a reduction of Z as W takes fewer distinct values than Z. If W can be expressed in the form $W = h(Z)$ for every sufficient statistic Z, then W represents the "strongest" possible reduction of the original data d_O that is still sufficient. To formalize this idea of reduction and to get around the problem of indeterminacy with regard to one-to-one functions of W, it is more convenient to use the idea of a partition as follows.

Let $\mathcal{D}_O$ be the sample space of D_O. A partition of $\mathcal{D}_O$ is a collection of mutually exclusive subsets of $\mathcal{D}_O$ such that their union is $\mathcal{D}_O$. With any statistic $W = g(D_O)$, we can induce a partition $\mathcal{P}_W$ of $\mathcal{D}_O$ such that each set in $\mathcal{P}_W$ is the inverse image of a value of W. Then two outcomes, d_{O_1} and d_{O_2}, belong to the same partition set of $\mathcal{P}_W$ if $g(d_{O_1}) = g(d_{O_2})$. Furthermore, any one-to-one function of W will induce the same partition as W, namely $\mathcal{P}_W$.

Definition   A statistic $W_1 = g_1(D_O)$ is said to be minimal sufficient for $\boldsymbol{\theta}$ if and only if for any sufficient statistic $W = g(D_O)$, each partition set of $\mathcal{P}_W$ is a subset of a partition set of $\mathcal{P}_{W_1}$.   □

We note that if W_1 is minimal sufficient, then every set in $\mathcal{P}_{W_1}$ must be a union of sets in $\mathcal{P}_W$. As each set A in $\mathcal{P}_{W_1}$ maps into a single value of W_1, say w_1, the union of sets in $\mathcal{P}_W$ that make up A represents a mapping of values of W into w_1. We can therefore define a function k such that $W_1 = k(W)$. The values of W correspond to the partition sets in $\mathcal{P}_W$ and the values of W_1 correspond to the partition sets in $\mathcal{P}_{W_1}$. Therefore an alternative definition of minimal sufficiency is: W_1 is minimal sufficient for $\boldsymbol{\theta}$ if, for every sufficient statistic W, there exists a function k such that $W_1 = k(W)$. If k is a one-to-one function, then W is also minimal sufficient.

As minimal sufficient statistics give us the greatest data reduction without losing information about $\boldsymbol{\theta}$, the following theorem is particularly useful in sampling. The first proof given generalizes the approach used by Cassel et al. [1977, p. 37] for conventional designs to include adaptive designs. An alternative proof is also given, generalizing to include adaptive designs the factorization approach described by Chaudhuri and Stenger [1992, p. 17]. The minimal sufficiency of the reduced data d_R for conventional and adaptive designs was noted in Basu [1969].

Theorem 2.3.   With any conventional or adaptive design satisfying (2.2), D_R is a minimal sufficient statistic for $\boldsymbol{\theta}$.

Proof: We note first of all that we talk about *a* minimal sufficient statistic as any one-to-one function of a minimal sufficient statistic will also be minimal sufficient. Suppose that the sample s_O gives rise to $D_O = d_O$ so that, from (2.2),

$$P_\theta(D_O = d_O) = p(s_O \mid \mathbf{y}_s) I_{\Theta_{d_R}}(\boldsymbol{\theta}), \quad \boldsymbol{\theta} \in \Theta, \tag{2.8}$$

where $I_{\Theta_{d_R}}(\boldsymbol{\theta}) = 1$ when $\boldsymbol{\theta}$ is consistent with d_R, that is, $\boldsymbol{\theta} \in \Theta_{d_R}$, and 0 otherwise. Let $W = g(D_O)$ be any sufficient statistic. As $P_\theta[W = g(d_O)] \geq P_\theta(D_O = d_O) > 0$ for $\boldsymbol{\theta} \in \Theta_{d_R}$, it follows from the definition of sufficiency that

$$P_\theta\big[D_O = d_O \mid W = g(d_O)\big] = a(d_O) \tag{2.9}$$

does not depend on $\boldsymbol{\theta}$ for $\boldsymbol{\theta} \in \Theta_{d_R}$. Furthermore, as

$$
\begin{aligned}
P_\theta\big[D_O = d_O \,|\, W = g(d_O)\big] &= \frac{P_\theta\big[D_O = d_O \text{ and } W = g(d_O)\big]}{P_\theta\big[W = g(d_O)\big]} \\
&= \frac{P_\theta(D_O = d_O)}{P_\theta\big[W = g(d_O)\big]} \\
&> 0 \quad \text{for } \boldsymbol{\theta} \in \Theta_{d_R},
\end{aligned}
\tag{2.10}
$$

we have $a(d_O) > 0$ for $\boldsymbol{\theta} \in \Theta_{d_R}$. Cross multiplying (2.10) leads to

$$
P_\theta(D_O = d_O) = a(d_O)P_\theta\big[W = g(d_O)\big] \text{ for } \boldsymbol{\theta} \in \Theta_{d_R}.
\tag{2.11}
$$

Now $D_R = r_d(D_O)$ is sufficient for D_O and this defines a partition $\mathcal{P}_R$, say. Let $\mathcal{P}_W$ be a partition defined by W. We shall now show that each partition set of $\mathcal{P}_W$ is a subset of a partition set of $\mathcal{P}_R$.

Consider a partition set A of $\mathcal{P}_W$ which contains just one element, say d_{O_1}. Then one and only one value of d_O, namely $d_O = d_{O_1}$, maps to $g(d_O)$. As the inverse image of $r_d(d_{O_1})$ contains at least the element d_{O_1}, A belongs to a partition set of $\mathcal{P}_R$. We can assume that g is not one-to-one, otherwise W is essentially the same as D_O. Then there exists a partition set B of $\mathcal{P}_W$ with at least two elements, say d_{O_1} and d_{O_2} satisfying $g(d_{O_1}) = g(d_{O_2})$. From (2.11) we have

$$
P_\theta(D_O = d_{O_1}) = a(d_{O_1})P_\theta\big[W = g(d_{O_1})\big]
$$

and

$$
\begin{aligned}
P_\theta(D_O = d_{O_2}) &= a(d_{O_2})P_\theta\big[W = g(d_{O_2})\big] \\
&= a(d_{O_2})P_\theta\big[W = g(d_{O_1})\big].
\end{aligned}
$$

Combining these two probabilities and using (2.8) gives

$$
p(s_{O_2} \,|\, \mathbf{y}_s)I_{\Theta_{d_{R2}}}(\boldsymbol{\theta}) = \frac{a(d_{O_2})}{a(d_{O_1})}p(s_{O_1} \,|\, \mathbf{y}_s)I_{\Theta_{d_{R1}}}(\boldsymbol{\theta}).
\tag{2.12}
$$

Since $p(s_{O_1} \,|\, \mathbf{y}_s)$ and $p(s_{O_2} \,|\, \mathbf{y}_s)$ are both positive, we therefore conclude that $I_{\Theta_{d_{R2}}}(\boldsymbol{\theta}) = 1$ if and only if $I_{\Theta_{d_{R1}}}(\boldsymbol{\theta}) = 1$ for all $\boldsymbol{\theta} \in \Theta$. Hence $\Theta_{d_{R1}} = \Theta_{d_{R2}}$ so that $r_d(d_{O_1}) = r_d(d_{O_2})$ by (2.1). This implies that d_{O_1} and d_{O_2} belong to the same partition set of $\mathcal{P}_R$ so that B belongs to a partition set of $\mathcal{P}_R$. Thus every partition set of $\mathcal{P}_W$ is a subset of a partition set of $\mathcal{P}_R$ and, by the definition of minimal sufficiency, D_R is a minimal sufficient statistic. $\qquad\square$

An alternative proof of the minimal sufficiency of D_R for any conventional or adaptive design can be based on the factorization (2.3) of the likelihood. This approach uses the theorem about minimal sufficiency which says that if $T = t(X)$ is a statistic such that for any two data values x and x^*, $t(x) = t(x^*)$ if and only if $f(x, \boldsymbol{\theta}) = k(x, x^*)f(x^*, \boldsymbol{\theta})$ for all $\boldsymbol{\theta}$, for some function $k(x, x^*) > 0$, then T is minimal sufficient (cf., Arnold [1990, p. 340]). In what follows we equate d_O with x and $f(x, \boldsymbol{\theta})$ with $P_\theta(D_O = d_O)$.

Alternative Proof of Theorem 2.3: Let $d_O = (s_O, \mathbf{y}_s)$ and $d_O^* = (s_O^*, \mathbf{y}_{s^*})$ be any two data values with $p(s_O \mid \mathbf{y}_s) > 0$ and $p(s_O^* \mid \mathbf{y}_{s^*}) > 0$, and let r_d be our usual reduction function.

Suppose first that $r_d(d_O) = r_d(d_O^*)$, that is, $d_R = d_R^* (= d$, say), $\mathbf{y}_s = \mathbf{y}_{s^*}$ and $\Theta_{d_R} = \Theta_{d_R^*} = \Theta_d$. Then, by (2.2),

$$f(d_O; \boldsymbol{\theta}) = p(s_O \mid \mathbf{y}_s)I_{\Theta_d}(\boldsymbol{\theta}) \qquad \text{and} \qquad f(d_O^*; \boldsymbol{\theta}) = p(s_O^* \mid \mathbf{y}_s)I_{\Theta_d}(\boldsymbol{\theta})$$

for all $\boldsymbol{\theta} \in \Theta$ so that

$$f(d_O; \boldsymbol{\theta}) = k(d_O, d_O^*)f(d_O^*; \boldsymbol{\theta}),$$

with $k(d_O, d_O^*) = p(s_O \mid \mathbf{y}_s)/p(s_O^* \mid \mathbf{y}_s)$.

Conversely, suppose that $f(d_O; \boldsymbol{\theta}) = k(d_O, d_O^*)f(d_O^*; \boldsymbol{\theta})$ for some function $k(d_O, d_O^*)$ not depending on $\boldsymbol{\theta}$. Then, substituting for f,

$$p(d_O \mid \mathbf{y}_s)I_{\Theta_d}(\boldsymbol{\theta}) = k(d_O, d_O^*)p(d_O^* \mid \mathbf{y}_{s^*})I_{\Theta_{d^*}}(\boldsymbol{\theta})$$

for all $\boldsymbol{\theta}$. Since $p(d_O \mid \mathbf{y}_s)$ and $p(d_O^* \mid \mathbf{y}_{s^*})$ are both positive, and the indicator function I takes on only the values 0 and 1, equality requires that the two indicator functions are 0 or 1 at the same time. Thus, $I_{\Theta_d}(\boldsymbol{\theta}) = I_{\Theta_{d^*}}(\boldsymbol{\theta})$ for all $\boldsymbol{\theta}$ so that $\Theta_d = \Theta_{d^*}$ and hence $r_d(d_O) = r_d(d_O^*)$.

We have proved that $r_d(d_O) = r_d(d_O^*)$ if and only if $f(d_O; \boldsymbol{\theta}) = k(d_O, d_O^*)f(d_O^*; \boldsymbol{\theta})$ for all $\boldsymbol{\theta}$ and so the reduction d_R is minimal sufficient for $\boldsymbol{\theta}$.   □

2.5   RAO–BLACKWELL THEOREM

In this section, we are going to compare various estimators and look at a technique for modifying a given estimator so that it utilizes the data more efficiently. The efficiency of an estimator $T = T(D_O)$ of some parameter $\phi = \phi(\boldsymbol{\theta})$ (e.g., μ or τ) can be defined in terms of its mean-square error, namely

$$\text{MSE}[T] = \text{E}\big[(T - \phi)^2\big]$$

$$= \text{E}\big[(T - \text{E}[T] + \text{E}[T] - \phi)^2\big]$$

$$= \text{var}[T] + \big(\text{E}[T] - \phi\big)^2.$$

When T is unbiased, MSE[T] is the variance of T. Since a minimal sufficient statistic gives us the best reduction of the data D_O, any estimator that depends on more than

D_R contains surplus information about θ (and ϕ). We can use the following theorem to "reduce" the estimator further and thus provide an estimator with a smaller MSE. In fact, we shall show that

$$T_R = E[T \mid D_R] = \eta(D_R), \quad \text{say,}$$

is generally an improvement on T as it depends only on D_R. The following theorem is more general than we need at present, and it is its corollaries that are more directly applicable in this book. The following result was developed independently by Rao [1945] and Blackwell [1947]. A generalization of the theorem from mean-square error (MSE) to convex loss functions is given by Lehmann [1983, p. 50].

Theorem 2.4   Rao–Blackwell theorem.   Let $T = T(D_O)$ be any (not necessarily unbiased) estimator of a parameter $\phi = \phi(\theta)$, and let W be sufficient for θ. Define

$$T_W = E[T \mid W] = \eta(W).$$

Then

1. T_W is an estimator.
2. $E[T_W] = E[T]$.
3. $\text{MSE}[T_W] \leq \text{MSE}[T]$ with strict inequality for all $\theta \in \Theta$ such that $P_\theta(T \neq T_W) > 0$.

Proof:

1. Since W is sufficient for θ, the conditional distribution of T given W does not depend on θ. Hence T_W does not depend on θ and can be regarded as an estimator.
2. By the usual properties of conditional expectation,

$$E[T_W] = E_W E[T \mid W] = E[T].$$

3. We first note that

$$E[T - T_W \mid W] = E[T \mid W] - T_W = 0. \tag{2.13}$$

Then

$$\begin{aligned} \text{MSE[T]} &= E\left[(T - \phi)^2\right] \\ &= E\left[(T - T_W + T_W - \phi)^2\right] \\ &= E\left[(T - T_W)^2\right] + \text{MSE}[T_W] \end{aligned} \tag{2.14}$$

since, by (2.13),

$$E[(T - T_W)(T_W - \phi)] = E_W E[(T - T_W)(T_W - \phi) \mid W]$$
$$= E_W(T_W - \phi)E[T - T_W \mid W]$$
$$= 0.$$

It now follows from (2.14) that

$$\text{MSE}[T] \geq \text{MSE}[T_W]$$

with equality if and only if $E[(T - T_W)^2] = 0$, that is, if and only if $T = T_W$ with probability 1. Strict inequality will occur if T is different from T_W over a set of nonzero probability.   □

Corollary 1.   Since D_R is sufficient for θ, it follows that the above theorem applies with $W = D_R$ and

$$T_W = T_R = E[T \mid D_R].$$

Corollary 2.   If T is unbiased, then, from (2.14), mean-squared errors become variances and

$$\text{var}[T_W] = \text{var}[T] - E_W E[(T - T_W)^2 \mid W]$$
$$= \text{var}[T] - E_W \{\text{var}[T \mid W]\}. \tag{2.15}$$

Example 2.3   Random Sampling with Replacement.   In random sampling with replacement, $\bar{y}$, the mean of all n of the y_i values (some of which may be repeat selections) and $\bar{y}_\nu$, the mean of the y_i for the ν distinct units in the sample, are both unbiased estimators of μ. Which one is preferred? Equation (1.13) states that $\text{var}[\bar{y}_\nu] \leq \text{var}[\bar{y}]$, and we now prove this.

We begin with $T = \bar{y}$ and try and improve on this estimator using the above theorem; that is, setting $W = D_R$ we find that

$$T_W = E[\bar{y} \mid D_R].$$

Now

$$\bar{y} = \frac{1}{n} \sum_{i=1}^{N} y_i n_i = \frac{1}{n} \sum_{i=1}^{\nu} y_i n_i,$$

where n_i is the number of times unit i occurs in the sample. Suppose we fix $D_R = d_R$. Then $n_i = 0$ if $i \notin s_R$, and there are ν of the n_i's ($i \in s_R$) that are positive. For $i \in s_R$

we find the conditional distribution of n_i given d_R by considering the occupancy problem of distributing n balls in ν cells in such a way that each of the cells ends up with at least one ball. We therefore have ν balls, one in each of ν cells, and the other $n - \nu$ are distributed at random among the cells. Of those other balls, $n_i - 1$ will fall into the ith cell so that $n_i - 1 \sim \mathrm{Bin}(n - \nu, 1/\nu)$. Hence, for $i \in s_R$,

$$\mathrm{E}[n_i \mid D_R = d_R] = \mathrm{E}[(n_i - 1) \mid \nu \text{ cells}] + 1$$

$$= \frac{n - \nu}{\nu} + 1 = \frac{n}{\nu}. \tag{2.16}$$

For $i \notin s_R$, $n_i = 0$ and $\mathrm{E}[n_i \mid D_R = d_R] = 0$. Hence

$$\mathrm{E}[\bar{y} \mid D_R = d_R] = \frac{1}{n} \sum_{i=1}^{\nu} y_i \mathrm{E}[n_i \mid \nu]$$

$$= \frac{1}{n} \sum_{i=1}^{\nu} y_i \frac{n}{\nu} = \bar{y}_\nu. \tag{2.17}$$

Since $\bar{y}$ and $\bar{y}_\nu$ are both unbiased, it follows from (2.15) with $W = D_R$ and $T = \bar{y}$ that $\mathrm{var}[\bar{y}_\nu] \leq \mathrm{var}[\bar{y}]$, with strict inequality if $\nu < n$. $\qquad\square$

We see that, given an estimator, we can usually improve on it in terms of MSE or variance if we have a sufficient statistic. The new estimator T_W is a function $\eta(W)$ of the sufficient statistic W. When is there no improvement? This will occur when the original estimator is a function of the minimal sufficient statistic. To see this, suppose that W_1 is minimal sufficient and $T = T(W_1)$ is a function of W_1. If W is sufficient, then, by the alternative definition of minimal sufficiency, there exists a function k such that $W_1 = k(W)$. Hence

$$T_W = \mathrm{E}\big[T(W_1) \mid W\big] = \mathrm{E}\{T[k(W)] \mid W\} = T \tag{2.18}$$

and we end up with the same estimator.

In the above example, $\bar{y}$ is not a function of D_R. Although $\bar{y}$ does not depend on the order of the units selected, it does depend on the multiplicity of the units in the sample. We know, then, that the Rao–Blackwell theorem will provide an improved estimator. This estimator is $\bar{y}_\nu$, which does not depend on the multiplicities.

For the adaptive designs in the following chapters, the Rao–Blackwell method will be repeatedly useful in finding practical unbiased estimators. The idea is to start with a simple though perhaps inefficient unbiased estimator and take its conditional expectation given a sufficient statistic.

We now consider the question of whether there exists an unbiased estimator of μ that has the smallest variance of all unbiased estimators for all μ, namely the uniformly minimum variance unbiased estimator (UMVUE). To do this we need the concept of completeness.

2.6 COMPLETE STATISTICS

A statistic $W = g(D_O)$ is said to be complete for $\boldsymbol{\theta}$ if for any function $h(W)$, $E[h(W)] = 0$ for all $\boldsymbol{\theta} \in \Theta$ implies that $h(W) = 0$ with probability 1 for all $\boldsymbol{\theta} \in \Theta$.

Theorem 2.5. Let T be an unbiased estimator of $\phi = \phi(\boldsymbol{\theta})$. If W_1 is sufficient and complete for $\boldsymbol{\theta}$, then:

1. There exists a unique unbiased estimator T_1 which is a function of W_1.
2. T_1 is a UMVUE of ϕ.
(3) W_1 is minimal sufficient for ϕ.

Proof:

1. Let $T_1 = E[T \mid W_1] = \eta(W_1)$, say. Then

$$E[T_1] = E_{W_1} E[T \mid W_1] = E[T] = \phi$$

and T_1 is unbiased. To prove uniqueness, suppose that T_1 and T_2 are two unbiased estimators of ϕ that are functions of W_1. Then, setting $h(W_1) = T_1(W_1) - T_2(W_1)$,

$$E\big[T_1(W_1) - T_2(W_1)\big] = 0, \quad \text{for all} \quad \boldsymbol{\theta} \in \Theta,$$

which implies that $T_1(W_1) = T_2(W_1)$ with probability 1 as W_1 is complete. Hence T_1 is a unique function of W_1 irrespective of the choice of T.
2. Given any unbiased T that is not a function of W_1, then, from Theorem 2.4, Corollary 2,

$$\text{var}[T_1] < \text{var}[T].$$

Since, by **1**, T_1 is the same for every T, it follows that T_1 is the UMVUE of ϕ.
3. Let W be any sufficient statistic. Since $T_1 = \eta(W_1)$ is unbiased, it follows from Theorem 2.4 that

$$T_W = E\big[\eta(W_1) \mid W\big]$$

is unbiased; and, from Theorem 2.4, Corollary 2, we have $\text{var}[T_W] \leq \text{var}[T_1]$. However, by **2**, T_1 is the UMVUE of ϕ, so that $T_W = T_1$ with probability 1. Hence

$$E\big[\eta(W_1) \mid W\big] = \eta(W_1) \tag{2.19}$$

with probability 1. We now show that this implies that W_1 is a function of W. $\quad\square$

Suppose that W_1 is not a function of W. Then for some w there exists a w_0 such that $0 < P(W_1 = w_0 \mid W = w) < 1$, which implies that

$$P(W_1 = w_1 \mid W = w) \neq 1 \qquad (2.20)$$

for all w_1. Furthermore, since $\eta(W_1)$ is a discrete random variable and not a constant, it will have a maximum value $\eta_{\max}$ and it will take at least one other value less than $\eta_{\max}$. Combining these two results gives

$$\begin{aligned}
\mathrm{E}\big[\eta(W_1) \mid W = w\big] &= \sum_{w_1} \eta(w_1) P(W_1 = w_1 \mid W = w) \\
&< \eta_{\max} \sum_{w_1} P(W_1 = w_1 \mid W = w) \\
&= \eta_{\max}.
\end{aligned}$$

Here the the inequality (2.20) implies that there will be at least two terms in the above summation and not just one equal to $\eta(w_1)$ for some w_1. Hence (2.19) does not hold for $\eta(W_1) = \eta_{\max}$, and we have a contradiction. It therefore follows that we must have $P(W_1 = w_1 \mid W = w) = 1$ for some w_1 and that W_1 is a function of W. Since W_1 is a function of every sufficient statistic W, W_1 is minimal sufficient. □

We have proved above that a complete sufficient statistic is also minimal. Unfortunately a minimal sufficient statistic is not necessarily complete, nor is a complete statistic necessarily sufficient. In particular, the minimal sufficient statistic for the fixed population sampling situation is not complete, as is shown in the following theorem (Cassel et al. [1977, p. 44]).

Theorem 2.6.   The statistic $D_R = r_d(D_O)$ is not complete.

Proof: We have to show that there exists some function $h(D_R)$ such that $\mathrm{E}[h(D_R)] = 0$ for all $\theta \in \Theta$, and $P_\theta[h(D_R) \neq 0] > 0$ for at least one θ.

Consider the label $k, k \in \{1, 2, \ldots, N\}$, with inclusion probability α_k ($0 < \alpha_k < 1$) of being in the reduced set of labels s_R. Let $\mathcal{S}_k$ be the set of all s_R that include this label and define

$$h(D_R) = \begin{cases} c/\alpha_k, & s_R \in \mathcal{S}_k, \\ -c/(1 - \alpha_k), & s_R \notin \mathcal{S}_k. \end{cases}$$

Since $P_\theta(s_R \in \mathcal{S}_k) = \alpha_k$, we have

$$\mathrm{E}\big[h(D_R)\big] = \frac{c}{\alpha_k}\alpha_k - \frac{c}{1 - \alpha_k}(1 - \alpha_k) = 0 \qquad (2.21)$$

for all $\theta \in \Theta$. However, for $c \neq 0$, $P_\theta[h(D_R) \neq 0] = 1$ for all $\theta \in \Theta$. Hence D_R is not complete. □

It should be noted that if $W = W(D_R)$ is unbiased for ϕ, then so is $W_1 = W(D_R) + h(D_R)$ for all $c, c \neq 0$, where $h(D_R)$ is defined in the above theorem. If such a W exists, then W_1 is a function of D_R and there exists infinitely many unbiased estimators of ϕ that are functions of the sufficient statistic D_R. This raises the question of whether there is just one unbiased estimator which has the smallest variance for all $\boldsymbol{\theta}$. The following theorem (based on results of Godambe [1955], Godambe and Joshi [1965], and Basu [1971]; and see Cassel et al. [1977, pp. 68–71], Lehmann [1983, pp. 211], and Chaudhuri and Stenger [1992, pp. 11–14]) proves that such a UMVUE of ϕ does not exist as it would have to have zero variance. It is helpful here to use the representation $\boldsymbol{\theta} = \{(1, y_1), (2, y_2), \ldots, (N, y_N)\}$ rather than $\boldsymbol{\theta} = (y_1, y_2, \ldots, y_N)'$.

Theorem 2.7. Consider any sampling design $p(s_O \mid \mathbf{y})$. Let $\boldsymbol{\theta}^* = \{(1, y_1^*), (2, y_2^*), \ldots, (N, y_N^*)\}$ be any preassigned value of $\boldsymbol{\theta}$. If $\phi = \phi(\boldsymbol{\theta})$ has an unbiased estimator, then there exists an unbiased estimator T_1 of ϕ with $\mathrm{var}_{\boldsymbol{\theta}^*}[T_1] = 0$.

Proof: Let

$$T = T(D_O)$$
$$= T\left[(i_1, y_{i_1}), (i_2, y_{i_2}), \ldots, (i_n, y_{i_n})\right]$$

be any unbiased estimator of ϕ, and define

$$T^* = T\left[(i_1, y_{i_1}^*), (i_2, y_{i_2}^*), \ldots, (i_n, y_{i_n}^*)\right].$$

Consider the estimator $T_1 = T - T^* + \phi(\boldsymbol{\theta}^*)$. Since $\mathrm{E}_{\boldsymbol{\theta}}[T] = \phi(\boldsymbol{\theta})$ and $\mathrm{E}_{\boldsymbol{\theta}}[T^*] = \phi(\boldsymbol{\theta}^*)$, we have $\mathrm{E}_{\boldsymbol{\theta}}[T_1] = \phi(\boldsymbol{\theta})$ and T_1 is an unbiased estimator of ϕ. When $\boldsymbol{\theta} = \boldsymbol{\theta}^*$, $T_1 = \phi(\boldsymbol{\theta}^*)$ which is a constant so that its variance is zero. $\square$

One of the consequences of the preceding theorem is that for some of the adaptive designs described in the following chapters we derive more than one unbiased estimator that is a function of the minimal sufficient statistic. Each of the estimators has practical appeal and yet neither is uniformly better than the other. The underlying cause of the nonuniqueness is the fact that the minimal sufficient statistic in the fixed population sampling situation, with data consisting of labels as well as y-values, is not complete.

2.7 IGNORING THE LABELS

An essential ingredient of Theorem 2.7 is the estimator T^*. This estimator will depend on the labels even if T doesn't. It is the use of the labels that precludes the existence of a UMVUE. We shall see below that if we are able to ignore the information from the labels, then a UMVUE exists for some sampling designs. When can we ignore the labels? There will, of course, be some occasions when there are no labels attached to the units. Otherwise it might be suggested that the labels can be ignored when there is no evidence of a relationship between the label i and the corresponding y_i-value. However, ignoring the labels when they are available is a controversial topic, and for a further discussion of various views held see Chaudhuri and Vos [1988, p. 88]. Most of the adaptive schemes that we shall consider are label dependent so that the following theory plays a minor role in this book. The practical issue is that even though an optimal estimator may exist within the class of estimators not using the labels, a better estimator may be readily available outside that class.

Suppose that we take a simple random sample of size n, and $s_R = \{i_1, i_2, \ldots, i_n\}$ is the unordered set of (distinct) labels. Let

$$\mathbf{y}_R = \{y_{i_1}, y_{i_2}, \ldots, y_{i_n}\}$$

be the unordered set of corresponding y-values (which may not be distinct) and define $\mathbf{y}_{rank} = (y_{(1)}, y_{(2)}, \ldots, y_{(n)})$ to be the set of y's ranked according to size, namely $y_{(1)} \leq y_{(2)} \leq \cdots \leq y_{(n)}$. The reduced set $\mathbf{y}_R$ and the order statistics $\mathbf{y}_{rank}$ are equivalent in the sense that the knowledge of one implies knowledge of the other. To prove the following theorem we have to restrict Θ to be the Cartesian product

$$\Theta = V \times V \times \cdots \times V, \quad V \subset \mathbf{R}$$
$$= \mathbb{R}^N(V), \quad \text{say,}$$

using the notation of Royall [1968]. It is assumed that s_R is unknown or knowledge of s_R is discarded from D_R: The labels are only used to implement a particular sampling design. The following theorem is based on Royall [1968] and Lehmann [1983, p. 209].

Theorem 2.8. The order statistics $\mathbf{y}_{rank}$ are complete for $\boldsymbol{\theta}$ when $\boldsymbol{\theta} \in \Theta = \mathbb{R}^N(V)$.

Proof: Let h be a function such that

$$\mathrm{E}\big[h(Y_{(1)}, Y_{(2)}, \ldots, Y_{(n)})\big] = 0 \qquad \text{for all } \boldsymbol{\theta} \in \Theta. \qquad (2.22)$$

We now show that $h(y_{(1)}, y_{(2)}, \ldots, y_{(n)}) = 0$ for all $y_{(1)} \leq y_{(2)} \leq \cdots \leq y_{(n)}$. This is done in stages. To begin with, let $\boldsymbol{\theta} = (a, a, \ldots, a)'$, $a \in V$. Then (2.22) implies that

$$0 = \sum_{s_R} p(s_R) h(y_{(1)}, y_{(2)}, \ldots, y_{(n)}) \qquad (2.23)$$
$$= h(a, a, \ldots, a) \sum_{s_R} p(s_R) \qquad \text{for all } a \in V.$$

Since $p(s_R) > 0$ for all s_R, we have $h(a, a, \ldots, a) = 0$ for all a. Next, suppose that $(N - 1)$ elements of $\boldsymbol{\theta}$ are a and the remaining element is b $(b > a)$. Since (2.23) has two kinds of terms, those corresponding to samples consisting of n a's and those in which the sample contains just one b, (2.23) now becomes

$$p_0 h(a, a, \ldots, a) + p_1 h(a, a, \ldots, b) = 0,$$

where p_0 and p_1 are nonzero probabilities. Hence $h(a, a, \ldots, a, b) = 0$ for all a and b $(a < b)$.

Now suppose $\boldsymbol{\theta}$ has $N - 2$ elements equal to a and two elements equal to b. Then (2.23) becomes

$$p_0 h(a, a, \ldots, a) + p_1 h(a, a, \ldots, a, b) + p_2 h(\ldots, a, b, b) = 0$$

and $h(a, a, \ldots, a, b, b) = 0$ for all a and b $(a < b)$. Proceeding inductively we see that $h(a, \ldots, a, b, \ldots, b) = 0$ for any r a's and $(n - r)$ b's, $r = 0, 1, \ldots, n$.

The next stage in the inductive argument is to consider $\boldsymbol{\theta}$ of the form $(a, \ldots, a, b, c)'$ with $a < b < c$, then $\boldsymbol{\theta}$ of the form $(a, \ldots, a, b, b, c)'$, and so on, showing successively that $h(a, \ldots, a, b, c), h(a, \ldots, a, b, b, c), \ldots$ are all zero. Using induction, we find that $h(a, \ldots, a, b, \ldots, b, c, \ldots, c) = 0$ for any r a's, s b's, and $(n - r - s)$ c's. Continuing in this fashion we find that $h(y_{(1)}, y_{(2)}, \ldots, y_{(n)}) = 0$ for all $y_{(1)} \leq y_{(2)} \leq \cdots, y_{(n)}$. Thus $\mathbf{y}_{rank}$ is complete. $\qquad\qquad\square$

As noted by Lehmann [1983], we have not made use of the assumption of simple random sampling. The result is therefore valid for other sampling schemes where the probabilities $p(s_R)$ are known and positive for all s_R, for example stratified random sampling. The result says that if the data do not include the labels, then the order statistics are complete and so minimum variance estimators exist. However, deliberately ignoring the labels is not advised (see discussion in Cassel et al. [1977, pp. 47–51]). With the adaptive designs described in this monograph, the estimators making use of the unit label information in the data perform in general far better than estimators not making use of the labels.

2.8 BIAS OF CONVENTIONAL ESTIMATORS WITH ADAPTIVE DESIGNS

Adaptive designs may introduce biases into standard estimators that are unbiased with conventional designs. For example, the sample mean is an unbiased estimator of the population mean under a variety of conventional designs including simple random sampling. But if the sample size of a random sample depends sequentially on observed values, the sample mean is no longer unbiased.

Example 2.4   Bias with an Adaptive Design.   Consider a population of three units for which the values of the variable of interest are 0, 0, and 12. To estimate the population mean, the following sampling design is used. One of the units is selected at random without replacement and its value observed. If the value is zero, no more units are selected; but if the value is greater than zero, one additional unit is selected at random. The three possible sets of observations obtained under the design, each having probability 1/3, are (0), (0), and (12, 0), giving respective sample means 0, 0, and 6. Thus, the expected value of the sample mean under this design is the average of three sample means, namely 2, whereas the population mean is 4. □

2.9 FINDING DESIGN-UNBIASED STRATEGIES

In developing adaptive sampling strategies that are design unbiased, the following four approaches have proven useful and are used in the following chapters of this book.

1. The Rao–Blackwell method may be used to obtain unbiased estimators for designs in which the form of the desired estimator is not immediately evident. Starting with any unbiased estimator, which may be quite inefficient, a better unbiased estimator is obtained by taking the conditional expectation of the initial estimator given a sufficient statistic. The starting unbiased estimator of the population mean may be the value of the first unit selected, the sample mean of an initial sample, a stratified sample mean of an initial sample, or an unbiased estimate (such as a Hansen–Hurwitz or Horvitz–Thompson estimate) based on an unequal probability initial sample.

2. Unbiased estimators may be obtained by taking into account unequal selection, inclusion, utilization, or intersection probabilities for different units. In conventional unequal probability sampling, unbiased estimators can be formed by dividing observed values by draw-by-draw selection probabilities, as in the Hansen–Hurwitz estimator, or by inclusion probabilities, as in the Horvitz–Thompson estimator. In adaptive designs such as adaptive cluster sampling described in Chapter 4, the selection and inclusion probabilities are not known for every unit in the sample. With such designs we shall find that unbiased estimators can be obtained using the probability that a unit's value is utilized in the estimator or the probability that the initial sample intersects a particular group (network) of units containing the unit.

3. For adaptive surveys in k phases, it is sometimes possible to form an unbiased estimator of the population mean or total from each phase, and then form a combined estimator as a weighted average of the k individual estimators. Such an estimator will be unbiased provided the k weights are fixed in advance, that is, do not depend adaptively on observed values. The fixed-weight adaptive allocation strategies described in Chapter 7 are examples of this method.

4. In some situations it is possible to construct special adaptive designs under which a conventional estimator is unbiased. An example of such a design is the adaptive allocation plan in which the sample size in each stratum depends on observations in other strata previously observed during the survey. In Chapter 7, we see that the conventional stratified estimator is unbiased with such a design. However, if the sample size in a stratum depends on initial observations within that stratum, the conventional stratified estimator is biased.

CHAPTER 3

Stochastic Population Sampling Theory

3.1 INTRODUCTION

In Chapter 2, sampling theory was examined with the population values $y_1, y_2, \ldots, y_N$ of the variable of interest viewed as fixed. However, in many sampling situations it is realistic and useful to consider the vector $\mathbf{y}$ as a realization of a random vector $\mathbf{Y} = (Y_1, Y_2, \ldots, Y_N)'$. For example, if y_i is the number of plants of a given species in the ith plot of the study region, then y_i is a realization of a random variable Y_i, the distribution of which depends on seed dispersion, germination probabilities, rainfall, and so on. In this situation, the population total, population mean, and other population quantities will also be random variables. Thus $Z = Z(\mathbf{Y}) = \sum_{i=1}^{N} Y_i$ is now a random variable with realization $z = \sum_{i=1}^{N} y_i$.

In what follows we shall assume that $\mathbf{Y}$ has a density function belonging to the family of density functions $\{f(\mathbf{y}; \boldsymbol{\phi}) : \boldsymbol{\phi} \in \Phi\}$, where $\boldsymbol{\phi}$ is generally unknown. In addition to inference about the unknown parameter $\boldsymbol{\phi}$, there is an emphasis on predicting the realized value of a population random variable like Z from the sample data rather than estimating $\boldsymbol{\phi}$. The notation will reflect the change in emphasis away from $\mathbf{y}$ as a fixed parameter $\boldsymbol{\theta}$ to $\mathbf{y}$ as a realization of a random variable $\mathbf{Y}$. For the theory developed in this chapter, the full model for the sampling situation incorporates both design and population model components.

This chapter is intended to give a general overview of some of the density functions and predictive methods useful for inference in sampling. The focus of attention is on which results apply to conventional designs but not adaptive designs, which apply to conventional and adaptive but not nonstandard designs, and which apply to designs in general.

3.2 PROBABILITY DENSITY OF OUTCOMES

Once again we let s_O denote the original sample, in the order selected and possibly containing repeat values. A given realization $\mathbf{y}$ of the population vector together with a sample s_O selected will be termed an "outcome." The set of possible outcomes

51

$\{(s_O, \mathbf{y}); s_O \in \mathcal{S}, \mathbf{y} \in \mathcal{Y}\}$ will be termed the *outcome space*. With a model $f(\mathbf{y}; \boldsymbol{\phi})$ for the population and a design $p(s_O \mid \mathbf{y}; \boldsymbol{\phi})$ for selecting the sample, the probability density function for outcomes is given by

$$f_{S_O, \mathbf{Y}}(s_O, \mathbf{y}; \boldsymbol{\phi}) = p(s_O \mid \mathbf{y}; \boldsymbol{\phi}) f(\mathbf{y}; \boldsymbol{\phi}). \qquad (3.1)$$

(In the following the subscripts on density functions will often be omitted for notational simplicity, with f denoting the density function of random variables indicated by the arguments.) For an adaptive design, the design component of (3.1) is $p(s_O \mid \mathbf{y}; \boldsymbol{\phi}) = p(s_O \mid \mathbf{y}_s)$. For a conventional design, the design component is $p(s_O \mid \mathbf{y}; \boldsymbol{\phi}) = p(s_O)$.

Note that an outcome $(s_O, \mathbf{y})$ is not completely observed, since the components of $\mathbf{y}$ not in the sample s_O are not observed. However, the outcome space is basic to the analysis of the sampling situation, since the various quantities we are interested in, such as the data, the reduced data, and reordered outcomes separating the observed from the unobserved components, are random variables defined on this space as functions of outcomes.

The population density $f(\mathbf{y}; \boldsymbol{\phi})$ can be continuous or discrete. When it is discrete, integrals in the following sections of this chapter are replaced by corresponding sums. The design $p(s_O \mid \mathbf{y}; \boldsymbol{\phi})$ is usually assumed to be discrete, though it can be continuous as in the selection of point sample sites from a spatial or temporal study region. In this chapter we assume that there are only a finite number of possible samples except in the section on spatial/temporal extensions.

In Section 2.2, we defined the ordered data as $d_O = (s_O, \mathbf{y}_O)$, with order usually corresponding to the order of selection, and the reduced data $d = (s, \mathbf{y}_s)$, with the reduced sample s consisting of the distinct units in the sample ordered from smallest to largest, and $\mathbf{y}_s$ ordered to correspond to s. The information in d_O is equivalent to that contained in the pair $(s_O, \mathbf{y}_s)$ so that, as mentioned in Section 2.2, it is often convenient to write $d_O = (s_O, \mathbf{y}_s)$. From a distributional point of view, using the notation $\mathbf{y}_s$ is more helpful here than $\mathbf{y}_O$ as we are only interested in the Y_i's associated with the *distinct* units.

When only the reduced data are relevant to the sampling and inference at hand, we may consider the reduced outcome space, consisting of all possible reduced outcomes $(s, \mathbf{y})$. The density for a reduced outcome is

$$f_{S, \mathbf{Y}}(s, \mathbf{y}; \boldsymbol{\phi}) = p(s \mid \mathbf{y}; \boldsymbol{\phi}) f(\mathbf{y}; \boldsymbol{\phi}),$$

where $p(s \mid \mathbf{y}; \boldsymbol{\phi}) = \sum_{\{s_O : r(s_O) = s\}} p(s_O \mid \mathbf{y}; \boldsymbol{\phi})$ and r is the reduction function.

Thus we see that the outcome space and its family of density functions are basic to the sampling model. Population characteristics that one would like to predict, such as the population total or mean, or the components of $\mathbf{y}$ not in the sample, are also random variables defined on the outcome space. All information for prediction and estimation in sampling are contained in the outcome space and its density, and all other density functions will be derived from (3.1).

Example 3.1   Outcome Space.   Consider a population of $N = 4$ units, with the possible values of the vector **y** given below:

y:	(0,0,0,1)	(0,0,1,1)	(0,0,1,0)	(0,1,0,0)	(1,1,0,0)	(1,0,0,0)

The population could be viewed as having very simple clustering tendencies, since whenever two positive values of y_i occur, they occur together on one side or the other of the population.

Consider the following design of fixed sample size $n = 2$. First, one unit is selected at random from the population. If $y_i = 0$ for that unit, then the second unit is selected at random from the two units on the other side of the population. (One side of the population consists of units 1 and 2 and the other side consists of units 3 and 4.) If, on the other hand, $y_i = 1$ for the first selected unit, then the other unit in the same side is selected. For example, suppose unit 1 is selected first. If $y_1 = 0$, then the second sample unit is selected at random from $\{3, 4\}$. If $y_1 = 1$, then unit 2 is the second selection. This design is adaptive as the selection of the second unit depends on the value observed for the first unit.

Since there are 12 possible ordered samples s_O of size $n = 2$, and 6 possible values of **y** as given above, the outcome space consists of 72 possible values of $(s_O, \mathbf{y})$. Table 3.1 gives the design probabilities $p(s_O|\mathbf{y})$ for each of the 72 possible outcomes.

For example, the probability of selecting the ordered sample (1,3) when the population values are $\mathbf{y} = (0, 0, 0, 1)$ is $p[(1, 3), (0, 0, 1, 1)] = (1/4)(1/2) = 1/8$, since the probability of selecting unit 1 as the first unit is 1/4 and then, given that $y_1 = 0$, the probability of selecting unit 3 second is 1/2. On the other hand, the probability

Table 3.1.   The Outcome Space for the Example.*

$s_O \setminus \mathbf{y}$:	(0,0,0,1)	(0,0,1,1)	(0,0,1,0)	(0,1,0,0)	(1,1,0,0)	(1,0,0,0)
(1,2)	0	0	0	0	1/4	1/4
(2,1)	0	0	0	1/4	1/4	0
(1,3)	1/8	1/8	1/8	1/8	0	0
(3,1)	1/8	0	0	1/8	1/8	1/8
(1,4)	1/8	1/8	1/8	1/8	0	0
(4,1)	0	0	1/8	1/8	1/8	1/8
(2,3)	1/8	1/8	1/8	0	0	1/8
(3,2)	1/8	0	0	1/8	1/8	1/8
(2,4)	1/8	1/8	1/8	0	0	1/8
(4,2)	0	0	1/8	1/8	1/8	1/8
(3,4)	0	1/4	1/4	0	0	0
(4,3)	1/4	1/4	0	0	0	0

*For each of the 72 possible outcomes, the design probability $p(s_O \mid \mathbf{y})$ under the adaptive design is given.

Table 3.2. A Specified Population Model.

y:	(0,0,0,1)	(0,0,1,1)	(0,0,1,0)	(0,1,0,0)	(1,1,0,0)	(1,0,0,0)
$f(\mathbf{y})$:	1/6	1/6	1/6	1/6	1/6	1/6

of selecting (3,1)—the same units in the other order—is zero when $\mathbf{y} = (0,0,1,1)$ since, given $y_3 = 1$, the probability of selecting unit 1 second is zero.

So far, no stochastic model has been assumed for $\mathbf{y}$. The design probabilities $p(s_O \mid \mathbf{y})$ add to unity down each column of the table. Now suppose we assume the simple model that each of the six possible values of the population vector $\mathbf{y}$ has equal probability $f(\mathbf{y}) = 1/6$. The population model is given in Table 3.2.

Because the population model is discrete, the outcome density (3.1) is a probability. Under the model of Table 3.2, the probability of an outcome $f(s_O, \mathbf{y}) = p(s_O \mid \mathbf{y})f(\mathbf{y})$ is obtained by multiplying the design probability of Table 3.1 by 1/6. The outcome probabilities for each of the 72 outcomes are then given in Table 3.3.

Instead of assuming one particular distribution as in Table 3.3 for the population model, we could assume a family of distributions indexed by a parameter. For example, suppose we assume one probability ϕ for each of the four values of $\mathbf{y}$ for which $\sum_{i=1}^{N} y_i = 1$, and another probability for each of the two values for which $\sum_{i=1}^{N} y_i = 2$. Because the probabilities have to sum to one over all possible values of $\mathbf{y}$, the parametric model will be as specified in Table 3.4, with $0 \leq \phi \leq 1/4$. Under the model in Table 3.4, the probability of an outcome such as $(s_O, \mathbf{y}) = [(1,3), (0,0,1,1)]$ would be $p(s_O \mid \mathbf{y})f(\mathbf{y}; \phi) = (1/8)[(1/2) - 2\phi]$.

Table 3.3. The Outcome Probabilities $f(s_O, \mathbf{y}) = p(s_O \mid \mathbf{y})f(\mathbf{y})$ with the Specified Model and the Adaptive Design.

s_O \ y:	(0,0,0,1)	(0,0,1,1)	(0,0,1,0)	(0,1,0,0)	(1,1,0,0)	(1,0,0,0)
(1,2)	0	0	0	0	1/24	1/24
(2,1)	0	0	0	1/24	1/24	0
(1,3)	1/48	1/48	1/48	1/48	0	0
(3,1)	1/48	0	0	1/48	1/48	1/48
(1,4)	1/48	1/48	1/48	1/48	0	0
(4,1)	0	0	1/48	1/48	1/48	1/48
(2,3)	1/48	1/48	1/48	0	0	1/48
(3,2)	1/48	0	0	1/48	1/48	1/48
(2,4)	1/48	1/48	1/48	0	0	1/48
(4,2)	0	0	1/48	1/48	1/48	1/48
(3,4)	0	1/24	1/24	0	0	0
(4,3)	1/24	1/24	0	0	0	0

Table 3.4. A Parametric Population Model.

y:	(0,0,0,1)	(0,0,1,1)	(0,0,1,0)	(0,1,0,0)	(1,1,0,0)	(1,0,0,0)
$f(y; \phi)$:	ϕ	$(1/2) - 2\phi$	ϕ	ϕ	$(1/2) - 2\phi$	ϕ

It is often convenient to work with the reduced outcome space, combining outcomes whose ordered samples s_O have the same reduced sample s. For example, the two ordered samples $(1,2)$ and $(2,1)$ have the same reduced sample $s = (1, 2)$, so that the first two rows of the original table would be combined. The design probability $p(s \mid y)$ is obtained by adding the probabilities $p(s_O \mid y)$ over all ordered samples s_O having the reduction s. The reduced outcome space has 36 possible outcomes, with design probabilities $p(s \mid y)$ given in Table 3.5.

The probabilities $f(s, y) = p(s \mid y)f(y)$ for the reduced outcomes under the model in which $f(y) = 1/6$ for each possible value of y are given in Table 3.6.

Table 3.5. The Design Probabilities $p(s \mid y)$ for the Reduced Outcome Space, With the Adaptive Design.

$s \setminus y$:	(0,0,0,1)	(0,0,1,1)	(0,0,1,0)	(0,1,0,0)	(1,1,0,0)	(1,0,0,0)
(1,2)	0	0	0	1/4	1/2	1/4
(1,3)	1/4	1/8	1/8	1/4	1/8	1/8
(1,4)	1/8	1/8	1/4	1/4	1/8	1/8
(2,3)	1/4	1/8	1/8	1/8	1/8	1/4
(2,4)	1/8	1/8	1/4	1/8	1/8	1/4
(3,4)	1/4	1/2	1/4	0	0	0

Table 3.6. The Outcome Probabilities $f(s, y) = p(s \mid y)f(y)$ for the Reduced Sample Space, With the Adaptive Design and the Model $f(y) = 1/6$ for Each Value of y.

$s \setminus y$:	(0,0,0,1)	(0,0,1,1)	(0,0,1,0)	(0,1,0,0)	(1,1,0,0)	(1,0,0,0)
(1,2)	0	0	0	1/24	1/12	1/24
(1,3)	1/24	1/48	1/48	1/24	1/48	1/48
(1,4)	1/48	1/48	1/24	1/24	1/48	1/48
(2,3)	1/24	1/48	1/48	1/48	1/48	1/24
(2,4)	1/48	1/48	1/24	1/48	1/48	1/24
(3,4)	1/24	1/12	1/24	0	0	0

□

3.3 JOINT DENSITY OF OBSERVED AND UNOBSERVED

It is often convenient to reorder the population vector $\mathbf{y}$ to put the components $\mathbf{y}_s$ in the sample first and the components $\mathbf{y}_{\bar{s}}$ not in the sample second. From the previous section we recall that $\mathbf{y}_s$ is the vector of y-values corresponding to s, the ν [$= \nu(s)$] ordered distinct elements of s_O. Since the sample selection of y-values depends on s, we can show this dependence by defining the ith element of $\mathbf{y}_s$, for $i = 1, 2, \ldots, \nu$, to be the s_ith element of $\mathbf{y}$ ($s_1 < s_2 < \cdots < s_\nu$). We similarly let $\bar{s}$ be the vector of labels not in the sample, ordered from smallest to largest, and let $\mathbf{y}_{\bar{s}}$ be the corresponding vector of y-values. We can define the jth element of $\mathbf{y}_{\bar{s}}$ to be the $\bar{s}_j$th element of $\mathbf{y}$ ($j = 1, 2, \ldots, N - \nu$). Thus, for any outcome $(s_O, \mathbf{y})$ there corresponds a unique reordered outcome $(s_O, \mathbf{y}_s, \mathbf{y}_{\bar{s}})$, and vice versa.

Formally, the reordering of the outcome is an invertible transformation h given by

$$h(s_O, \mathbf{y}) = (s_O, \mathbf{y}_s, \mathbf{y}_{\bar{s}}).$$

The inverse transformation may be written

$$g(s_O, \mathbf{y}_s, \mathbf{y}_{\bar{s}}) = (s_O, \mathbf{y}).$$

In fact, the permutation h consists of multiplying by a permutation (orthogonal) matrix. The Jacobian of the transformation, as for any permutation, has absolute value 1. Therefore, the density of the permuted vector $(s_O, \mathbf{y}_s, \mathbf{y}_{\bar{s}})$ is

$$f_{(S_O, \mathbf{Y}_S, \mathbf{Y}_{\bar{S}})}(s_O, \mathbf{y}_s, \mathbf{y}_{\bar{s}}; \boldsymbol{\phi}) = f_{(S_O, \mathbf{Y})}\big[g(s_O, \mathbf{y}_s, \mathbf{y}_{\bar{s}}); \boldsymbol{\phi}\big].$$

Now g has two components, so we may write $g = (g_1, g_2)$, where $g_1(s_O, \mathbf{y}_s, \mathbf{y}_{\bar{s}}) = s_O$ and $g_2(s_O, \mathbf{y}_s, \mathbf{y}_{\bar{s}}) = \mathbf{y}$. The second component g_2 of the inverse transformation takes the reordered outcome $(s_O, \mathbf{y}_s, \mathbf{y}_{\bar{s}})$ and reconstructs the original order of the vector $\mathbf{y}$. In fact, this reordering depends on s_O only through s, so that for the vector $\mathbf{y}$ corresponding to the reordered outcome $(s_O, \mathbf{y}_s, \mathbf{y}_{\bar{s}})$ we may write $\mathbf{y}(s, \mathbf{y}_s, \mathbf{y}_{\bar{s}})$ for $g_2(s_O, \mathbf{y}_s, \mathbf{y}_{\bar{s}})$.

With this notation the density of the reordered outcome is

$$
\begin{aligned}
f_{(S_O, \mathbf{Y}_S, \mathbf{Y}_{\bar{S}})}(s_O, \mathbf{y}_s, \mathbf{y}_{\bar{s}}; \boldsymbol{\phi}) &= f_{(S_O, \mathbf{Y})}\big[s_O, \mathbf{y}(s, \mathbf{y}_s, \mathbf{y}_{\bar{s}}); \boldsymbol{\phi}\big] \\
&= p\big[s_O \mid \mathbf{y}(s, \mathbf{y}_s, \mathbf{y}_{\bar{s}}); \boldsymbol{\phi}\big] f\big[\mathbf{y}(s, \mathbf{y}_s, \mathbf{y}_{\bar{s}}); \boldsymbol{\phi}\big]
\end{aligned}
\tag{3.2}
$$

where the last density f is just the population model evaluated at the value of $\mathbf{y}$ (in original order) corresponding to the reordered outcome. Where no confusion arises we can write more simply

$$f_{(S_O, \mathbf{Y}_S, \mathbf{Y}_{\bar{S}})}(s_O, \mathbf{y}_s, \mathbf{y}_{\bar{s}}; \boldsymbol{\phi}) = p(s_O \mid \mathbf{y}; \boldsymbol{\phi}) f(\mathbf{y}; \boldsymbol{\phi}),$$

in which it is understood that $\mathbf{y}$ is the original vector corresponding to the reordered outcome $(s_O, \mathbf{y}_s, \mathbf{y}_{\bar{s}})$.

If the design is conventional or adaptive, that is, the design depends on $\mathbf{y}$ only through $\mathbf{y}_s$ (and not on $\mathbf{y}_{\bar{s}}$ or $\boldsymbol{\phi}$), we have

$$f_{(S_O, \mathbf{Y}_s, \mathbf{Y}_{\bar{s}})}(s_O, \mathbf{y}_s, \mathbf{y}_{\bar{s}}; \boldsymbol{\phi}) = p(s_O \mid \mathbf{y}_s) f\left[\mathbf{y}(s, \mathbf{y}_s, \mathbf{y}_{\bar{s}}); \boldsymbol{\phi}\right] \qquad (3.3)$$

and the design probability contains no unknowns. By conditioning the left-hand side of (3.3) on $\mathbf{y}_s$ and $\mathbf{y}_{\bar{s}}$, we see that we can also write

$$f\left[\mathbf{y}(s, \mathbf{y}_s, \mathbf{y}_{\bar{s}}); \boldsymbol{\phi}\right] = f_s(\mathbf{y}_s, \mathbf{y}_{\bar{s}}; \boldsymbol{\phi}), \qquad (3.4)$$

the subscript s indicating that the joint density of $\mathbf{y}_s$ and $\mathbf{y}_{\bar{s}}$ depends on s.

Note that in (3.2) the joint density of observed and unobserved depends on the population model f only through the *reduced* sample s. The original sample s_O appears only in the design probability.

Example 3.2   Reordered Outcome.   The reordering of an outcome $(s_O, \mathbf{y})$ in Example 3.1 into observed and unobserved components $(s_O, \mathbf{y}_s, \mathbf{y}_{\bar{s}})$ transforms for example the outcome $(s_O, \mathbf{y}) = [(1, 3), (0, 0, 1, 1)]$ into the reordered outcome $(s_O, \mathbf{y}_s, \mathbf{y}_{\bar{s}}) = [(1, 3), (0, 1), (0, 1)]$. Either way, the probability (3.3) of the outcome under the adaptive design and the model $f(\mathbf{y}) \equiv 1/6$ is

$$f_{(S_O, \mathbf{Y}_s, \mathbf{Y}_{\bar{s}})}\left[(1, 3), (0, 1), (0, 1)\right] = p\left[(1, 3) \mid (0, 1)\right] f\left[(0, 0, 1, 1)\right]$$
$$= (1/8)(1/6) = 1/48. \qquad \square$$

3.4   PROBABILITY DENSITY OF THE DATA

Since $d_O = (s_O, \mathbf{y}_s)$ is the observed data, it follows from (3.2) that the density of the corresponding random variable D_O is

$$f_{D_O}(s_O, \mathbf{y}_s; \boldsymbol{\phi}) = \int f(s_O, \mathbf{y}_s, \mathbf{y}_{\bar{s}}; \boldsymbol{\phi}) \, d\, \mathbf{y}_{\bar{s}}$$

$$= \int p\left(s_O \mid \mathbf{y}(s, \mathbf{y}_s, \mathbf{y}_{\bar{s}}); \boldsymbol{\phi}\right) f\left[\mathbf{y}(s, \mathbf{y}_s, \mathbf{y}_{\bar{s}}); \boldsymbol{\phi}\right] d\, \mathbf{y}_{\bar{s}}, \qquad (3.5)$$

where $d\, \mathbf{y}_{\bar{s}} = \prod_{i \in \bar{s}} dy_i$ and the integration is over all components $\bar{s}$ of $\mathbf{y}$ not in the sample.

With a conventional or adaptive design, $p(s_O \mid \mathbf{y}; \boldsymbol{\phi})$ can depend on $\mathbf{y}$ only through $\mathbf{y}_s$, and we have

$$f_{D_O}(s_O, \mathbf{y}_s; \boldsymbol{\phi}) = p(s_O \mid \mathbf{y}_s) \int f\left[\mathbf{y}(s, \mathbf{y}_s, \mathbf{y}_{\bar{s}}); \boldsymbol{\phi}\right] d\, \mathbf{y}_{\bar{s}}.$$

The marginal density of $\mathbf{y}_s$ for *fixed coordinates* $s = (s_1, s_2, \ldots, s_\nu)$ is

$$f_s(\mathbf{y}_s; \boldsymbol{\phi}) = \int f\left[\mathbf{y}(s, \mathbf{y}_s, \mathbf{y}_{\bar{s}}); \boldsymbol{\phi}\right] d\, \mathbf{y}_{\bar{s}} = \int f_s(\mathbf{y}_s, \mathbf{y}_{\bar{s}}; \boldsymbol{\phi}) d\, \mathbf{y}_{\bar{s}}, \qquad (3.6)$$

which integrates the population density $f(\mathbf{y}; \boldsymbol{\phi})$ over all $\mathbf{y}$ with the given values $\mathbf{y}_s$ for the components given by s. With a conventional or adaptive design, the probability density for the data value d_O can thus be written

$$f_{D_O}(s_O, \mathbf{y}_s; \boldsymbol{\phi}) = p(s_O \mid \mathbf{y}_s) f_s(\mathbf{y}_s; \boldsymbol{\phi}). \tag{3.7}$$

Note that f_s depends only on the reduced sample rather than on the original ordered sample. For instance, if $\mathbf{Y}$ has a multivariate normal distribution with dimension N, then $\mathbf{Y}_s$ for fixed s has a multivariate normal distribution with dimension ν and parameters consisting of the respective elements of the mean vector and covariance matrix of $\mathbf{Y}$.

From (3.7), the likelihood for $\boldsymbol{\phi}$ given the data d_O is $L(\boldsymbol{\phi} \mid d_O) = p(s_O \mid \mathbf{y}_s) f_s(\mathbf{y}_s; \boldsymbol{\phi})$. Hence, for an adaptive or conventional design, the ratio of likelihoods for two values of $\boldsymbol{\phi}$ is $\lambda(d_O) = f_s(\mathbf{y}_s; \boldsymbol{\phi}_1)/f_s(\mathbf{y}_s; \boldsymbol{\phi}_2)$, which does not depend on the design p. With a nonstandard design, on the other hand, the ratio of likelihoods given by (3.5) does depend on the design.

Example 3.3   Density of the Data.   The density of a data value in Example 3.1, with the adaptive design and the model $f(\mathbf{y}) \equiv 1/6$, is obtained by adding the probabilities of outcomes having that data value. Because of the discrete model, the integral in (3.5) is replaced by the sum over all outcomes having the given data value. For example, the ordered data value $(s_O, \mathbf{y}_O) = [(1,3), (0,0)]$ is obtained from the two ordered outcomes $[(1,3), (0,0,0,1)]$ and $[(1,3), (0,1,0,0)]$, each having probability $(1/8)(1/6) = 1/48$, so the probability of the data value is $f_{D_O}[(1,3), (0,0)] = 1/48 + 1/48 = 1/24$. The density of the associated reduced data value $(s, \mathbf{y}_s) = [(1,3), (0,0)]$ is obtained by adding the probabilities of the two reduced outcomes in

Table 3.7. Data Values and Their Probabilities, With the Adaptive Design and the Model $f(\mathbf{y}) = 1/6$ for each y.*

$s \setminus$ y:	(0,0,0,1)	(0,0,1,1)	(0,0,1,0)	(0,1,0,0)	(1,1,0,0)	(1,0,0,0)
(1,2)	$\mathbf{y}_s = (0,0)$ $f_D(s, \mathbf{y}_s) = 0$			(0,1) 1/24	(1,1) 1/12	(1,0) 1/24
(1,3)	(0,0) 1/12	(0,1) 1/24		(0,0)	(1,0) 1/24	
(1,4)	(0,1) 1/24		(0,0) 1/12		(1,0) 1/24	
(2,3)	(0,0) 1/12	(0,1) 1/24		(1,0) 1/24		(0,0)
(2,4)	(0,1) 1/24		(0,0) 1/12	(1,0) 1/24		(0,0)
(3,4)	(0,1) 1/24	(1,1) 1/12	(1,0) 1/24	(0,0) 0		

*Sets of outcomes having the same data value are combined as blocks when contiguous.

Table 3.6, namely $f_D[(1,3),(0,0)] = 1/24 + 1/24 = 1/12$. Alternatively it is given by adding the probabilities of the four ordered outcomes in Table 3.3 giving the reduced data value.

In Table 3.7, we block off regions of the outcome space giving the same data. The upper entry in each block is the value $\mathbf{y}_s$ of the reduced data. The lower entry is the probability of that data value, obtained by adding the outcome probabilities for all outcomes in Table 3.6 giving that data value. Blocks without the probability entry indicate that it is given in a block to the left having the same data value. For example, the data $(s, \mathbf{y}_s) = [(1,3),(0,0)]$ are given by two noncontiguous outcomes in the second row of the table, so the probability $1/2$ (which is the sum of two probabilities) is listed only in the first of those two outcomes. □

3.5 CONDITIONAL DENSITY OF UNOBSERVED VALUES GIVEN THE DATA

Given the data $d_O = (s_O, \mathbf{y}_s)$, the conditional density of the components $\mathbf{y}_{\bar{s}}$ not in the sample is

$$f(\mathbf{y}_{\bar{s}} \mid s_O, \mathbf{y}_s; \boldsymbol{\phi}) = \frac{f(s_O, \mathbf{y}_s, \mathbf{y}_{\bar{s}}; \boldsymbol{\phi})}{f(s_O, \mathbf{y}_s; \boldsymbol{\phi})}.$$

Using (3.2) and (3.5),

$$f(\mathbf{y}_{\bar{s}} \mid s_O, \mathbf{y}_s; \boldsymbol{\phi}) = \frac{p(s_O \mid \mathbf{y}; \boldsymbol{\phi})f(\mathbf{y}; \boldsymbol{\phi})}{\int p(s_O \mid \mathbf{y}; \boldsymbol{\phi})f(\mathbf{y}; \boldsymbol{\phi}) \, d \, \mathbf{y}_{\bar{s}}}, \tag{3.8}$$

where $\mathbf{y} = \mathbf{y}(s, \mathbf{y}_s, \mathbf{y}_{\bar{s}})$.

If the design is conventional or adaptive, the conditional density of the unobserved values given the data is, by (3.7),

$$f(\mathbf{y}_{\bar{s}} \mid s_O, \mathbf{y}_s; \boldsymbol{\phi}) = \frac{p(s_O \mid \mathbf{y}_s)f(\mathbf{y}; \boldsymbol{\phi})}{p(s_O \mid \mathbf{y}_s)f_s(\mathbf{y}_s; \boldsymbol{\phi})},$$

so that the design term cancels out giving

$$f(\mathbf{y}_{\bar{s}} \mid s_O, \mathbf{y}_s; \boldsymbol{\phi}) = \frac{f(\mathbf{y}; \boldsymbol{\phi})}{f_s(\mathbf{y}_s; \boldsymbol{\phi})} = \frac{f_s(\mathbf{y}_s, \mathbf{y}_{\bar{s}}; \boldsymbol{\phi})}{f_s(\mathbf{y}_s; \boldsymbol{\phi})}, \tag{3.9}$$

where $\mathbf{y} = \mathbf{y}(s, \mathbf{y}_s, \mathbf{y}_{\bar{s}})$.

Thus, for a conventional or adaptive design, the conditional density given the data depends on $\boldsymbol{\phi}$ but not on the design. Also, for a conventional or adaptive design, the conditional distribution of the unobserved values $\mathbf{y}_{\bar{s}}$ given the data $(s_O, \mathbf{y}_O)$ depends on the data only through the reduced data $(s, \mathbf{y}_s)$. With a nonstandard design, in which the design $p(s_O \mid \mathbf{y}; \boldsymbol{\phi})$ depends on y-values of units not in the sample, the conditional density does depend on the design.

Example 3.4 Conditional Density Given the Data.   For the adaptive design and model $f(\mathbf{y}) \equiv 1/6$ of Example 3.1, given the reduced data value $(s, \mathbf{y}_s) = [(1, 3), (0, 0)]$, the conditional probability that the unobserved values are $\mathbf{y}_{\bar{s}} = (1, 0)$ is obtained by dividing the probability of the reduced outcome $(s, \mathbf{y}) = [(1, 3), (0, 1, 0, 0)]$, which gives the right observed and unobserved components, by the probability of the given data value. Using the probabilities of Tables 3.6 and 3.7, this is $(1/24)/(1/24 + 1/24) = 1/2$. This probability would also equal 1/2 for any other adaptive or conventional design as (3.9) tells us that, in this case, the probability does not depend on the design. Equivalently, with the discrete model, the conditional probability of the $\mathbf{y}$ given the data $d = (s, \mathbf{y}_s)$ is the conditional probability of $\mathbf{y}_{\bar{s}}$ given the data. That is, the conditional probability that $\mathbf{Y}$ is (0,1,0,0), given data with reduced value $(s, \mathbf{y}_s) = [(1, 3), (0, 0)]$, is 1/2.   □

3.6 CONDITIONAL DENSITY OF Y GIVEN THE SAMPLE

The conditional density of $\mathbf{y}$ given the sample s_O is, by (3.1),

$$f(\mathbf{y} \mid s_O; \boldsymbol{\phi}) = \frac{f(s_O, \mathbf{y}; \boldsymbol{\phi})}{f(s_O; \boldsymbol{\phi})}$$

$$= \frac{p(s_O \mid \mathbf{y}; \boldsymbol{\phi})f(\mathbf{y}; \boldsymbol{\phi})}{\int p(s_O \mid \mathbf{y})f(\mathbf{y}; \boldsymbol{\phi})\, d\,\mathbf{y}}, \tag{3.10}$$

where $d\,\mathbf{y} = \prod_{i=1}^{N} dy_i$. This density depends on the design, except when the design does not depend on any y-values. For a conventional design, the design is independent of the population values and $f(\mathbf{y} \mid s_O; \boldsymbol{\phi}) = f(\mathbf{y}; \boldsymbol{\phi})$. For an adaptive or nonstandard design, on the other hand, the conditional density given the sample depends on the design used as well as on the population model.

The conditional density given the reduced sample s (or equivalently s_R), rather than the ordered sample, is obtained from (3.10) by summing both numerator and denominator over all ordered samples giving the reduced sample s. This means that we can simply replace s_O by s in the above equations. The conditional density given s also depends on the design unless the design does not depend on any y-values.

Example 3.5 Conditional Density Given the Sample.   With the adaptive design and the model $f(\mathbf{y}) \equiv 1/6$ of Example 3.1, the conditional density of $\mathbf{y}$ or of $\mathbf{y}_{\bar{s}}$ given the sample does depend on the design. For example, the conditional probability under the adaptive design of the value $\mathbf{y} = (0, 0, 1, 1)$ given the ordered sample $s_O = (1, 3)$ is, from Table 3.3, $(1/48)/(4/48) = 1/4$. The probability of the same value of $\mathbf{y}$ given only the reduced sample $s = (1, 3)$ is, from Table 3.6, $(1/48)/(8/48) = 1/8$. With a simple random sampling design, on the other hand, the probability of the sample $s_O = (1, 3)$, like that of every other ordered sample, would be $p(s_O) = 1/12$ and the probability under our model of each outcome would be $p(s_O)f(\mathbf{y}) = (1/12)(1/6) = 1/72$. Under simple random sampling, the probability of the value $\mathbf{y} = (0, 0, 1, 1)$

given the ordered sample $s_O = (1, 3)$ is $(1/72)/(6/72) = 1/6$, as it would be with any other conventional design as well. The probability of the same value of $\mathbf{y}$ given only the reduced sample is still $1/6$. □

3.7 DESIGN EXPECTATION (GIVEN THE REALIZATION)

The design expectation of an estimator $\hat{Z} = \hat{z}(D_O)$ is the conditional expectation given the realization $\mathbf{y}$ of the population process, namely

$$E[\hat{Z} \mid \mathbf{y}] = \sum_{s_O \in S} \hat{z}(s_O, \mathbf{y}_s) p(s_O \mid \mathbf{y}; \boldsymbol{\phi}). \qquad (3.11)$$

An estimator $\hat{Z}$ is said to be *design unbiased* for a population quantity $z(\mathbf{y})$ if $E[\hat{Z} \mid \mathbf{y}] = z(\mathbf{y})$ for all $\mathbf{y} \in \mathcal{Y}$. For such an estimator, the unconditional expectation is

$$E[\hat{Z}] = E_{\mathbf{y}}\{E[\hat{Z} \mid \mathbf{y}]\}$$
$$= E[z(\mathbf{Y})] = E[Z],$$

so that a design-unbiased estimator is always unconditionally unbiased, no matter what the true population model is. The subscript in the operator $E_{\mathbf{y}}$ is included as a reminder that the outer expectation is over all possible values of $\mathbf{y}$ with respect to the assumed population model.

The design mean-square error of $\hat{Z}$ is

$$E\{[\hat{Z} - z(\mathbf{y})]^2 \mid \mathbf{y}\} = \sum_{s_O \in S} [\hat{z}(s_O, \mathbf{y}_s) - z(\mathbf{y})]^2 p(s_O \mid \mathbf{y}; \boldsymbol{\phi}).$$

For a design-unbiased estimator, the design mean-square error is the conditional variance:

$$E\{[\hat{Z} - z(\mathbf{y})]^2 \mid \mathbf{y}\} = E\left(\{\hat{Z} - E[\hat{Z} \mid \mathbf{y}]\}^2 \mid \mathbf{y}\right)$$
$$= \mathrm{var}[\hat{Z} \mid \mathbf{y}].$$

The unconditional mean-square error can be written

$$E\{[\hat{Z} - z(\mathbf{Y})]^2\} = E_{\mathbf{y}}\left(E\{[\hat{Z} - z(\mathbf{y})]^2 \mid \mathbf{y}\}\right)$$
$$= E_{\mathbf{y}}\{\mathrm{var}[\hat{z}(D_O)|\mathbf{y}]\}$$
$$= \int \sum_{s_O \in S} [\hat{z}(s_O, \mathbf{y}_s) - z(\mathbf{y})]^2 p(s_O \mid \mathbf{y}; \boldsymbol{\phi}) f(\mathbf{y}; \boldsymbol{\phi}) \, d\mathbf{y}.$$

Since the sum has a finite number of terms, the order of integration and summation can be interchanged. In conclusion we note that the above results can be applied to the reduced data D by simply replacing s_O by s.

Table 3.8. Values of the Expansion Estimator $N\bar{y}$ for Each Outcome of the Population in Example 3.1.

$s \setminus y$:	(0,0,0,1)	(0,0,1,1)	(0,0,1,0)	(0,1,0,0)	(1,1,0,0)	(1,0,0,0)
(1,2)	0	0	0	2	4	2
(1,3)	0	2	2	0	2	2
(1,4)	2	2	0	0	2	2
(2,3)	0	2	2	2	2	0
(2,4)	2	2	0	2	2	0
(3,4)	2	4	2	0	0	0

Example 3.6 Design Expectation.   Consider the population of Example 3.1, with $N = 4$ units. Let $\bar{y}$ denote the sample mean of the 2 units in the sample, so that $N\bar{y}$ is the expansion estimator of the population total $z(\mathbf{y}) = \sum_{i=1}^{N} y_i$. The values of $N\bar{y}$ for each possible (reduced) outcome are given in Table 3.8.

With simple random sampling of the 2 units, the design probability does not depend on $\mathbf{y}$ and is $p(s) = 1/6$ for each possible sample. With simple random sampling, the expansion estimator is design unbiased for the population total no matter what the value of $\mathbf{y}$. For example, if $\mathbf{y} = (0, 0, 1, 1)$ the population total is $z(\mathbf{y}) = 2$ and the design expectation of $N\bar{y}$ is, using (3.11), $E[N\bar{y} \mid \mathbf{y} = (0, 0, 1, 1)] = 0(1/6) + 2(4/6) + 4(1/6) = 2$, which is the correct population value. $\square$

With the adaptive design of Example 3.1, however, using the design probabilities of Table 3.5, the design expectation is $E[N\bar{y} \mid \mathbf{y} = (0, 0, 1, 1)] = 0(0) + 2(4/8) + 4(1/2) = 3$, so that the expansion estimator is not design unbiased with the adaptive design.

One estimator of the population total that would be design unbiased with the adaptive design of Example 3.1 would be Ny_{O1}, where y_{O1} is the value for the first unit selected. This would not be an efficient estimator, however, since it depends on the ordered data and uses only one of the two observed values. In later chapters, however, we shall find such simple if inefficient unbiased estimators to be useful starting points in finding more efficient design-unbiased estimators with adaptive designs.

3.8   MODEL EXPECTATION GIVEN THE SAMPLE

For conventional designs, model expectation is usually defined to be the conditional expectation given the sample s_O or s (cf. Särndal et al. [1992, p. 534]). With an adaptive design, the conditional density given the sample depends on the design, by Section 3.6, so that conditional expectations given the sample also depend on the design as well as the model. Two important consequences of this difference are that (1) finding unbiased strategies under an assumed population model with adaptive designs usually involves consideration of the model and the design in combination, and (2) many of the standard results in the sampling literature about unbiased, best unbiased, and best linear unbiased estimation or prediction under assumed models apply only to the class of conventional designs.

For conventional and adaptive sampling, the conditional expectation of a predictor $\hat{Z} = \hat{z}(D_O)$ given the sample is

$$
\begin{aligned}
\mathrm{E}[\hat{Z} \mid s_O] &= \int \hat{z}(s_O, \mathbf{y}_s) f(d_O \mid s_O; \boldsymbol{\phi}) \, d\, \mathbf{y}_s \\
&= \frac{\int \hat{z}(s_O, \mathbf{y}_s) f(s_O, \mathbf{y}_s; \boldsymbol{\phi}) \, d\, \mathbf{y}_s}{f(s_O; \boldsymbol{\phi})} \\
&= \frac{\int \hat{z}(s_O, \mathbf{y}_s) p(s_O \mid \mathbf{y}_s) f_s(\mathbf{y}_s; \boldsymbol{\phi}) \, d\, \mathbf{y}_s}{\int p(s_O \mid \mathbf{t}_s) f(\mathbf{t}; \boldsymbol{\phi}) \, d\, \mathbf{t}},
\end{aligned} \tag{3.12}
$$

by (3.7), where $f_s(\mathbf{y}_s; \boldsymbol{\phi})$ is given by (3.6), and $d\,\mathbf{y}_s = \prod_{i \in s} dy_i$. Similarly, the prediction mean-square error is given by

$$
\mathrm{E}\{[\hat{Z} - z(\mathbf{Y})]^2 \mid s_O\} = \frac{\int [\hat{z}(s_O, \mathbf{y}_s) - z(\mathbf{y})]^2 p(s_O \mid \mathbf{y}_s) f_s(\mathbf{y}_s; \boldsymbol{\phi}) \, d\, \mathbf{y}_s}{\int p(s_O \mid \mathbf{t}_s) f(\mathbf{t}; \boldsymbol{\phi}) \, d\, \mathbf{t}}. \tag{3.13}
$$

If the design is conventional, then $p(s_O \mid \mathbf{y}_s) = p(s_O)$, and this term cancels out of the numerator and denominator of the above equations. Using (3.6) we then obtain

$$
\begin{aligned}
\mathrm{E}[\hat{Z} \mid s_O] &= \int \hat{z}(s_O, \mathbf{y}_s) f_s(\mathbf{y}_s; \boldsymbol{\phi}) \, d\, \mathbf{y}_s \\
&= \int \int \hat{z}(s_O, \mathbf{y}_s) f[\mathbf{y}(s_O, \mathbf{y}_s, \mathbf{y}_{\bar{s}}); \boldsymbol{\phi}] \, d\, \mathbf{y}_{\bar{s}} d\, \mathbf{y}_s \\
&= \int \hat{z}(s_O, \mathbf{y}_s) f(\mathbf{y}; \boldsymbol{\phi}) \, d\, \mathbf{y}
\end{aligned}
$$

and

$$
\mathrm{E}\{[\hat{Z} - z(\mathbf{Y})]^2 \mid s_O\} = \int [\hat{z}(s_O, \mathbf{y}_s) - z(\mathbf{y})]^2 f(\mathbf{y}; \boldsymbol{\phi}) \, d\, \mathbf{y}. \tag{3.14}
$$

A predictor $\hat{z}(D_O)$ of a population quantity $z(\mathbf{Y})$ is called "model unbiased given the sample" if

$$
\mathrm{E}[\hat{Z} \mid s_O] = \mathrm{E}[z(\mathbf{Y}) \mid s_O]
$$

for all $s_O \in \mathcal{S}$. When this is the case we also have $\mathrm{E}[\hat{Z}] = \mathrm{E}[z(\mathbf{Y})]$ and the prediction mean-square error is

$$
\mathrm{E}\left\{ [\hat{Z} - z(\mathbf{Y})]^2 \mid s_O \right\} = \mathrm{var}\left\{ [\hat{z}(D_O) - z(\mathbf{Y})] \mid s_O \right\}.
$$

The above also holds for discrete random variables with integrals replaced by sums. For expectations conditional on the reduced sample, s_O may be replaced by s in the above expressions.

With adaptive designs it is more useful to work with expectations that take into account both the design and the model. In subsequent chapters, we shall therefore use the term "model unbiased" to describe adaptive strategies that are *unconditionally* unbiased, with the unbiasedness depending on an assumed model but also on the design.

We conclude from the preceding discussion that conventional and adaptive sampling need to be treated differently when it comes to model expectation. We now give two examples where an estimate which is unbiased with a conventional design is biased with an adaptive design.

Example 3.7   Model Expectation with Conventional and Adaptive Designs. Consider the expansion estimator and population of Example 3.6. With any conventional design, expectations given the sample depend only on the model, not on the design. With the model $f(\mathbf{y}) \equiv 1/6$ of Table 3.2, the conditional expectation of the population total Z given any sample s is $E[Z \mid s] = 1(4/6) + 2(2/6) = 1.33$. The conditional expectation of the expansion estimator $N\bar{Y}$, given for example the sample $s = (1, 2)$ and using (3.12) and the values in the first row of Table 3.8, is $E[N\bar{Y} \mid s = (1, 2)] = 0(3/6) + 2(2/6) + 4(1/6) = 1.33$. Because of symmetries in the assumed model, we get the same result for any other sample. Note that the expansion estimator is model unbiased for the population total under the assumed model with *any* conventional design, and not just with simple random sampling. For example the design could be purposive selection of the sample (1,2).

With the adaptive design of Example 3.1, under the same model, the expansion estimator is not model unbiased. Using (3.12) and the outcome probabilities of Table 3.6 at the end of Section 3.2, the conditional expectation of $N\bar{Y}$ given the sample (1,2) is

$$E\left[N\bar{Y} \mid s = (1, 2)\right] = \left[0(0) + 2(2/24) + 4(1/12)\right]/(4/24) = 3,$$

while under the same design and model

$$E\left[Z \mid s = (1, 2)\right] = \left[1(2/24) + 2(1/12)\right]/(4/24) = 1.5.$$

Unconditionally, summing with respect to all the probabilities in Table 3.6 (i.e, multiplying corresponding elements in Tables 3.6 and 3.8 and adding), the expectation of $N\bar{Y}$ under the adaptive design and the model assumed can be shown to be $E[N\bar{Y}] = 1.67$. This does not equal $E[Z] = 1.33$, so that the expansion estimator is not unconditionally unbiased with the adaptive design. Thus, the search for model-unbiased procedures in adaptive sampling, such as those described in later parts of this book, needs to take both design and model into account.   □

Example 3.8   Model Bias with an Adaptive Design.   Suppose a population has N units and that the associated variables of interest $Y_1, \ldots, Y_N$ are independent, identically distributed Bernoulli random variables with parameter $p = 1/2$; namely

$$P(Y_i = 1) = 1 - P(Y_i = 0) = 1/2.$$

Let $Z = z(\mathbf{Y}) = (1/N) \sum_{i=1}^{N} Y_i$ be the population proportion, so that under the assumed model, $E[Z] = 1/2$. Because of the assumed model, with any conventional design or purposive selection of a sample of $n \leq N$ units, the sample proportion $\hat{Z} = \hat{z}(\mathbf{Y}) = (1/\nu) \sum_{i=1}^{\nu} Y_{s_i}$ is an unbiased predictor; that is,

$$E[\hat{Z} \mid s_O] = \frac{1}{\nu} \sum_{i=1}^{\nu} E[Y_{s_i}] = 1/2 = E[Z \mid s_O].$$

In this case, $E[\hat{Z}] = E[Z] = 1/2$ also.

Now consider the same predictor with the following sequential design. One unit is selected at random from the population. If its associated value y_{s_1} is zero, sampling stops. If $y_{s_1} = 1$, select one more unit at random from the remaining units. Let $\hat{Z}$ be the sample proportion defined above. The possible samples are $s_O = (i)$ with $\hat{Z} = 0$, and $s_O = (i, j)$ with $\hat{Z} = 1/2(1 + Y_{s_2})$. Hence $E[\hat{Z} \mid s_O = (i)] = 0$ and $E[\hat{Z} \mid s_O = (i, j)] = 1/2(1 + E[Y_{s_2}]) = 3/4$. Neither of these values is equal to $1/2$. Thus, the predictor that was model unbiased conditional on the sample with any conventional design is not conditionally unbiased for any of the samples with the adaptive design. Further, $\hat{Z}$ is not unconditionally unbiased with the adaptive design since $E[\hat{Z}] = 0(1/2) + (3/4)(1/2) = 3/8$. □

3.9 PREDICTIVE OR BAYES DENSITIES

Suppose that in the population model $f(\mathbf{y}; \boldsymbol{\phi})$ the value of the parameter $\boldsymbol{\phi}$ is assumed known, so that the stochastic distribution of the population is known exactly. Inference about $\mathbf{Y}_{\bar{s}}$ or any function $Z = z(\mathbf{Y}_{\bar{s}})$ is then a problem in pure prediction and can be based on the conditional density $f(\mathbf{y}_{\bar{s}} \mid d_O; \boldsymbol{\phi})$ [cf. (3.8)] of the unobserved variables given the data.

In the Bayesian setup for sampling, either a specific distribution for the population vector $\mathbf{Y}$ is assumed as above or else $\boldsymbol{\phi}$ is viewed as a random vector with known prior density $\pi(\boldsymbol{\phi})$. Predictive inference about $\mathbf{Y}_{\bar{s}}$ is then based on the predictive density

$$f(\mathbf{y}_{\bar{s}} \mid d_O) = \int f(\mathbf{y}_{\bar{s}} \mid d_O; \boldsymbol{\phi}) \pi(\boldsymbol{\phi} \mid d_O) \, d\boldsymbol{\phi}, \tag{3.15}$$

where

$$\pi(\boldsymbol{\phi} \mid d_O) = \frac{f(d_O; \boldsymbol{\phi}) \pi(\boldsymbol{\phi})}{\int f(d_O; \boldsymbol{\phi}) \pi(\boldsymbol{\phi}) \, d\boldsymbol{\phi}} \tag{3.16}$$

is the posterior density of $\boldsymbol{\phi}$ given the data d_O.

For a conventional or adaptive design, the posterior density of $\boldsymbol{\phi}$, can be found by substituting (3.7) into (3.16) to give

$$\pi(\boldsymbol{\phi} \mid d_O) = \frac{p(s_O \mid \mathbf{y}_s) f_s(\mathbf{y}_s; \boldsymbol{\phi}) \pi(\boldsymbol{\phi})}{\int p(s_O \mid \mathbf{y}_s) f_s(\mathbf{y}_s; \boldsymbol{\phi}) \pi(\boldsymbol{\phi}) d\boldsymbol{\phi}}$$

$$= \frac{f_s(\mathbf{y}_s; \boldsymbol{\phi}) \pi(\boldsymbol{\phi})}{\int f_s(\mathbf{y}_s; \boldsymbol{\phi}) \pi(\boldsymbol{\phi}) d\boldsymbol{\phi}}$$

so that Bayes inference about $\boldsymbol{\phi}$ does not depend on the design if the design is conventional or adaptive. Also if the design is adaptive or conventional, the densities conditional on the data depend on the data only through the reduced data $d = (s, \mathbf{y}_s)$.

The predictive density of $\mathbf{Y}_{\bar{s}}$ for a conventional or adaptive design can also be found by substituting (3.9) into (3.15) and using $\pi(\boldsymbol{\phi} \mid d_O)$, namely

$$f(\mathbf{y}_{\bar{s}} \mid d_O) = \int \frac{f_s(\mathbf{y}_s, \mathbf{y}_{\bar{s}}; \boldsymbol{\phi})}{f_s(\mathbf{y}_s; \boldsymbol{\phi})} \frac{f_s(\mathbf{y}_s; \boldsymbol{\phi}) \pi(\boldsymbol{\phi})}{\left[\int f_s(\mathbf{y}_s; \boldsymbol{\phi}) \pi(\boldsymbol{\phi}) d\boldsymbol{\phi}\right]} d\boldsymbol{\phi}$$

$$= \frac{\int f_s(\mathbf{y}_s, \mathbf{y}_{\bar{s}}; \boldsymbol{\phi}) \pi(\boldsymbol{\phi}) d\boldsymbol{\phi}}{\int f_s(\mathbf{y}_s; \boldsymbol{\phi}) \pi(\boldsymbol{\phi}) d\boldsymbol{\phi}}$$

so that again Bayes predictive inference about $\mathbf{Y}_{\bar{s}}$ does not depend on the design if the design is conventional or adaptive. Bayes predictive inference does depend on the design if the design is nonstandard.

The pure prediction problem, in which $\boldsymbol{\phi}$ is assumed known, and the Bayes prediction problem, in which a known prior density for $\boldsymbol{\phi}$ is assumed, are formally the same since predictive inference in each case uses a known conditional distribution given the data. In Example 3.1, the specified model of Table 3.2 is an example of a population in which the stochastic distribution of the population is known exactly. The inference problem would also be one of pure prediction if a prior distribution were assumed for the parameter $\boldsymbol{\phi}$ in the model of Table 3.4.

3.10   PREDICTIVE EXPECTATION (GIVEN THE DATA)

If the population model $f(\mathbf{y}; \boldsymbol{\phi})$ is known—that is, $\boldsymbol{\phi}$ is known—then the unbiased predictor of $Z = z(\mathbf{Y})$ with lowest mean-square error is known to be the conditional expectation $\hat{Z} = \mathrm{E}[Z \mid d_O]$. If $\boldsymbol{\phi}$ is a random variable with known prior distribution $\pi(\boldsymbol{\phi})$, then $\hat{Z} = \mathrm{E}[Z \mid d_O]$ is the Bayes predictor of Z. Therefore to find $\hat{Z}$ we need the conditional distribution of $\mathbf{Y}$ given d_O. Assuming $\boldsymbol{\phi}$ known, then for any conventional or adaptive design, this conditional distribution of the unobserved components $\mathbf{y}_s$ is given by (3.9) as

$$f(\mathbf{y}_{\bar{s}} \mid d_O; \boldsymbol{\phi}) = \frac{f(\mathbf{y}; \boldsymbol{\phi})}{f_s(\mathbf{y}_s; \boldsymbol{\phi})},$$

where $\mathbf{y} = \mathbf{y}(s, \mathbf{y}_s, \mathbf{y}_{\bar{s}})$. Conditional on the data d_O, we fix s and $\mathbf{y}_s$ and only $\mathbf{y}_{\bar{s}}$ can vary. Then, expressing $\mathbf{y} = \mathbf{y}(s, \mathbf{y}_s, \mathbf{y}_{\bar{s}})$,

$$E[z(\mathbf{Y}) \mid d_O] = \int z(\mathbf{y}) f(\mathbf{y}_{\bar{s}} \mid d_O; \boldsymbol{\phi}) \, d\,\mathbf{y}_{\bar{s}}$$

$$= \frac{\int z(\mathbf{y}) f(\mathbf{y}; \boldsymbol{\phi}) \, d\,\mathbf{y}_{\bar{s}}}{f_s(\mathbf{y}_s; \boldsymbol{\phi})}$$

$$= \frac{\int z(\mathbf{y}) f(\mathbf{y}; \boldsymbol{\phi}) \, d\,\mathbf{y}_{\bar{s}}}{\int f(\mathbf{y}; \boldsymbol{\phi}) \, d\,\mathbf{y}_{\bar{s}}}. \tag{3.17}$$

The conditional mean-square error given the data is

$$E\big[(Z - E[Z \mid d_O])^2 \mid d_O\big] = \mathrm{var}[Z \mid d_O]$$

$$= \frac{\int \{z(\mathbf{y}) - \hat{z}\}^2 f(\mathbf{y}; \boldsymbol{\phi}) \, d\,\mathbf{y}_{\bar{s}}}{f_s(\mathbf{y}_s; \boldsymbol{\phi})}.$$

If $\boldsymbol{\phi}$ has a prior distribution $\pi(\boldsymbol{\phi})$, then, repeating the argument in Section 3.9, we find that we simply multiply each of the above integrals by $\pi(\boldsymbol{\phi})$ and integrate each with respect to $\boldsymbol{\phi}$.

Since for adaptive and conventional designs conditional densities given the data depend on the data only through the reduced data d, the expressions in this section given d_O equal the corresponding expressions given d.

Example 3.9 Predictive Expectation and Mean-Square Error. With the population of Example 3.1 and the specified model $f(\mathbf{y}) \equiv 1/6$ of Table 3.2, consider predicting the population total $Z = \sum_{i=1}^{N} Y_i$. Under the assumed model with $f(\mathbf{y}) = 1/6$ for all values of $\mathbf{y}$, the best unbiased estimator of Z is the conditional expectation $\hat{Z} = E(Z \mid D)$, given the observed data value d. The mean-square prediction error given the data is $E[(\hat{Z} - Z)^2 \mid d]$. Because they are conditional on the data d, the predictor $\hat{Z}$ and its conditional mean-square prediction error do not depend on the design used but only on the population model, provided the design is conventional or adaptive. For example, with the data $d = (s, \mathbf{y}_s) = [(1,3), (0, 1)]$, only two values $\mathbf{y} = (0, 0, 1, 1)$ and $\mathbf{y} = (0, 0, 1, 0)$ have positive joint probability with the data value. Those two values of $\mathbf{y}$ have population totals $z(0, 0, 1, 1) = 2$ and $z(0, 0, 1, 0) = 1$. With the integrals in (3.17) replaced by summations over all outcomes having the given data value $d = (s, \mathbf{y}_s)$, the conditional expectation, is $\hat{Z} = E\{[z(\mathbf{Y})|(s, \mathbf{y}_s) = [(1,3), (0, 1)]\} = [2(1/6) + 1(1/6)]/(1/6 + 1/6) = 2(1/2) + 1(1/2) = 1.5$, in which $(1/2)$ is the conditional probability of either of the values of $\mathbf{y}$ given d. Similarly, the conditional mean-square prediction error is $E[(\hat{Z} - Z)^2 \mid d)] = (1.5 - 2)^2(1/2) + (1.5 - 1)^2(1/2) = .25$. Table 3.9, which is independent of the design used, lists the best unbiased predictor together with its conditional mean-square prediction error for every possible data value with the given model.

Table 3.9. Values of the Predictor $\hat{Z} = E[Z \mid d]$ of the Population Total Z and of Conditional Mean-Square Prediction Error $E[(\hat{Z} - Z)^2 \mid d)]$ for Each Data Value.*

y:	(0,0,0,1)	(0,0,1,1)	(0,0,1,0)	(0,1,0,0)	(1,1,0,0)	(1,0,0,0)
$z(\mathbf{y})$:	1	2	1	1	2	1
$s \setminus f(\mathbf{y})$:	1/6	1/6	1/6	1/6	1/6	1/6
(1,2)	$\hat{z} = 1.333$			1	2	1
	$E[(\hat{Z} - Z)^2 \mid d] = .222$			0	0	0
(1,3)	1	1.5		1	1.5	
	0	.25		0	.25	
(1,4)	1.5		1		1.5	
	.25		0		.25	
(2,3)	1	1.5		1.5		1
	0	.25		.25		0
(2,4)	1.5		1	1.5		1
	.25		0	.25		0
(3,4)	1	2	1	1.333		
	0	0	0	.222		

*This table applies to any conventional or adaptive design under the model $f(\mathbf{y}) = 1/6$, for all $\mathbf{y}$.

Example 3.10   Prediction with a Conventional Design.   The predictors and their mean-square errors of Example 3.9 are functions of the data only and are not functions of the design used. However, in planning a survey it is of interest to use a good procedure for selecting the sample, so that the expected mean-square prediction error, over all data values that might be obtained, is as low as possible. Suppose any conventional design is to be used with the population of Example 3.9. Then expectations conditional on the sample do not depend on the design (Section 3.8). For example, the conditional expected mean-square prediction error given the sample $s = (0, 1)$, using (3.14) with s instead of s_O, or more directly from

$$E\left[(\hat{Z} - Z)^2 \mid s\right] = E\left\{E[(\hat{Z} - Z)^2 \mid d] \mid s\right\} \tag{3.18}$$

using Table 3.9, is $E[(\hat{Z} - Z)^2 \mid s = (1,2)] = .222(3/6) + 0(3/6) = .111$. Table 3.10 gives the conditional mean-square prediction error $E[(\hat{Z} - Z)^2 \mid s]$ for each of the possible samples, with a conventional design.

The unconditional mean-square error under the conventional design is then $E[(\hat{Z} - Z)^2] = \sum_s E[(\hat{Z} - Z)^2 \mid s]p(s)$. With simple random sampling, each sample has probability $p(s) = 1/6$ and so the mean square prediction error is the average of the six conditional ones, giving $E[(\hat{Z} - Z)^2] = .1482$. The optimal conventional design from the model point of view, however, would assign positive probability only to the samples (1,2) and (3,4), since they give lower mean-square error than the other samples. For example suppose one of those two samples was selected at random, so that $p(s) = 1/2$ for $s = (1, 2)$ and $s = (3, 4)$, and $p(s) = 0$ for all other samples. The

Table 3.10. Expected Mean-Square Prediction Error $E[(\hat{Z} - Z)^2 | s]$ Given the Sample, for Each Possible Sample, Under Any Conventional Design.

s	$E[(\hat{Z} - Z)^2|s]$
(1,2)	.1111
(1,3)	.1667
(1,4)	.1667
(2,3)	.1667
(2,4)	.1667
(3,4)	.1111

mean square prediction error under this design would be $E[(\hat{Z} - Z)^2] = .1111$. The same value would be obtained by the purposive design which selects the sample (1,2) with probability one.

Example 3.11   Prediction with an Adaptive Design.   For adaptive designs, in contrast to conventional designs as in Example 3.10, the expected mean-square error given the sample depends on the design used. At the inference stage, the best unbiased predictor $\hat{Z}$ and its mean-square error of Table 3.9 can be computed from the data without taking the design used into account. But in the planning stage of the survey, one would like to choose a procedure for selecting the sample that will give the lowest possible expected value of the mean-square error over all possible data values that might be obtained. For the adaptive design of Example 3.1, the mean-square error conditional on the sample is obtained using (3.13) [or (3.18) and Table 3.9] and the data probabilities in Table 3.7, divided by the row total of the probabilities. For example, the conditional mean-square prediction error given $s = (1, 3)$ is $E[(\hat{Z} - Z)^2 | s = (1, 3)] = [0(1/12) + .25(2/24)]/(1/12 + 2/24) = 0.125$. The values for each of the possible samples are listed in Table 3.11.

Table 3.11.   Expected Mean Square Prediction Error $E[(\hat{Z} - Z)^2 | s]$ Given the Sample, for Each Possible Sample, Under the Adaptive Design of this Example.

s	$E[(\hat{Z} - Z)^2 | s]$
(1,2)	0
(1,3)	.125
(1,4)	.125
(2,3)	.125
(2,4)	.125
(3,4)	0

The unconditional mean-square prediction error with the adaptive design is obtained by multiplying the value $E[(\hat{Z} - Z)^2 \mid d]$ for each data value from Table 3.8 by the probability in Table 3.7 of the data value under the adaptive design, and summing over all the possible data values. Alternatively, the conditional values in Table 3.10 can be multiplied by the marginal design probability $p(s) = \sum_{\mathbf{y}} p(s \mid \mathbf{y})$ for each row and summed. Using the first method, the unconditional mean-square error is

$$E[(\hat{Z} - Z)^2] = \sum_{d} E[(\hat{Z} - Z)^2 \mid d] f_D(d)$$

$$= .2222(0) + 0(16/24) + .25(8/24) = 0.0833.$$

Thus, under the assumed population model, the adaptive design gives lower expected mean-square error than the optimal conventional design. Further discussion of optimal sampling strategies, with some results on when the optimal strategy is conventional and when it is adaptive, is found in Chapter 10.　　　　　　　□

3.11　PREDICTIVE SUFFICIENCY

Suppose a random vector $\mathbf{X}$ is observed and one wishes to predict the value of a random variable or vector $\mathbf{Z}$. Let $f(\mathbf{x}, \mathbf{z}; \boldsymbol{\phi})$ be the joint density of $\mathbf{X}$ and $\mathbf{Z}$ depending on a parameter $\boldsymbol{\phi}$, with $\boldsymbol{\phi} \in \Phi$. The relevant information for predicting $\mathbf{Z}$ should be contained in the joint likelihood function $L_{\mathbf{x}}(\mathbf{z}, \boldsymbol{\phi}) = f(\mathbf{x}, \mathbf{z}; \boldsymbol{\phi})$ (Berger and Wolpert [1988], Bjørnstad [1990]).

The concept of predictive sufficiency was introduced by Lauritzen [1974] who used the term "totally sufficient" for a statistic sufficient for prediction of any combination of unknown random variables from an observed set of random variables, in the presence of unknown parameters. The concept was applied to the survey sampling situation by Bolfarine and Zacks [1992]. A variety of predictive likelihood functions that eliminate the "nuisance parameter" $\boldsymbol{\phi}$ have been discussed in Lauritzen [1974], Hinkley [1979], Lejeune and Faulkenberry [1982], Butler [1986, 1990], and others; reviews of the topic are found in Bjørnstad [1990] and Bolfarine and Zacks [1992]. The following definition of predictive sufficiency is based on Lauritzen [1974] and Bolfarine and Zacks [1992, p. 31].

Definition　　Suppose that the random vectors $\mathbf{X}$ and $\mathbf{Z}$ have joint density $f(\mathbf{x}, \mathbf{z}; \boldsymbol{\phi})$. A statistic $\mathbf{T} = \mathbf{g}(\mathbf{X})$ is said to be *predictive sufficient* for the random vector $\mathbf{Z}$ if, whenever $f(\mathbf{t}; \boldsymbol{\phi}) > 0$,

(1) The conditional distribution of $\mathbf{X}$ given $\mathbf{T}$ does not depend on $\boldsymbol{\phi}$, and
(2) $\mathbf{X}$ and $\mathbf{Z}$ are conditionally independent, given $\mathbf{T}$.　　　　　　　□

The first requirement in the definition is that $\mathbf{T}$ be sufficient for the parameter $\boldsymbol{\phi}$, and the second requirement is that $\mathbf{Y}$ and $\mathbf{Z}$ be conditionally independent given $\mathbf{T} = \mathbf{t}$. Describing $\mathbf{T}$ as being predictive sufficient "for $\mathbf{Z}$" is consistent with the usage of other writers such as Bolfarine and Zacks [1992] and emphasizes the interest in the prediction of unknowns in sampling. However, the concept is of sufficiency for all unknowns, both $\mathbf{Z}$ and $\boldsymbol{\phi}$; hence Lauritzen's use of the term "totally sufficient."

Equivalently, predictive sufficiency can be characterized with a single condition, as described below.

Theorem 3.1. The statistic $\mathbf{T} = g(\mathbf{X})$ is predictive sufficient for $\mathbf{Z}$ if and only if

$$f(\mathbf{x}, \mathbf{z}; \boldsymbol{\phi}) = h(\mathbf{x})f(\mathbf{t}, \mathbf{z}; \boldsymbol{\phi}) \qquad (3.19)$$

for some function $h(\mathbf{x})$ not depending on $\mathbf{z}$ or $\boldsymbol{\phi}$.

Proof: Suppose $\mathbf{T}$ is predictive sufficient for $\mathbf{Z}$. Then, from (1) of the definition, $\mathbf{T}$ is sufficient for the parameter $\boldsymbol{\phi}$ and we can write $f(\mathbf{x}; \boldsymbol{\phi}) = h(\mathbf{x})f(\mathbf{t}; \boldsymbol{\phi})$ by the Fisher–Neyman criterion. By (2) of the definition and since $\mathbf{t}$ is a function of $\mathbf{x}$,

$$f(\mathbf{z} \mid \mathbf{x}; \boldsymbol{\phi}) = f(\mathbf{z} \mid \mathbf{x}, \mathbf{t}; \boldsymbol{\phi})$$
$$= f(\mathbf{z} \mid \mathbf{t}; \boldsymbol{\phi}).$$

Thus,

$$f(\mathbf{z} \mid \mathbf{z}; \boldsymbol{\phi}) = f(\mathbf{x}; \boldsymbol{\phi})f(\mathbf{z} \mid \mathbf{x}; \boldsymbol{\phi})$$
$$= h(\mathbf{x})f(\mathbf{t}; \boldsymbol{\phi})f(\mathbf{z} \mid \mathbf{t}; \boldsymbol{\phi})$$
$$= h(\mathbf{x})f(\mathbf{t}, \mathbf{z}; \boldsymbol{\phi}).$$

Conversely, suppose (3.19) holds. Then the marginal density of $\mathbf{x}$ is

$$f(\mathbf{x}; \boldsymbol{\phi}) = \int f(\mathbf{x}, \mathbf{z}; \boldsymbol{\phi})\, d\mathbf{z}$$
$$= h(\mathbf{x}) \int f(\mathbf{t}, \mathbf{z}; \boldsymbol{\phi})\, d\mathbf{z}$$
$$= h(\mathbf{x})f(\mathbf{t}; \boldsymbol{\phi}),$$

showing (1) of the definition holds.

Since $\mathbf{t}$ is a function of $\mathbf{x}$,

$$f(\mathbf{z} \mid \mathbf{x}, \mathbf{t}; \boldsymbol{\phi}) = f(\mathbf{z} \mid \mathbf{x}; \boldsymbol{\phi})$$
$$= \frac{f(\mathbf{x}, \mathbf{z}; \boldsymbol{\phi})}{f(\mathbf{x}; \boldsymbol{\phi})}$$
$$= \frac{h(\mathbf{x})f(\mathbf{t}, \mathbf{z}; \boldsymbol{\phi})}{h(\mathbf{x})f(\mathbf{t}; \boldsymbol{\phi})}$$
$$= f(\mathbf{z} \mid \mathbf{t}; \boldsymbol{\phi})$$

so that $\mathbf{Z}$ is conditionally independent of $\mathbf{X}$ given $\mathbf{T}$, showing (2) of the definition holds. $\quad\square$

A factorization theorem for predictive sufficiency follows.

Theorem 3.2 Factorization Theorem. The statistic $\mathbf{T} = \mathbf{g}(\mathbf{X})$ is predictive sufficient for $\mathbf{Z}$ if and only if the joint density of $\mathbf{X}$ and $\mathbf{Z}$ can be written

$$f(\mathbf{x}, \mathbf{z}; \boldsymbol{\phi}) = a(\mathbf{x})b(\mathbf{t}, \mathbf{z}, \boldsymbol{\phi})$$

for some functions a and b.

Proof: Suppose $\mathbf{T}$ is predictive sufficient. Then, since $\mathbf{t} = \mathbf{g}(\mathbf{x})$ is a function of $\mathbf{x}$, we have by (3.19) that

$$f(\mathbf{x}, \mathbf{z}; \boldsymbol{\phi}) = h(\mathbf{x})f(\mathbf{t}, \mathbf{z}; \boldsymbol{\phi}),$$

which factors into $a(\mathbf{x}) = h(\mathbf{x})$ and $b(\mathbf{t}, \mathbf{z}, \boldsymbol{\phi}) = f(\mathbf{t}, \mathbf{z}; \boldsymbol{\phi}))$.

Conversely, suppose that the factorization holds, i.e., there exist functions a and b such that $f(\mathbf{x}, \mathbf{z}; \boldsymbol{\phi}) = a(\mathbf{x})b(\mathbf{t}, \mathbf{z}, \boldsymbol{\phi})$. The marginal density of $\mathbf{x}$ is

$$\begin{aligned} f(\mathbf{x}; \boldsymbol{\phi}) &= \int f(\mathbf{x}, \mathbf{z}; \boldsymbol{\phi})\, d\mathbf{z} \\ &= a(\mathbf{x}) \int b(\mathbf{t}, \mathbf{z}; \boldsymbol{\phi})\, d\mathbf{z} \\ &= a(\mathbf{x})k(\mathbf{t}, \boldsymbol{\phi}), \end{aligned}$$

where the function k depends only on $\mathbf{t}$ and $\boldsymbol{\phi}$. Thus, by the ordinary factorization theorem (cf., Lehmann [1986]), $\mathbf{t}$ is sufficient for the parameter $\boldsymbol{\phi}$, showing (1) of the definition holds.

Since $\mathbf{t}$ is a function of $\mathbf{x}$,

$$\begin{aligned} f(\mathbf{z} \,|\, \mathbf{x}, \mathbf{t}; \boldsymbol{\phi}) &= f(\mathbf{z} \,|\, \mathbf{x}; \boldsymbol{\phi}) \\ &= \frac{f(\mathbf{x}, \mathbf{z}; \boldsymbol{\phi})}{f(\mathbf{x}; \boldsymbol{\phi})} \\ &= \frac{a(\mathbf{x})b(\mathbf{t}, \mathbf{z}, \boldsymbol{\phi})}{a(\mathbf{x})k(\mathbf{t}, \boldsymbol{\phi})} \\ &= g(\mathbf{t}, \mathbf{z}, \boldsymbol{\phi}), \end{aligned}$$

where the function $g = b/k$ does not depend on $\mathbf{x}$ so that $\mathbf{Z}$ is conditionally independent of $\mathbf{X}$, given $\mathbf{T}$, showing that (2) of the definition holds.

$\quad\square$

Definition   A statistic $\mathbf{T} = \mathbf{g}(\mathbf{X})$ is *minimal* predictive sufficient for $\mathbf{Z}$ if $\mathbf{T}$ is a function of every other predictive sufficient statistic for $\mathbf{Z}$.   □

Suppose $\mathbf{T}$ is classically minimal sufficient for the parameter $\boldsymbol{\phi}$ and that $\mathbf{X}$ and $\mathbf{Z}$ are conditionally independent given $\mathbf{T} = \mathbf{t}$ (i.e., $\mathbf{T}$ is also predictive sufficient). Then $\mathbf{T}$ is minimal predictive sufficient for $\mathbf{Z}$. For suppose $\mathbf{T}_1$ is any other predictive sufficient statistic for $\mathbf{Z}$, then by (1) of the definition, $\mathbf{T}_1$ is sufficient for $\boldsymbol{\phi}$. But since $\mathbf{T}$ is minimal sufficient for $\boldsymbol{\phi}$, $\mathbf{T}$ is a function of $\mathbf{T}_1$. Hence $\mathbf{T}$ is a function of every predictive sufficient statistic for $\mathbf{Z}$, so that $\mathbf{T}$ is minimal predictive sufficient.

The next theorem sometimes provides a useful way to determine whether a statistic is minimal predictive sufficient.

Theorem 3.3.   Suppose that $\mathbf{T} = \mathbf{g}(\mathbf{X})$ is a statistic such that for two sample points $\mathbf{x}$ and $\mathbf{x}^*$, $\mathbf{g}(\mathbf{x}) = \mathbf{g}(\mathbf{x}^*)$ if and only if $f(\mathbf{x}, \mathbf{z}; \boldsymbol{\phi}) = k(\mathbf{x}, \mathbf{x}^*)f(\mathbf{x}^*, \mathbf{z}; \boldsymbol{\phi})$, for all values of $\mathbf{z}$ and $\boldsymbol{\phi}$ and some function k. Then $\mathbf{T}$ is minimal predictive sufficient for $\mathbf{Z}$.

Proof: Note that the function k may depend on the sample points $\mathbf{x}$ and $\mathbf{x}^*$ but does not depend on the unknowns $\mathbf{z}$ or $\boldsymbol{\phi}$. Suppose $\mathbf{g}(\mathbf{x}) = \mathbf{g}(\mathbf{x}^*)$ if and only if $f(\mathbf{x}, \mathbf{z}; \boldsymbol{\phi}) = k(\mathbf{x}, \mathbf{x}^*)f(\mathbf{x}^*, \mathbf{z}; \boldsymbol{\phi})$, for all values of $\mathbf{z}$ and $\boldsymbol{\phi}$. For each $\mathbf{t}$ in the range of $\mathbf{T}$, let $A_{\mathbf{t}} = \{\mathbf{x} : \mathbf{g}(\mathbf{x}) = \mathbf{t}\}$. Then the sets $A_{\mathbf{t}}$ form a partition $\mathcal{P}_T$ of $\mathcal{D}_O$, the sample space of D_O. Let $\mathbf{u}_{\mathbf{t}}$ ($= \mathbf{h}(\mathbf{t})$, say) be a particular point in $A_{\mathbf{t}}$, that is, $\mathbf{g}(\mathbf{u}_{\mathbf{t}}) = \mathbf{t}$. Now for any $\mathbf{x}$, $\mathbf{x} \in A_{\mathbf{t}}$ for some $\mathbf{t}$, where $\mathbf{t} = \mathbf{g}(\mathbf{x}) = \mathbf{g}(\mathbf{u}_{\mathbf{t}})$ and

$$f(\mathbf{x}, \mathbf{z}; \boldsymbol{\phi}) = k(\mathbf{x}, \mathbf{u}_{\mathbf{t}})f(\mathbf{u}_{\mathbf{t}}, \mathbf{z}; \boldsymbol{\phi})$$

$$= k\big(\mathbf{x}, \mathbf{h}(\mathbf{t})\big)f\big(\mathbf{h}(\mathbf{t}), \mathbf{z}; \boldsymbol{\phi}\big)$$

$$= a(\mathbf{x})b(\mathbf{t}, \mathbf{z}, \boldsymbol{\phi}),$$

so that by the factorization theorem (Theorem 3.2), $\mathbf{T}$ is predictive sufficient.

To show that $\mathbf{T}$ is minimal predictive sufficient, let $\mathbf{T}_1 = \mathbf{g}_1(\mathbf{X})$ be any other predictive sufficient statistic for $\mathbf{Z}$ with corresponding partition $\mathcal{P}_{T_1}$. Then, by the factorization theorem, there exist functions a_1 and b_1 such that $f(\mathbf{x}, \mathbf{z}; \boldsymbol{\phi}) = a_1(\mathbf{x})b_1(\mathbf{t}_1, \mathbf{z}, \boldsymbol{\phi})$. Arguing as in Theorem 2.3, we can assume that at least one set $B_{\mathbf{t}_1}$ of $\mathcal{P}_{T_1}$ has at least two elements, $\mathbf{x}$ and $\mathbf{x}^*$, such that $\mathbf{g}_1(\mathbf{x}) = \mathbf{g}_1(\mathbf{x}^*)$ ($= \mathbf{t}_1$, say). Then $b_1(\mathbf{t}_1, \mathbf{z}, \boldsymbol{\phi})$ is the same for $\mathbf{x}$ and $\mathbf{x}^*$ so that $f(\mathbf{x}, \mathbf{z}; \boldsymbol{\phi}) = [a_1(\mathbf{x})/a_1(\mathbf{x}^*)]f(\mathbf{x}^*, \mathbf{z}; \boldsymbol{\phi})$. Since the ratio $a_1(\mathbf{x})/a_1(\mathbf{x}^*)$ does not depend on $\mathbf{z}$ or $\boldsymbol{\phi}$, the assumption of the theorem implies that $\mathbf{g}(\mathbf{x}) = \mathbf{g}(\mathbf{x}^*)$ ($= \mathbf{t}_2$, say). Thus $\mathbf{x}$ and $\mathbf{x}^*$ belong to $A_{\mathbf{t}_2}$ and $B_{\mathbf{t}_1} \subset A_{\mathbf{t}_2}$. Thus every partition set of $\mathcal{P}_{T_1}$ is a partition set of $\mathcal{P}_T$, and T is a function of T_1. Hence $\mathbf{T}$ is minimal predictive sufficient for $\mathbf{Z}$.   □

We note that in endeavoring to apply the preceding theorem to situations, we must have $\mathbf{z}$ the same for both $\mathbf{x}$ and $\mathbf{x}^*$.

A statistic $\mathbf{T}$ is *complete*, that is, has a complete family of distributions, if no two different functions of $\mathbf{T}$ have the same expectation. This is equivalent to saying that $\mathbf{T}$ is complete if, for any function $\mathbf{h}(\mathbf{T})$, $E[\mathbf{h}(\mathbf{T}); \boldsymbol{\phi}] = \mathbf{0}$ for all $\boldsymbol{\phi} \in \Phi$ implies that $\mathbf{h}(\mathbf{T}) = \mathbf{0}$ with probability one for all $\boldsymbol{\phi} \in \Phi$.

3.12   PREDICTIVE SUFFICIENCY IN SAMPLING

We now wish to examine the application of predictive sufficiency to the finite population sampling situation with the model $f(\mathbf{y}; \boldsymbol{\phi})$ for the population vector $\mathbf{Y}$. In particular we shall focus on the prediction of $\mathbf{Y}_{\bar{s}}$, the vector of y-values corresponding to $\bar{s}$, the ordered set of unit labels not included in the sample. In applying the above theory we equate $\mathbf{x}$ with $d_O = (s_O, \mathbf{y}_s)$ and $\mathbf{z}$ with $\mathbf{y}_{\bar{s}}$. The information relevant to inference about $\mathbf{Y}_{\bar{s}}$ (and $\boldsymbol{\phi}$) is then contained in the joint density of D_O and $\mathbf{Y}_{\bar{s}}$ (Berger and Wolpert [1988]).

Theorem 3.4.   Let $\mathbf{t} = d = (s, \mathbf{y}_s)$ be the reduced data set. Then, for conventional or adaptive sampling, $\mathbf{T}$ is predictive sufficient for $\mathbf{Y}_{\bar{s}}$.

Proof: For conventional or adaptive sampling it follows from (3.2) that

$$f(s_O, \mathbf{y}_s, \mathbf{y}_{\bar{s}}; \boldsymbol{\phi}) = p(s_O \mid \mathbf{y}_s)f\big[\mathbf{y}(s, \mathbf{y}_s, \mathbf{y}_{\bar{s}}); \boldsymbol{\phi}\big],$$

where $f[\mathbf{y}(s, \mathbf{y}_s, \mathbf{y}_{\bar{s}}); \boldsymbol{\phi}] = b(\mathbf{t}, \mathbf{y}_{\bar{s}}, \boldsymbol{\phi})$ as $\mathbf{t}$ is a function of s and $\mathbf{y}_s$. Also setting $a(d_O) = p(s_O \mid \mathbf{y}_s)$ we have that

$$f(d_O, \mathbf{y}_{\bar{s}}; \boldsymbol{\phi}) = f(s_O, \mathbf{y}_s, \mathbf{y}_{\bar{s}}; \boldsymbol{\phi})$$
$$= a(d_O)b(\mathbf{t}, \mathbf{y}_{\bar{s}}; \boldsymbol{\phi}).$$

Hence, by Theorem 3.2, we see that $\mathbf{T}$ is predictive sufficient for $\mathbf{Y}_{\bar{s}}$ irrespective of the population model.   □

Although $\mathbf{T}$ is predictive sufficient it is not necessarily minimal predictive sufficient.

Suppose that for a statistic $\mathbf{T}_1$, the population model component f further factorizes into

$$f\big[\mathbf{y}(s, \mathbf{y}_s, \mathbf{y}_{\bar{s}}); \boldsymbol{\phi}\big] = a_s(\mathbf{y}_s)b_s(\mathbf{t}_1, \mathbf{y}_{\bar{s}}, \boldsymbol{\phi}),$$

then $(s, \mathbf{t}_1)$ is predictive sufficient for $\mathbf{y}_{\bar{s}}$. If, in addition, the function b_s is the same for each sample s, then $\mathbf{t}_1$ alone (i.e., without s) is predictive sufficient for $\mathbf{y}_{\bar{s}}$.

Now suppose that instead of inference about $\mathbf{Y}_{\bar{s}}$ we wish to focus specifically on a population characteristic $Z = z(\mathbf{Y})$ such as the population total. We assume Z to have the property that there exists a vector $\mathbf{v} = \mathbf{v}(\mathbf{y}_{\bar{s}})$ of variables for which there is an invertible function between $\mathbf{y}_{\bar{s}}$ and $(\mathbf{v}, z)$: its Jacobian will be a function of $\mathbf{v}$ and z, $J(\mathbf{v}, z)$, say. This implies that there is an invertible function between $(s_O, \mathbf{y}_s, \mathbf{y}_{\bar{s}})$ and $(s_O, \mathbf{y}_s, \mathbf{v}, z)$, but the Jacobian is still the same, as s_O and $\mathbf{y}_s$ have an identity transformation. Then, for a conventional or adaptive design, the joint distribution of $(s_O, \mathbf{y}_s, \mathbf{v}, z)$ is

$$f_1(s_O, \mathbf{y}_s, \mathbf{v}, z; \boldsymbol{\phi}) = f\big[s_O, \mathbf{y}_s, \mathbf{y}_{\bar{s}}(\mathbf{v}, z); \boldsymbol{\phi}\big]|J(\mathbf{v}, z)|$$
$$= p(s_O \mid \mathbf{y}_s)f_s\big[\mathbf{y}_s, \mathbf{y}_{\bar{s}}(\mathbf{v}, z); \boldsymbol{\phi}\big]|J(\mathbf{v}, z)|,$$

by (3.3) and (3.4). The joint density of the data and Z is then

$$f(s_O, \mathbf{y}_s, z; \boldsymbol{\phi}) = \int f_1(s_O, \mathbf{y}_s, \mathbf{v}, z; \boldsymbol{\phi}) \, d\,\mathbf{v}$$

$$= p(s_O \mid \mathbf{y}_s) \int f_s \left[\mathbf{y}_s, \mathbf{y}_{\bar{s}}(\mathbf{v}, z); \boldsymbol{\phi} \right] |J(\mathbf{v}, z)| \, d\,\mathbf{v}$$

$$= a(s_O, \mathbf{y}_s) b(s, \mathbf{y}_s, z, \boldsymbol{\phi})$$

$$= a(d_O) b(\mathbf{t}, z, \boldsymbol{\phi}).$$

Hence, by Theorem 3.2, the reduced data $T = D_R$ are sufficient for Z. We note that when Z is the population total it can be readily shown that such a function $\mathbf{v} = \mathbf{v}(\mathbf{y}_{\bar{s}})$ exists.

With $Z = z(\mathbf{Y})$ a function of $\mathbf{y}$ not depending on the sample s, for example with Z the population total, we can use Theorem 3.3 to try and prove minimal predictive sufficiency. Since

$$f(s_O, \mathbf{y}_s, z; \boldsymbol{\phi}) = p(s_O \mid \mathbf{y}_s) f_s(\mathbf{y}_s, z; \boldsymbol{\phi}) = a(d_O) f_s(\mathbf{y}_s, z; \boldsymbol{\phi}),$$

we see that T is minimal predictive sufficient for Z if, for two values $d_O = (s_O, \mathbf{y}_s)$ and $d_O^* = (s_O^*, \mathbf{y}_{s^*})$, $T(d_O) = T(d_O^*)$ if and only if there is a function k not depending on the unknowns such that

$$f_s(\mathbf{y}_s, z; \boldsymbol{\phi}) = k(d_O, d_O^*) f_{s^*}(\mathbf{y}_{s^*}, z; \boldsymbol{\phi}),$$

[as $a(d_O)$ and $a(d_O^*)$ are incorporated into $k(d_O, d_O^*)$].

With a quantity Z such as the vector $\mathbf{Y}_{\bar{s}}$ of unknowns or the total of the unknowns, Z depends on the sample as well as on $\mathbf{Y}$. Even so, Theorem 3.3 can still be used. Interestingly, even though two data values d_O and d_O^* might involve two different samples, even two different reduced samples, a vector such as $Z = \mathbf{Y}_{\bar{s}}$ is just a random vector of $N - \nu$ components that conceivably could take the same value with d_O and d_O^* even though different components of $\mathbf{Y}$ are involved.

We now give a number of examples. For ease of exposition it is sometimes helpful to have a different notation for $\mathbf{y}_s$, namely $\mathbf{v}$ [$= \mathbf{v}(s, \mathbf{y}_s)$], where $v_i = y_{s_i}$ ($i = 1, 2, \ldots, \nu$) is the ith element of $\mathbf{y}_s$.

Example 3.12   Fixed Population Model.   The fixed population situation can be viewed as having a degenerate population model with point mass at the parameter value $\boldsymbol{\phi} = \boldsymbol{\theta} = (\theta_1, \ldots, \theta_N)'$, so that $P(Y_1 = \theta_1, \ldots, Y_N = \theta_N) = 1$ and the probability of any other value of $\mathbf{Y}$ is zero (Rao [1977]). With adaptive or conventional sampling, to predict any function $Z = z(\mathbf{Y})$, the relevant density is

$$f(s_O, \mathbf{y}_s, \mathbf{y}_{\bar{s}}; \boldsymbol{\theta}) = p(s_O \mid \mathbf{y}_s) f_s(\mathbf{y}_s, \mathbf{y}_{\bar{s}}; \boldsymbol{\theta}).$$

Since $P(\mathbf{Y} = \boldsymbol{\theta}) = 1$, the factor $f_s(\mathbf{y}_s, \mathbf{y}_{\bar{s}}; \boldsymbol{\theta})$ is the indicator function $I(s, \mathbf{y}_s, \mathbf{y}_{\bar{s}}, \boldsymbol{\theta})$ which equals 1 if the ith component of $\mathbf{y}_s$ equals the s_ith component of $\boldsymbol{\theta}$ and the jth component of $\mathbf{y}_{\bar{s}}$ equals the $\bar{s}_j$th component of $\boldsymbol{\theta}$ (for $i = 1, \ldots, \nu$ and $j = 1, \ldots, N - \nu$), and equals zero otherwise.

The data are connected to the unknowns only through $I(s, \mathbf{y}_s, \mathbf{y}_{\bar{s}}, \boldsymbol{\theta})$, and the minimal sufficient statistic consists of the reduced data vector $d = (s, \mathbf{y}_s)$.

Example 3.13 I.I.D. Normal Model. Suppose that the population values $Y_1, \ldots, Y_N$ are independent and identically distributed (i.i.d.) random variables from a normal distribution with unknown mean μ and variance σ^2 and that we wish to predict the population total $Z = z(\mathbf{Y}) = \sum_{i=1}^{N} Y_i$. With conventional or adaptive sampling, the joint density of the data and Z is

$$f(s_O, \mathbf{y}_s, z; \mu, \sigma^2) = p(s_O \mid \mathbf{y}_s) f_s(\mathbf{y}_s, z; \mu, \sigma^2)$$
$$= p(s_O \mid \mathbf{y}_s) f_s(z \mid \mathbf{y}_s; \mu, \sigma^2) f_s(\mathbf{y}_s; \mu, \sigma^2).$$

Introducing $\mathbf{v} = \mathbf{y}_s$, let t_1 and t_2 be respectively the sample mean and sample variance of the values for the ν distinct units in the sample; that is, $t_1 = \sum_{i=1}^{\nu} (v_i/\nu) = \bar{v}$ and $t_2 = \sum_{i=1}^{\nu} (v_i - \bar{v})^2 / (\nu - 1)$. Since Z is the sum of N i.i.d. normal random variables, with ν of them given, the conditional distribution of Z is normal with mean $(N - \nu)\mu + \nu t_1$ and variance $(N - \nu)\sigma^2$. Thus the conditional density of Z given $\mathbf{y}_s$ is a function of z, t_1, $\nu(s)$, μ, and σ^2. The last factor $f_s(\mathbf{y}_s; \mu, \sigma^2)$ is the joint density of $\nu(s)$ independent normal random variables and can be expressed as follows:

$$f_s(\mathbf{y}_s; \mu, \sigma^2) = \frac{1}{(2\pi\sigma^2)^{\nu/2}} \exp\left\{ -\frac{1}{2\sigma^2} \left[\sum_{i=1}^{\nu} (v_i - \mu)^2 \right] \right\}$$

$$= \frac{1}{(2\pi\sigma^2)^{\nu/2}} \exp\left\{ -\frac{1}{2\sigma^2} \left[\sum_{i=1}^{\nu} (v_i - \bar{v})^2 + \nu(\bar{v} - \mu)^2 \right] \right\}$$

$$= h\big(t_1, t_2, \nu(s), \mu, \sigma^2\big).$$

Thus, setting $a(s_O, \mathbf{y}_s) = p(s_O \mid \mathbf{y}_s)$, the joint density can be written

$$f(s_O, \mathbf{y}_s, z; \mu, \sigma^2) = a(s_O, \mathbf{y}_s) b(t_1, t_2, \nu(s), z, \mu, \sigma^2).$$

and $(t_1, t_2, \nu(s))'$ is predictive sufficient for z by the factorization theorem. Suppose, in addition that the design has fixed effective sample size so that the number of distinct units in every possible sample is n. Because of the fixed sample size and the i.i.d. variables, the densities f_s are the same for every possible sample so that the subscript s in f_s can be dropped. In this case (t_1, t_2) is the minimal sufficient statistic so that all predictive inference about the population total may be based on the sample mean and variance.

Example 3.14 Multivariate Normal Linear Model. Suppose that the population vector $\mathbf{Y}$ has an N-dimensional multivariate normal distribution with mean vector $\mathbf{X}\boldsymbol{\beta}$ and variance-covariance matrix $\sigma^2\boldsymbol{\Sigma}$, where $\mathbf{X}$ and $\boldsymbol{\Sigma}$ are known matrices, $\boldsymbol{\beta}$ is a vector of p unknown parameters, and σ^2 is unknown. For adaptive or conventional sampling, the joint density for predicting the population total $Z = \sum_{i=1}^{N} Y_i$ is

$$f(s_O, \mathbf{y}_s, z; \boldsymbol{\beta}, \sigma^2) = p(s_O \mid \mathbf{y}_s) f_s(z \mid \mathbf{y}_s; \boldsymbol{\beta}, \sigma^2) f_s(\mathbf{y}_s; \boldsymbol{\beta}, \sigma^2).$$

Let $\mathbf{X}_s$ denote the matrix of rows from $\mathbf{X}$ corresponding to the $\nu(s)$ distinct units in the reduced sample s. Let $\boldsymbol{\Sigma}_{ss}$ denote the covariance matrix of the sample values $\mathbf{Y}_s$ $(= \mathbf{V})$; the elements of $\boldsymbol{\Sigma}_{ss}$ are selected from $\boldsymbol{\Sigma}$ according to the units in s. Let $\boldsymbol{\Sigma}_{s\bar{s}}$ be the covariance matrix of $\mathbf{Y}_s$ and $\mathbf{Y}_{\bar{s}}$ and let $\mathbf{1}_N$ denote an appropriate vector of N 1's. Then the marginal density $f_s(\mathbf{y}_s; \boldsymbol{\beta}, \sigma^2)$ of the values $\mathbf{Y}_s$ in the reduced sample is ν-dimensional multivariate normal with mean $\mathbf{X}_s\boldsymbol{\beta}$ and covariance matrix $\sigma^2 \boldsymbol{\Sigma}_{ss}$. If s were fixed, the sufficient statistics for that family of distributions would be $\mathbf{X}_s'\boldsymbol{\Sigma}_{ss}^{-1}\mathbf{Y}_s$ and $\mathbf{Y}_s'\boldsymbol{\Sigma}_{ss}^{-1}\mathbf{Y}_s$. However as s is not fixed, the marginal density also depends on s through $\mathbf{X}_s'\boldsymbol{\Sigma}_{ss}^{-1}\mathbf{X}_s$.

The population total Z is a normal random variable with mean $\mathbf{1}_N'\mathbf{X}\boldsymbol{\beta}$ and variance $\mathbf{1}_N'\boldsymbol{\Sigma}\mathbf{1}_N$. For a given sample s, Z and $\mathbf{Y}_s$ have a joint multivariate normal distribution. To completely specify that distribution, we need to know the vector of covariances of Z with $\mathbf{Y}_s$, namely

$$\begin{aligned} \text{cov}[Z, \mathbf{Y}_s] &= \text{cov}[\mathbf{1}_\nu'\mathbf{Y}_s + \mathbf{1}_{N-\nu}'\mathbf{Y}_{\bar{s}}, \mathbf{Y}_s] \\ &= (\mathbf{1}_\nu'\boldsymbol{\Sigma}_{ss} + \mathbf{1}_{N-\nu}'\boldsymbol{\Sigma}_{\bar{s}s})\sigma^2 \\ &= \mathbf{c}'\sigma^2, \quad \text{say.} \end{aligned}$$

For the coordinates given by s, we can use this joint distribution to find the conditional density of Z given $\mathbf{y}_s$. From Seber [1977, p. 36] this is the univariate normal with mean $\mathbf{1}_N'\mathbf{X}\boldsymbol{\beta} + \mathbf{c}'\boldsymbol{\Sigma}_{ss}^{-1}(\mathbf{y}_s - \mathbf{X}_s\boldsymbol{\beta}) = \mathbf{c}'\boldsymbol{\Sigma}_{ss}^{-1}\mathbf{Y}_s + (\mathbf{X}'\mathbf{1}_N - \mathbf{X}_s'\boldsymbol{\Sigma}_{ss}^{-1}\mathbf{c})'\boldsymbol{\beta}$ and variance $(\mathbf{1}_N'\boldsymbol{\Sigma}\mathbf{1}_N - \mathbf{c}'\boldsymbol{\Sigma}_{ss}^{-1}\mathbf{c})\sigma^2$. Thus $f_s(z \mid \mathbf{y}_s; \boldsymbol{\beta}, \sigma^2)$ depends not only on the statistic $\mathbf{c}'\boldsymbol{\Sigma}_{ss}^{-1}\mathbf{Y}_s$ and the unknowns z, $\boldsymbol{\beta}$, and σ^2, but it also depends on s through the additional statistics $\mathbf{c}'\boldsymbol{\Sigma}_{ss}^{-1}\mathbf{X}_s$ and $\mathbf{c}'\boldsymbol{\Sigma}_{ss}^{-1}\mathbf{c}$. Thus setting $a(s_O, \mathbf{y}_s) = p(s_O \mid \mathbf{y}_s)$ we find that

$$f(s_O, \mathbf{y}_s, z; \boldsymbol{\beta}, \sigma^2) = a(s_O, \mathbf{y}_s)b(\mathbf{t}, z, \boldsymbol{\beta}, \sigma^2)$$

where

$$\mathbf{t} = \left(\mathbf{X}_s'\boldsymbol{\Sigma}_{ss}^{-1}\mathbf{y}_s, \mathbf{y}_s'\boldsymbol{\Sigma}_{ss}^{-1}\mathbf{y}_s, \mathbf{c}'\boldsymbol{\Sigma}_{ss}^{-1}\mathbf{y}_s, \mathbf{X}_s'\boldsymbol{\Sigma}_{ss}^{-1}\mathbf{X}_s, \mathbf{c}'\boldsymbol{\Sigma}_{ss}^{-1}\mathbf{X}_s, \mathbf{c}'\boldsymbol{\Sigma}_{ss}^{-1}\mathbf{c} \right)'.$$

Hence $\mathbf{T}$ is predictive sufficient for the population total. □

3.13 PREDICTIVE LIKELIHOODS

Because the reduced data $d = (s, \mathbf{y}_s)$ are predictive sufficient with any conventional or adaptive design, and because the data d_O including order of selection and repeat selections affect the likelihood only through the occurrence of s_O in the design part, we shall for notational simplicity use only the reduced data in this section. From the likelihood point of view the joint density $f(d, z; \boldsymbol{\phi})$, assuming it exists, contains the relevant information for predictive inference about the random variable $Z = z(\mathbf{Y})$. The unknown parameter $\boldsymbol{\phi}$ may be of interest in itself as an object of inference, or it may be simply a "nuisance" parameter to be dealt with in the prediction of Z. To obtain a likelihood function for z based on the data d and not depending on the parameter $\boldsymbol{\phi}$, several methods have been developed. Because of the unknown $\boldsymbol{\phi}$, no unique definition exists for the likelihood of z.

The joint density can be factored into

$$f(d, z; \boldsymbol{\phi}) = f(z \mid d; \boldsymbol{\phi}) f(d; \boldsymbol{\phi}).$$

For perspective, assume first that the value of $\boldsymbol{\phi}$ is known. Reasonable choices for a point predictor include the conditional mean

$$\hat{Z} = \mathrm{E}\big[Z \mid D = d; \boldsymbol{\phi}\big]$$
$$= \int z f(z \mid d; \boldsymbol{\phi}) \, dz,$$

or the value of z that maximizes the conditional density given the data; that is,

$$\tilde{Z} = \arg \sup_{z} f(z \mid d; \boldsymbol{\phi}).$$

The problem is that $\boldsymbol{\phi}$ is in general not known. Perhaps the simplest solution then is to form an estimative likelihood function for z by substituting an estimate, such as a maximum likelihood estimate, of $\boldsymbol{\phi}$ into $f(z \mid d; \boldsymbol{\phi})$. Let

$$\hat{\boldsymbol{\phi}} = \arg \sup_{\boldsymbol{\phi}} f(d; \boldsymbol{\phi})$$

be a maximum likelihood estimator (m.l.e.) of $\boldsymbol{\phi}$ (i.e., maximizes $f(d; \boldsymbol{\phi})$). The *estimative predictive likelihood* of z is defined as

$$L_e(z \mid d) = f(z \mid d; \hat{\boldsymbol{\phi}}). \tag{3.20}$$

A value $\hat{Z}$ that maximizes (3.20) is called an *estimative maximum likelihood predictor* of Z (cf. Bolfarine and Zacks [1992, p. 101]).

For many familiar prediction problems, the estimative approach seems to lead to reasonable estimates. Examples in sampling include standard regression predictors and the "kriging" predictors of spatial statistics. However, by treating the estimated parameter value as if it were the true value, the estimative likelihood L_e may under-assess the uncertainty in some problems (cf. Bjørnstad [1990]).

Another approach to predictive likelihood is to find the values $\hat{z}$ and $\hat{\phi}$ that simultaneously maximize the joint density $f(d, z; \phi)$ with the given data d. Let

$$\hat{\phi}_z = \arg\sup_{\phi} f(d, z; \phi).$$

The *profile predictive likelihood* of z is defined as

$$L_p(z \mid d) = f(d, z; \hat{\phi}_z). \tag{3.21}$$

Additionally, the profile predictive likelihood may be normalized by dividing by a constant $k(d)$ that makes the likelihood integrate over z to 1. A value $\hat{Z}$ that maximizes (3.21) is called a *maximum profile likelihood predictor* of Z (Lejeune and Faulkenberry [1982], Bolfarine and Zacks [1992, p. 102]). The ordinary m.l.e. $\hat{\phi}$ is the value of ϕ that gives the highest density for the data actually observed, while the profile maximum likelihood estimator $\hat{\phi}$ maximizes the joint density of both the observed d and the hypothesized z.

Other approaches to predictive likelihood include the sufficiency-based predictive likelihoods of Lauritzen [1974], Hinkley [1979], and Butler [1986, 1990]. These likelihoods involve the conditional joint distribution of Z and D given a statistic $t(D, Z)$ that is sufficient for ϕ. Because of the sufficiency, the joint conditional distribution does not involve ϕ. For standard normal models, the estimative, profile, and sufficiency-based approaches all lead to the same maximum likelihood predictors, though the sufficiency-based approach may lead to greater computational difficulties (Bolfarine and Zacks [1992]). For a general review see Bjørnstad [1990]).

For normal models, the maximum likelihood predictors tend to coincide with the best linear unbiased predictors under conventional designs (see Bolfarine and Zacks [1992] and the examples below). The maximum likelihood predictors do not depend on the design used for any conventional or adaptive design. Although Hinkley [1979] gives an example in which maximum likelihood prediction depends on a stopping rule, the stopping rule in his example involves not only the observed variable (i.e., the design) but the variable to be predicted (i.e., the unobserved Z). In the sampling situation, a stopping rule involving a variable Z to be predicted would be part of the population model, not the sampling design.

Example 3.15   Prediction of the Unobserved Population Vector. Here $\mathbf{z} = \mathbf{y}_{\bar{s}}$, and we consider adaptive or conventional designs. Then

$$
\begin{aligned}
f(d, \mathbf{z}; \phi) &= f(s, \mathbf{y}_s, \mathbf{y}_{\bar{s}}; \phi) \\
&= p(s \mid \mathbf{y}_s) f_s(\mathbf{y}_s, \mathbf{y}_{\bar{s}}; \phi) \\
&= p(s \mid \mathbf{y}_s) f_s(\mathbf{y}_{\bar{s}} \mid \mathbf{y}_s; \phi) f_s(\mathbf{y}_s; \phi).
\end{aligned}
$$

Now

$$f(d; \phi) = p(s \mid \mathbf{y}_s) f_s(\mathbf{y}_s; \phi),$$

and the m.l.e. $\hat{\boldsymbol{\phi}}$ is the value of $\boldsymbol{\phi}$ that maximizes $f_s(\mathbf{y}_s; \boldsymbol{\phi})$. The estimative predictive likelihood is then

$$L_e(\mathbf{z} \mid d) = f(\mathbf{z} \mid d; \hat{\boldsymbol{\phi}})$$

$$= \frac{f(d, \mathbf{z}; \hat{\boldsymbol{\phi}})}{f(d; \hat{\boldsymbol{\phi}})}$$

$$= f_s(\mathbf{y}_{\bar{s}} \mid \mathbf{y}_s; \hat{\boldsymbol{\phi}}).$$

The value of $\mathbf{y}_{\bar{s}}$ maximizing the conditional estimative density $f_s(\mathbf{y}_{\bar{s}} \mid \mathbf{y}_s; \hat{\boldsymbol{\phi}})$ is the estimative maximum likelihood predictor of $\mathbf{y}$.

Example 3.16   I.I.D. Normal Model.   This is a continuation of Example 3.13 in the previous section. Given a conventional or adaptive sample, the joint density of the data and population total Z is

$$f(s, \mathbf{y}_s, z; \mu, \sigma^2) = p(s \mid \mathbf{y}_s) f_s(z \mid \mathbf{y}_s; \mu, \sigma^2) f_s(\mathbf{y}_s; \mu, \sigma^2).$$

The theory of Example 3.15 applies here with $\mathbf{y}_{\bar{s}}$ replaced by z. Thus since $f_s(\mathbf{y}_s; \mu, \sigma^2)$, the marginal density function of $\mathbf{Y}_s \ (= \mathbf{V})$, is v-dimensional multivariate normal, the m.l.e.'s of μ and σ^2 are $\hat{\mu} = \bar{v} = \sum_{i=1}^{v} v_i/v$ and $\hat{\sigma}^2 = \sum_{i=1}^{v} (v_i - \bar{v})^2/v$, respectively. To get the predictive likelihood, we require the conditional density $f_s(z \mid \mathbf{y}_s; \mu, \sigma^2)$, which is normal with mean $(N - v)\mu + v\bar{v}$ and variance $(N - v)\sigma^2$. For any given values of μ and σ^2, this density is maximized at its mode, that is, by setting z equal to the conditional mean. Thus replacing μ and σ^2 by their maximum likelihood estimates, the maximum estimative likelihood predictor for the population total is $\hat{Z} = (N - v)\bar{v} + v\bar{v} = N\bar{v}$, which is known as the "expansion" predictor.

Since the conditional distribution of Z given $\mathbf{y}_s$ with $\hat{\mu}$ and $\hat{\sigma}^2$ assumed to be the true parameters is normal with mean $\hat{Z}$ and variance $(N - v)\hat{\sigma}^2$, a $(1 - \alpha)$ prediction interval for Z based on quantiles of this estimative predictive likelihood is

$$\hat{Z} \pm t\hat{\sigma}\sqrt{N - v},$$

where t is the upper $\alpha/2$ point of the standard normal distribution and $v = v(s)$ is a function of s.

The maximum profile likelihood predictor can be shown to also be $\hat{Z}$ (Bjørnstad [1990, p. 249], Bolfarine and Zacks [1992]). A $(1 - \alpha)$ prediction interval for Z based on quantiles of the profile predictive likelihood is

$$\hat{Z} \pm t_v \hat{\sigma} \left[\frac{N(N - v)}{v} \right]^{1/2},$$

where t_v is the upper $\alpha/2$ point of the t-distribution with v degrees of freedom. For comparison, the classical $(1 - \alpha)$ prediction interval for Z based on pivotal methods is

$$\hat{Z} \pm t_{\nu-1}\hat{\sigma}\left[\frac{N(N-\nu)}{(\nu-1)}\right]^{1/2},$$

where $t_{\nu-1}$ is the upper $\alpha/2$ point of the t-distribution with $(\nu-1)$ degrees of freedom.

Example 3.17   Multivariate Normal Linear Model.   This is a continuation of Example 3.14 in the previous section. Given a conventional or adaptive sample, the joint density of the data and population total Z is

$$f(s, \mathbf{y}_s, z; \boldsymbol{\beta}, \sigma^2) = p(s \mid \mathbf{y}_s)f_s(z \mid \mathbf{y}_s; \boldsymbol{\beta}, \sigma^2)f_s(\mathbf{y}_s; \boldsymbol{\beta}, \sigma^2).$$

The marginal density $f_s(\mathbf{y}_s; \boldsymbol{\beta}, \sigma^2)$ is ν-dimensional multivariate normal with mean $\mathbf{X}_s\boldsymbol{\beta}$ and covariance $\sigma^2\boldsymbol{\Sigma}_{ss}$ (with $\boldsymbol{\Sigma}_{ss}$ known). The values of $\boldsymbol{\beta}$ and σ^2 that maximize the marginal density are the maximum likelihood estimates

$$\hat{\boldsymbol{\beta}} = (\mathbf{X}_s'\boldsymbol{\Sigma}_{ss}^{-1}\mathbf{X}_s)^{-1}\mathbf{X}_s'\boldsymbol{\Sigma}_{ss}^{-1}\mathbf{Y}_s$$

and

$$\hat{\sigma}^2 = (\mathbf{Y}_s - \mathbf{X}\hat{\boldsymbol{\beta}})'\boldsymbol{\Sigma}_{ss}^{-1}(\mathbf{Y}_s - \mathbf{X}\hat{\boldsymbol{\beta}})/\nu.$$

From Example 3.14, the conditional density of Z given $\mathbf{y}_s$ is univariate normal with mean $\mathbf{c}'\boldsymbol{\Sigma}_{ss}^{-1}\mathbf{Y}_s + (\mathbf{X}'\mathbf{1}_N - \mathbf{X}_s'\boldsymbol{\Sigma}_{ss}^{-1}\mathbf{c})'\boldsymbol{\beta}$ and variance $(\mathbf{1}_N'\boldsymbol{\Sigma}\mathbf{1}_N - \mathbf{c}'\boldsymbol{\Sigma}_{ss}^{-1}\mathbf{c})\sigma^2$. Because a normal density has its mode at its mean, for any fixed values of the parameters $\boldsymbol{\beta}$ and σ^2, the value of z that maximizes $f_s(z \mid \mathbf{y}_s; \boldsymbol{\beta}, \sigma^2)$ is the conditional mean. Thus the maximum estimative likelihood predictor of the population total is

$$\hat{Z} = \mathbf{c}'\boldsymbol{\Sigma}_{ss}^{-1}\mathbf{Y}_s + (\mathbf{X}'\mathbf{1}_N - \mathbf{X}_s'\boldsymbol{\Sigma}_{ss}^{-1}\mathbf{c})'\hat{\boldsymbol{\beta}}.$$

In addition, $\hat{Z}$ will be recognized as the usual predictor used in regression estimation and in spatial "kriging" (cf. Cressie [1993, pp. 173–174]). Whereas best linear unbiasedness holds under the model only with a conventional design, the maximum likelihood property holds with adaptive designs as well.

Example 3.18   Model for a Stratified Population.   A stratified population is modeled as independent normal random variables with means and variances depending on stratum membership, so that for stratum h with N_h units, Y_{hi} has a normal (μ_h, σ_h^2) distribution $(h = 1, \ldots, L, i = 1, \ldots, N_h)$. A stratified sample is selected by any design of the type $p(s \mid \mathbf{y}; \boldsymbol{\phi}) = p(s \mid \mathbf{y}_s)$. Then, as a special case of Example 3.7, the maximum estimative likelihood predictor of the population total can be shown to be

$$\hat{Z} = \sum_{h=1}^{L} N_h\bar{y}_h.$$

Thus under the assumed model the standard predictor $\hat{Z}$ is a maximum likelihood predictor for the population total, irrespective of whether the design is conventional or adaptive.

Example 3.19   Size-Biased Sampling.   In size-biased sampling, objects such as groups of animals are observed with probability related to a variable of interest such as group size (Cook and Martin [1974], Godambe and Rajarshi [1989]), Drummer and McDonald [1987], Quang [1991], Otto and Pollock [1990], and Quinn [1981]). Let y_i be the size of the ith group. Following Godambe and Rajarshi [1989], the population model assumed is of the form

$$f(\mathbf{y}; \boldsymbol{\phi}) = \prod_{i=1}^{N} f_i(y_i; \boldsymbol{\phi});$$

that is, sizes of groups are generated independently. After $\mathbf{y}$ is realized, a Bernoulli experiment with chance of success $w_i(y_i)$ determines which of the N groups are observed. The probability of observing a specific sample s of the groups is

$$p(s \mid \mathbf{y}) = \prod_{i \in s} w_i(y_i) \prod_{i \in \bar{s}} [1 - w_i(y_i)],$$

so that the design depends on y-values for the units outside the sample as well as those within the sample. The joint density of the data and the unobserved values is thus

$$f(s, \mathbf{y}; \boldsymbol{\phi}) = p(s \mid \mathbf{y}) f(\mathbf{y}; \boldsymbol{\phi})$$
$$= \prod_{i \in s} w_i(y_i) \prod_{i \in \bar{s}} [1 - w_i(y_i)] \prod_{i \in s} f_i(y_i; \boldsymbol{\phi}).$$

The unknowns are linked with the data in both model and design components, so that inference based on likelihood methods will involve the selection probabilities. In fact, maximum likelihood for this situation involves some peculiarities (Godambe and Rajarshi [1989]). Interestingly, whereas the probability of selecting s as the sample depends on y-values of units outside the sample, the probability that a unit i is included in the sample depends only on y_i. (See also the discussion of nonresponse models at the end of this chapter.)

Example 3.20   Nonstandard Design.   Suppose an environmental regulatory agency wishes to assess the fuel efficiency of a certain model of car. So the agency requests that the automobile manufacturer loan them $n = 10$ cars for testing. The model for the population of N cars produced is that fuel efficiency (in miles per gallon, say, or kilometers per liter) is that $Y_1, \ldots, Y_N$ are i.i.d. normal with unknown mean μ and unknown variance σ^2.

It turns out that the manufacturer is able to test with the same type of testing equipment as the agency the fuel efficiency of each car as it comes off the assembly line. Of the 1000 cars of that model produced during the test week, the company selects the 10 cars with the highest fuel efficiency and gives them to the agency for testing.

Now, with any conventional design such as a simple random sample or even a purposively selected sample not depending on the variable of interest, the sample mean of the ten cars tested by the agency would produce a maximum likelihood estimate of the population mean for that model. Under the assumed model and with a conventionally selected sample, the sample mean has a normal distribution, with mean μ and variance $\sigma^2/10$. With the informative design used by the manufacturer, however, the sample mean is the mean of the largest 10 order statistics of 1000 i.i.d. normal random variables and has quite a different distribution and a different likelihood.

Interestingly, had the manufacturer used any regular adaptive design, for example adding cars to the sample until five of them gave a fuel efficiency reading above a specified constant c, then likelihood-based estimation (but not unbiased estimation) would be unaffected by the design and the sample mean would be the maximum likelihood estimate of the population mean.

With the nonstandard design, the design involves unknowns as well as the data. Effective inference in such a case requires knowledge of the design used. □

3.14 BEST UNBIASED PREDICTORS

Suppose that a random vector $\mathbf{X}$ is observed with the objective of predicting the value of a random variable Z and that the joint density of $\mathbf{X}$ and Z is $f(\mathbf{x}, z; \boldsymbol{\phi})$, where $\boldsymbol{\phi} \in \Phi$. Suppose $\mathbf{T} = \mathbf{g}(\mathbf{X})$ is a predictive sufficient statistic for Z.

Let $\hat{Z}$ be any unbiased predictor of Z; that is, $\mathrm{E}[\hat{Z}] = \mathrm{E}[Z]$. By the extension of the Rao–Blackwell Theorem to prediction (Bolfarine and Zacks [1992]), a better unbiased predictor of Z can be obtained by taking the conditional expectation of $\hat{Z}$ given $\mathbf{T} = \mathbf{t}$. Writing $\hat{Z}_{RB} = \mathrm{E}[\hat{Z} \mid \mathbf{t}]$, then

$$\mathrm{E}[\hat{Z}_{RB}] = \mathrm{E}\{\mathrm{E}[\hat{Z} \mid \mathbf{t}]\}$$
$$= \mathrm{E}[\hat{Z}]$$
$$= \mathrm{E}[Z]$$

and

$$\mathrm{E}[(\hat{Z} - Z)^2] = \mathrm{E}\{\mathrm{E}[(\hat{Z} - \hat{Z}_{RB} + \hat{Z}_{RB} - Z)^2 \mid \mathbf{t}]\}$$
$$= \mathrm{E}\{\mathrm{E}[(\hat{Z} - \hat{Z}_{RB})^2 \mid \mathbf{t}]\} + \mathrm{E}\{\mathrm{E}[(\hat{Z}_{RB} - Z)^2 \mid \mathbf{t}]\}$$
$$\geq \mathrm{E}\{\mathrm{E}[(\hat{Z}_{RB} - Z)^2 \mid \mathbf{t}]\}$$
$$= \mathrm{E}[(\hat{Z}_{RB} - Z)^2],$$

since

$$\mathrm{E}[(\hat{Z} - \hat{Z}_{RB})(\hat{Z}_{RB} - Z) \mid \mathbf{t}] = (\hat{Z}_{RB} - Z)\mathrm{E}[(\hat{Z} - \hat{Z}_{RB}) \mid \mathbf{t}] = 0.$$

Suppose further that the sufficient statistic $\mathbf{T}$ is complete. Then any two functions of $\mathbf{t}$ with the same expectation are equal to each other with probability 1, so that $\hat{Z}_{RB}$ is the (essentially) unique function of $\mathbf{t}$ with expectation $E[Z]$. Thus, by the extension of the Rao–Blackwell and Lehman–Scheffé theorems to prediction (Bolfarine and Zacks [1992]), any function of a complete sufficient $\mathbf{T}$ with expectation $E[Z]$ is a best unbiased predictor of Z.

In applying the above theory to sampling from finite populations, we focus on model expectations given the sample. Recall that with a conventional design, model expectations given the sample may depend on the sample selected but do not depend on the design used to obtain the sample. However, with an adaptive design, conditional expectations given the sample depend on the design as well as on the model. Also, if a predictor is conditionally unbiased given the sample, it will be unconditionally unbiased as well. In the following examples, each of the above expectations is replaced by expectation conditional on s. The results illustrated on best unbiased prediction under assumed population models hold for conventional designs but in general not for adaptive designs.

Example 3.21   I.I.D. Normal Model—Conventional Design.   Suppose Y_1, $\ldots, Y_N$ are i.i.d. normal with unknown mean μ and unknown variance σ^2, and that the sample s is selected by any conventional design $p(s)$ not depending on any y-values. Then expectations given s do not depend on the design. Hence, if $Z = \sum_{i=1}^{N} Y_i$ is the population total and $\hat{Z} = N\bar{V}$, where $\bar{V}$ is the sample mean of the Y-values $\mathbf{Y}_s$ in the reduced sample, then $E[\hat{Z} \mid s] = E[Z \mid s] = N\mu$. Thus $\hat{Z}$ is an unbiased predictor of Z. Note that $\hat{Z}$ is in general not unbiased with an adaptive design such as the adaptive cluster sampling described in Chapter 4.

In Example 3.13, it was shown that the minimal predictive sufficient statistic with this model is $(t_1, t_2, \nu(s))'$. When sample size is not fixed, this statistic is not complete, so that more than one unbiased predictor may be based on it. For example, a second unbiased predictor of Z is $\tilde{Z} = (\nu/E[\nu])\hat{Z}$. Thus, when the effective sample size under the design is random, no uniformly best unbiased predictor of the population mean or total exists. However, $\hat{Z}$ is the best *conditionally* unbiased predictor of Z, given $\nu(s)$, provided the design is conventional.

Example 3.22   Multivariate Normal Linear Model.   Suppose that the population vector $\mathbf{Y}$ has an N-dimensional multivariate normal distribution with mean vector $\boldsymbol{\beta}\mathbf{X}$ and variance-covariance matrix $\sigma^2\boldsymbol{\Sigma}$, where $\mathbf{X}$ is a known matrix of full rank and $\boldsymbol{\Sigma}$ is a known positive definite matrix, $\boldsymbol{\beta}$ is a vector of p unknown parameters, and σ^2 is unknown. In Example 3.14, the statistic

$$\mathbf{T} = \left(\mathbf{X}_s'\boldsymbol{\Sigma}_{ss}^{-1}\mathbf{Y}_s, \mathbf{Y}_s'\boldsymbol{\Sigma}_{ss}^{-1}\mathbf{Y}_s, \mathbf{c}'\boldsymbol{\Sigma}_{ss}^{-1}\mathbf{Y}_s, \mathbf{X}_s'\boldsymbol{\Sigma}_{ss}^{-1}\mathbf{X}_s, \mathbf{c}'\boldsymbol{\Sigma}_{ss}^{-1}\mathbf{X}_s, \mathbf{c}'\boldsymbol{\Sigma}_{ss}^{-1}\mathbf{c}\right)'$$

was shown to be predictive sufficient for the population total Z. Given the sample s, the components of $\mathbf{T}$ involving only x-values (and s) are fixed, while the remaining random components involving Y-values have a complete distribution.

The usual predictor of the population total based on normal theory and with the sample given (Tam [1987], Bolfarine and Zacks [1992]) can be written (cf. Example 3.7)

$$\hat{Z} = \mathbf{c}'\mathbf{\Sigma}_{ss}^{-1}\mathbf{Y}_s + (\mathbf{X}'\mathbf{1}_N - \mathbf{X}_s'\mathbf{\Sigma}_{ss}^{-1}\mathbf{c})'\hat{\boldsymbol{\beta}}.$$

Conditional on the sample s, the x-values are fixed and the sufficient statistic is complete, by the usual normal theory. Thus we can say that $\hat{Z}$ is the best *conditionally* unbiased predictor of Z, given s, provided the design is conventional. The same estimator would not be unbiased in general with an adaptive design, because then the conditional expectation given s depends on the design.

Example 3.23   Model for a Stratified Population.   Suppose a stratified population with L strata is modeled as independent random variables Y_{hi}, $h = 1,\ldots L$, $i = 1,\ldots,N_h$, with Y_{hi} having a normal (μ_h, σ_h^2) distribution. A sample is selected by any conventional design. Then, as a special case of the previous example, the best unbiased predictor of the population total is

$$\hat{Z} = \sum_{h=1}^{L} N_h \bar{y}_h.$$

Thus, under the assumed model, the best conditionally unbiased predictor, given the sample, is the standard stratified sampling estimator of the population total. Under the assumed model, this predictor is best conditionally unbiased under any conventional design, including stratified random sampling or even purposive selection of the specific sample s. However, it is not in general unbiased with an adaptive allocation design. A specific adaptive allocation design under which this predictor is unbiased is described in Chapter 7. On the other hand, the standard predictor $\hat{Z}$ is a maximum likelihood predictor for the population total under the assumed normal model irrespective of whether the design is conventional or adaptive.

3.15   HOMOGENEOUS LINEAR ESTIMATORS

A homogeneous linear estimator is an estimator of the form

$$\hat{z}(d) = \sum_{i=1}^{N} y_i w_{i,s,\mathbf{y}_s}.$$

For the ith unit in the population, the weight $w_{i,s,\mathbf{y}_s}$ may depend on the sample selected as well as on i. The weight is zero for any unit not in the sample so that $w_{i,s,\mathbf{y}_s} = 0$ for $i \notin s$ and

$$\hat{z}(d) = \sum_{i \in s} y_i w_{i,s,\mathbf{y}_s}.$$

For a conventional design the weight does not depend on $\mathbf{y}$, so that $w_{i,s,y_s} = w_{i,s}$. However, with an adaptive design, the weight w_{i,s,y_s} may depend on the population vector $\mathbf{y}$ through $\mathbf{y}_s$.

For a homogeneous linear estimator the design expectation is

$$
E[\hat{z}(D) \mid \mathbf{y}] = \sum_{s \in S} \sum_{i=1}^{N} y_i w_{i,s,y_s} p(s \mid \mathbf{y}; \boldsymbol{\phi})
$$

$$
= \sum_{i=1}^{N} y_i \sum_{s \in S} w_{i,s,y_s} p(s \mid \mathbf{y}; \boldsymbol{\phi}).
$$

Now, since $w_{i,s,y_s} = 0$ for $i \notin s$, the last sum on the right can be written, for each fixed i,

$$
\sum_{s \in S} w_{i,s,y_s} p(s \mid \mathbf{y}; \boldsymbol{\phi}) = \sum_{s \in S: i \in s} w_{i,s,y_s} p(s \mid \mathbf{y}; \boldsymbol{\phi}).
$$

If the weight does not depend on the sample, so that $w_{i,s,y_s} = w_i$, the above sum is

$$
\sum_{s \in S: i \in s} w_i p(s \mid \mathbf{y}; \boldsymbol{\phi}) = w_i \sum_{s \in S: i \in s} p(s \mid \mathbf{y}; \boldsymbol{\phi})
$$

$$
= w_i \pi_i,
$$

where π_i is the probability that unit i is included in the sample. Note that this probability depends on $\mathbf{y}$.

For conventional designs, the Horvitz–Thompson estimator uses

$$
w_{i,s} = \frac{1}{\pi_i}
$$

so that the weight does not depend on the sample s. For adaptive cluster sampling, the weight is of the form

$$
w_{i,s,y_s} = \frac{I_{i,s,y_s}}{\alpha_i},
$$

where I_{i,s,y_s} is a indicator function depending, in a complicated manner, on the criteria for determining whether unit i is to be included in the estimator or not.

Linear predictors such as regression and kriging predictors are also of the form

$$
\hat{z}(d) = \sum_{i \in s} y_i w_{i,s}.
$$

This has model expectation

$$
E[\hat{z}(D) \mid s] = \int \sum_{i \in s} y_i w_{i,s} f(\mathbf{y} \mid s; \boldsymbol{\phi}) d\mathbf{y}
$$

$$= \sum_{i \in s} w_{i,s} \int y_i f(\mathbf{y} \mid s; \boldsymbol{\phi}) \, d\mathbf{y}$$

$$= \sum_{i \in s} w_{i,s} \mathrm{E}[Y_i \mid s].$$

We note from Section 3.6 that for an adaptive design, $f(\mathbf{y} \mid s; \boldsymbol{\phi})$ (and hence $\mathrm{E}[Y_i \mid s]$) may depend on the design $p(s \mid \mathbf{y}_s)$. Thus, even from a model-based perspective, a conventional estimator such as the sample mean which may be model unbiased with any conventional design, will not in general be model unbiased with an adaptive design.

If the design is conventional, it follows from Section 3.6 that $f(\mathbf{y} \mid s; \boldsymbol{\phi}) = f(\mathbf{y}; \boldsymbol{\phi})$ and

$$\mathrm{E}[\hat{z}(D) \mid s] = \sum_{i \in s} w_{i,s} \mathrm{E}[Y_i],$$

where $\mathrm{E}[Y_i]$ is now a constant not depending on the design.

3.16 BEST LINEAR UNBIASED PREDICTION

Suppose that the population vector $\mathbf{Y}$ has an N-dimensional multivariate distribution with mean vector $\boldsymbol{\beta}\mathbf{X}$ and variance-covariance matrix $\sigma^2\boldsymbol{\Sigma}$, where $\mathbf{X}$ is a known matrix of full rank and $\boldsymbol{\Sigma}$ is a known positive definite matrix, $\boldsymbol{\beta}$ is a vector of p unknown parameters, and σ^2 is unknown. No specific distribution such as the normal is assumed.

Given the (reduced) sample s, the problem is to find a linear predictor of the population total Z of the form $\hat{Z} = \mathbf{a}'\mathbf{Y}_s$, subject to $\mathrm{E}[\hat{Z} \mid s] = \mathrm{E}[Z \mid s]$, which minimizes the mean-square error $\mathrm{E}[(\hat{Z} - Z)^2 \mid s]$. The solution for any conventional design, usually obtained using Lagrange's method, is given by

$$\hat{Z} = \mathbf{c}'\boldsymbol{\Sigma}_{ss}^{-1}\mathbf{Y}_s + (\mathbf{X}'\mathbf{1}_N - \mathbf{X}_s'\boldsymbol{\Sigma}_{ss}^{-1}\mathbf{c})'\hat{\boldsymbol{\beta}}$$

$$= \mathbf{c}'\boldsymbol{\Sigma}_{ss}^{-1}\mathbf{Y}_s + (\mathbf{X}'\mathbf{1}_N - \mathbf{X}_s'\boldsymbol{\Sigma}_{ss}^{-1}\mathbf{c})'(\mathbf{X}_s'\boldsymbol{\Sigma}_{ss}^{-1}\mathbf{X}_s)^{-1}\mathbf{X}_s'\boldsymbol{\Sigma}_{ss}^{-1}\mathbf{Y}_s$$

$$(= \mathbf{a}'\mathbf{Y}_s, \quad \text{say}),$$

where $\mathbf{c}' = \mathrm{cov}[Z, \mathbf{Y}_s]/\sigma^2$ is given in Example 3.14 in Section 3.12. (cf. Royall [1976], Tam [1987], Cressie [1993, p. 174], and Bolfarine and Zacks [1992, p. 25]). This means that the predictor, which is best unbiased (conditional on the sample) under the normal assumption, is best linear unbiased with only the assumed mean and covariance structure. This result holds for any conventional design, including even unequal probability designs and standard purposive selection procedures. With adaptive designs, on the other hand, the linear predictor $\hat{Z}$ is not in general unbiased.

Thus, for example, with conventional designs the expansion predictor $\hat{Z} = N\bar{V}$, where $\bar{V}$ is the mean of the elements of $\mathbf{Y}_s$, is the best linear conditionally unbiased predictor of the population total, given the sample size, under the assumption that $Y_1, \ldots, Y_N$ are i.i.d. random variables (cf. Example 3.16). The stratified sampling predictor $\hat{Z} = \sum_{h=1}^{L} N_h \bar{V}_h$ is the best linear predictor of the population total (conditional on the sample) under the assumption that the Y's are independent random variables, with means and variances depending on strata (cf. Example 3.23). With adaptive designs, conditional expectations given the sample s depend on the design, so that in general the same predictors are not unbiased.

3.17 SPATIAL–TEMPORAL EXTENSIONS

In spatial, environmental, and ecological sampling situations the problems go beyond the classical sampling framework in several ways. The variable of interest is defined over a continuous region of space and time rather than a finite set of units. The "units" or sites on which observations are obtained may be points, lines, sets, or filtering operations depending on detectability or weighting functions reflective of the observational method. The natural generalization of classical sampling to the environmental setting has a spatial–temporal pattern or process for the population and a collection of detectability or weighting functions or other "units" from which the sample is selected. A fixed spatial pattern corresponds to the fixed population approach in classical sampling, whereas a spatial stochastic process corresponds to the model-based approach. The index set of the stochastic process corresponds to the population units of finite population sampling. The collection of points, sets, detectability functions, or weighting functions correspond to the sampling units.

With a variable of interest that is continuous in space and time, consider a spatial–temporal stochastic process

$$\{Z(s), s \in \mathcal{S}\}.$$

The index set $\mathcal{S}$ is a region in space and time, an element s is a location or a point in space or space-time, and the random variable $Z(s)$ represents the value of the variable of interest at s. The population total T is the integral of the process over a study region R, namely

$$T = \int_R z(v) \, dv.$$

For a discrete variable, consider a spatial–temporal point process

$$\{N(A), A \in \mathcal{A}\}.$$

The set A is a subset of the (spatial–temporal) study region, the index set $\mathcal{A}$ is a collection of such subsets, and $N(A)$ is the number of the objects of interest in A. A realization of the process gives a spatial pattern $\{z(s), s \in \mathcal{S}\}$ or $\{n(A), A \in \mathcal{A}\}$.

In extending classical sampling theory to accommodate the spatial–temporal setting of ecological resources, the following three connections are made. (1) The unit labels $\{1, 2, \ldots, N\}$ of finite population sampling become the index set $\mathcal{S}$ on which the spatial–temporal stochastic process is defined. (2) The sampling units—which may be complex primary or systematic units—selected in classical sampling are replaced by the set $\mathcal{T}$, which indexes the possible sites, units, detectability functions, or weighting functions used in observing the spatial–temporal pattern or process. For example, in line transect surveys of waterfowl, the set $\mathcal{T}$ is a baseline from which transect locations are selected at random or with probability proportional to transect length. (3) The classical design-based, fixed population approach extends to the fixed spatial pattern $\{z(s), s \in \mathcal{S}\}$ in which no model is involved or in which a design-based analysis proceeds conditional on the specific realization of a model. With these three connections, the population model approach of the finite population sampling literature can be extended to the full spatial–temporal stochastic model $\{Z(s), s \in \mathcal{S}\}$.

3.18 NONSAMPLING ERRORS

Sources of nonsampling errors in surveys include measurement errors, frame errors, nonresponse, and data entry errors (see Biemer et al. [1991], Lessler and Kalsbeek [1992], and Rubin [1987]). Nonsampling errors can introduce biases, which must be adjusted for, and additional variance terms, which must be estimated.

In many surveys the variable of interest y_i for a unit i in the sample is not observed without error. Rather, the observed or measured value is $y_i^{(m)}$, which may differ from y_i due to measurement error or other cause. With measurement error, the data consist of $(s_O, \mathbf{y}_s^{(m)})$. In analyzing surveys with measurement error it is sometimes mathematically useful to consider the hypothetical values $\mathbf{y}_{\bar{s}}^{(m)}$ for units not in the sample. The density of an outcome is

$$f(s_O, \mathbf{y}^{(m)}, \mathbf{y}; \boldsymbol{\phi}) = p(s_O \mid \mathbf{y}^{(m)}, \mathbf{y}; \boldsymbol{\phi}) f(\mathbf{y}^{(m)}, \mathbf{y}; \boldsymbol{\phi})$$
$$= p(s_O \mid \mathbf{y}^{(m)}, \mathbf{y}; \boldsymbol{\phi}) f_m(\mathbf{y}^{(m)} \mid \mathbf{y}; \boldsymbol{\phi}) f(\mathbf{y}; \boldsymbol{\phi}).$$

For an adaptive design, the selection probability depends only on observed values in the sample, so

$$p(s_O \mid \mathbf{y}^{(m)}, \mathbf{y}; \boldsymbol{\phi}) = p(s_O \mid \mathbf{y}_{\bar{s}}^{(m)}).$$

The conditional density $f_m(\mathbf{y}^{(m)} \mid \mathbf{y}; \boldsymbol{\phi})$ is the measurement error model. An example of a measurement error model is

$$f_m(\mathbf{y}_s^{(m)} \mid \mathbf{y}; \boldsymbol{\phi}) = \prod_{i=1}^{N} f_{mi}(y_i^{(m)} \mid y_i; \boldsymbol{\phi}).$$

With such a model the errors are conditionally independent given the realized true y-values, and the error for unit i depends on y only for that unit. A model commonly assumed has $Y_i^{(m)} = Y_i + \epsilon_i$, where the ϵ_i are independent random variables with variances σ_i^2 (see Cochran [1977, p. 377] and and Lessler and Kalsbeek [1992, p. 247]). Additionally it may be assumed that conditional on $Y_i = y_i$, $Y_i^{(m)}$ has a normal distribution with mean y_i and variance σ_i^2. With repeated measurements on the ith unit we write $y_{it}^{(m)}$ for the tth measurement.

Imperfect detectability is a common source of nonsampling error in ecological surveys and other surveys in which objects are counted. With y_i representing the number of animals actually present in the ith unit and $y_i^{(m)}$ the number detected by the observer, a simple model is that, conditional on y_i, $Y_i^{(m)}$ has a binomial distribution in which the number of trials is y_i and the probability of success for each trial is a constant probability of detection g. A slightly different notation is used in Chapter 9. A more flexible model is given by $Y_i^{(m)} = \sum_{j=1}^{M_i} Y_{ij}Z_{ij}(y_{ij})$, where y_{ij} may be a continuous variable representing the size of the jth object in the ith sampling unit, and the Z_{ij} are independent Bernoulli random variables whose probabilities of success (detection) may depend on the sizes.

Nonresponse occurs in households when no one is at home in a selected household or the occupants decline to fill out a questionnaire. In ecological surveys, nonresponse arises when some selected plots are not observed because of unfavorable weather or difficult terrain. Very often the nonresponse tendencies are related to values of the variable of interest, including often values not in the data, so that careful modeling and analysis of nonresponse is necessary for effective inference. In surveys with nonresponse, the data consist of the sample s_O containing distinct units given by s, a vector r_s of 1's and 0's indicating the units of the sample for which a response was obtained, and the values $\mathbf{y}_{s_r}$ of the variable of interest of the units s_r in the sample for which response was obtained. Let R denote a random vector of length N taking values 1 and 0 indicating hypothetical response or not for every unit in the sample. The density for an outcome is

$$f(s_O, \mathbf{r}, \mathbf{y}; \boldsymbol{\phi}) = p(s_O \mid \mathbf{r}, \mathbf{y}; \boldsymbol{\phi})f_r(\mathbf{r} \mid \mathbf{y}; \boldsymbol{\phi})f(\mathbf{y}; \boldsymbol{\phi}).$$

The conditional density $f_r(\mathbf{r} \mid \mathbf{y}; \boldsymbol{\phi})$ is the nonresponse model. For an adaptive design, the probability of selecting the sample depends only on observed values in the sample, so that

$$p(s_O \mid \mathbf{r}, \mathbf{y}; \boldsymbol{\phi}) = p(s_O \mid \mathbf{r}_s, \mathbf{y}_{s_r}).$$

With such a design, one could for example add neighboring households to the sample whenever nonresponse occurred at a selected household. The situation in the size-biased sampling can be modeled as a nonresponse situation but differs from the classical nonresponse case in that data indicating that a unit had not "responded" is not available.

As an example of a difficult nonresponse situation, surveys of environmental contaminants and of endangered species often rely on access permission from landowners. Suppose that, in order to avoid development restrictions and/or costs to remedy the situation, landowners are less likely to give permission when they know the contamination level is high or that the endangered species uses the land as habitat.

The sample of sites at which measurements are made are then not representative of the study region as a whole, and it may be necessary to model the nonresponse relationships to produce realistic estimates.

With conventional sampling designs, nonsampling errors mainly affect the properties of estimators and predictors. With adaptive designs, nonsampling errors can in addition affect what sample is selected. In this monograph, nonsampling errors arising from imperfect detectability of objects are treated in some detail for adaptive sampling in Chapter 9.

CHAPTER 4

Adaptive Cluster Sampling

4.1 INTRODUCTION

Adaptive cluster sampling was motivated by the problem of sampling rare, clustered populations. In adaptive cluster sampling, an initial sample of units is selected and, whenever the value of the variable of interest satisfies a specified condition, neighboring units are added to the sample. The condition for extra sampling might be presence of the rare animal or plant species, high abundance of a spatially clustered species, detection of "hot spots" in an environmental pollution study, observed high concentration of mineral ore or fossil fuel, infection with a rare disease or positive value of a related indicator variable in an epidemiology study, or observation of a rare characteristic of interest in a household survey. The neighborhood of a unit may be defined by spatial proximity or, in the case of human populations, by social or genetic links or other connections.

Using the fixed population approach of Chapter 2, we describe in this chapter a variety of adaptive cluster sampling strategies, with initial samples selected by various sampling schemes such as simple random sampling with or without replacement, systematic sampling, unequal probability sampling, simple Latin square $+1$ sampling, cluster sampling, and stratified sampling. Computations are illustrated in some examples. Comparisons of adaptive cluster sampling and conventional designs are derived theoretically for the simplest strategy and are carried out computationally with examples for that strategy and others. In the next chapter, we examine the factors influencing the relative efficiency of adaptive cluster sampling to simple random sampling, including within-network variation, rarity of the population, sample size and cost issues, and the use of auxiliary information such as a "rapid assessment" variables.

4.2 INITIAL RANDOM SAMPLE WITHOUT REPLACEMENT

In Example 1.1 of Section 1.6, we described the following adaptive cluster sampling scheme. For every unit we define a neighborhood that consists of the unit and a set of "neighboring" units. A sample of size n_1 is taken using simple random sampling (*srs*). If the *y*-value of a sampled unit satisfies a certain condition C, for example $y_i > c$, then the rest of the unit's neighborhood is added to the sample. If any other units in that neighborhood satisfy C, then their neighborhoods are also added to the sample. This process is continued until a cluster of units is obtained that contains a "boundary" of units called *edge* units that do not satisfy C. The final sample then consists of n_1 (not necessarily distinct) clusters, one for each unit selected in the initial sample.

We note that if a unit selected in the initial sample does not satisfy C, there is no augmentation and we have a cluster of size 1. Although the units in the initial sample are distinct on account of *srs*, repeat selections can occur in the final sample when a cluster includes more than one unit of the initial sample. For example, if two non-edge units in the same cluster are selected in the initial sample, then that cluster occurs twice in the final sample.

Neighborhoods can be defined to have a variety of patterns, and the units (plots) in a neighborhood do not have to be contiguous. However, the neighborhood relationship is assumed to be symmetric in the sense that if unit j is in the neighborhood of unit i, then i is also in the neighborhood of j. These neighborhoods do not depend on the population of *y*-values.

Although the cluster is the natural sample group, it is not the most useful entity to use for theoretical developments because of the double role that edge units can play. If an edge unit is selected in the initial sample, it forms a cluster of size 1. If it is not selected in the initial sample, then it can still be selected by being a member of any cluster for which it is an edge unit. We therefore introduce the idea of the network A_i for unit i, which is defined to be the cluster generated by unit i but with its edge units removed. A selection of any unit in A_i leads to the selection of all of A_i. If unit i is the only unit in a cluster satisfying C, then A_i consists of just unit i and forms a network of size 1. We also define any unit that does not satisfy C to be a network of size 1, as its selection does not lead to the inclusion of any other units. This means that all clusters of size 1 are also networks of size 1. Note that any edge unit is also a network of size 1. Thus any cluster consisting of more than one unit can be split into a network and further networks of size 1 (one for each edge unit). In contrast to having clusters which may overlap on their edge units, the distinct networks are *disjoint* and form a *partition* of the N units.

4.2.1 An Estimator Using Initial Intersection Probabilities

We shall now develop an estimator (Thompson [1990]) based on a modification of a Horvitz–Thompson estimator and compare it with the mean of the initial sample:

$$\bar{y}_1 = \frac{1}{n_1} \sum_{i=1}^{n_1} y_i. \tag{4.1}$$

We note that unit i will be included in the final sample either if any unit of A_i (including i itself) is selected as part of the initial sample or if any unit of a network of which unit i is an edge unit is selected. Let m_i denote the number of units in A_i, and let a_i denote the total number of units in networks of which unit i is an edge unit. If unit i satisfies C, then $a_i = 0$, while if unit i does not satisfy C, then $m_i = 1$. The probability that unit i is included in the sample is

$$\pi_i = 1 - \left[\binom{N - m_i - a_i}{n_1} \middle/ \binom{N}{n_1} \right]. \tag{4.2}$$

If we knew π_i for all the sampled units, we could use the Horvitz–Thompson estimator of (1.18), namely

$$\hat{\mu}_{HT} = \frac{1}{N} \sum_{i=1}^{\nu} \frac{y_i}{\pi_i}$$

$$= \frac{1}{N} \sum_{i=1}^{N} \frac{y_i I_i}{\pi_i}, \tag{4.3}$$

where $y_1, y_2, \ldots, y_\nu$ represent the y-values of the ν distinct units in the final sample, and I_i takes the value 1 when unit i is included in the sample and 0 otherwise. Unfortunately, although the m_i are known in (4.2) for all the units in the sample, some of the a_i are unknown. For example, if unit i is an edge unit for some cluster in the sample, then all the clusters it belongs to won't generally be sampled so that a_i will be unknown. To get around this problem, we drop a_i from π_i and consider the "partial" inclusion probability

$$\pi_i' = 1 - \left[\binom{N - m_i}{n_1} \middle/ \binom{N}{n_1} \right]. \tag{4.4}$$

This amounts to ignoring the edge units of clusters in the estimation process (though they are formally included in the final sample as part of the sufficient statistic D_R described in Section 2.4). Thus observations that do not satisfy the condition are ignored if they are not included in the initial sample. We now use the sample of n_1 networks (some of which may be the same), rather than the n_1 clusters, for estimating μ. The probability π_i' can then be interpreted as the probability that unit i is utilized in the estimator, or equivalently as the probability that the initial sample intersects A_i, the network for unit i.

The unbiased estimator based on the the initial intersection probabilities takes the form

$$\hat{\mu} = \frac{1}{N} \sum_{i=1}^{N} \frac{y_i I_i'}{\pi_i'}, \tag{4.5}$$

where I_i' takes the value 1 (with probability π_i') if the initial sample intersects A_i, and 0 otherwise.

In evaluating the properties of $\hat{\mu}$, it is more convenient to write the sum (4.5) in terms of the distinct networks, as the intersection probability π_i' is the same ($= \alpha_k$, say) for each unit i in the kth network. Hence

$$\hat{\mu} = \frac{1}{N} \sum_{k=1}^{K} \frac{y_k^* J_k}{\alpha_k}$$

$$= \frac{1}{N} \sum_{k=1}^{\kappa} \frac{y_k^*}{\alpha_k}, \qquad (4.6)$$

where y_k^* is the sum of the y-values for the kth network, K is the total number of distinct networks in the population, κ is the number of distinct networks in the sample, and J_k takes a value of 1 (with probability α_k) if the initial sample intersects the kth network, and 0 otherwise. If there are x_k units in the kth network, then

$$\alpha_k = 1 - \left[\binom{N - x_k}{n_1} \Big/ \binom{N}{n_1} \right]. \qquad (4.7)$$

Also, letting $p_{jk} = P(j$th and kth networks not intersected), then

$$p_{jk} = P(J_j \neq 1 \cap J_k \neq 1)$$

$$= \binom{N - x_j - x_k}{n_1} \Big/ \binom{N}{n_1},$$

so that from (1.30) and (1.31) the probability that networks j and k are both intersected is

$$\alpha_{jk} = \alpha_j + \alpha_k - (1 - p_{jk})$$

$$= 1 - \left[\binom{N - x_j}{n_1} + \binom{N - x_k}{n_1} - \binom{N - x_j - x_k}{n_1} \right] \Big/ \binom{N}{n_1}. \qquad (4.8)$$

To obtain the expected value and variance of $\hat{\mu}$, we set $z_k = y_k^*/\alpha_k$, $\pi_k = \alpha_k$ and $\pi_{jk} = \alpha_{jk}$ and obtain from Equations (1.4) and (1.5)

$$E[\hat{\mu}] = \frac{1}{N} \sum_{k=1}^{K} z_k \alpha_k$$

$$= \frac{1}{N} \sum_{k=1}^{K} y_k^*$$

$$= \frac{1}{N} \sum_{i=1}^{N} y_i$$

$$= \frac{\tau}{N}$$

$$= \mu \qquad (4.9)$$

and

$$\text{var}[\hat{\mu}] = \frac{1}{N^2} \left[\sum_{j=1}^{K} \sum_{k=1}^{K} y_j^* y_k^* \left(\frac{\alpha_{jk} - \alpha_j \alpha_k}{\alpha_j \alpha_k} \right) \right]. \tag{4.10}$$

From (1.6) an unbiased estimator of the above variance is

$$\widehat{\text{var}}[\hat{\mu}] = \frac{1}{N^2} \left[\sum_{j=1}^{K} \sum_{k=1}^{K} y_j^* y_k^* \left(\frac{\alpha_{jk} - \alpha_j \alpha_k}{\alpha_{jk} \alpha_j \alpha_k} \right) J_j J_k \right]$$

$$= \frac{1}{N^2} \left[\sum_{j=1}^{K} \sum_{k=1}^{K} \frac{y_j^* y_k^*}{\alpha_{jk}} \left(\frac{\alpha_{jk}}{\alpha_j \alpha_k} - 1 \right) \right], \tag{4.11}$$

where α_{jj} is interpreted as α_j. Examples of the computations are given in Section 4.6.1.

Smith et al. [1995] describe a simulation study which includes the use of adaptive cluster sampling with an initial simple random sample using the estimator $\hat{\mu}$ for estimating densities of wintering waterfowl. The relative efficiency of the adaptive strategy to conventional simple random sampling depended on the species studied (some species having more clustered distributions than others), on the unit sizes used, and on the condition chosen for the adaptive addition of units. For each species, the right combination of unit and condition would give the adaptive strategy higher efficiency than the conventional strategy, whereas for some species the conventional strategy was the more efficient for certain combinations of unit size and condition. Ramsey and Sjamsoe'oed [1995] discuss the use of adaptive cluster sampling with an initial simple random sample to investigate the relationship between animals' activities and habitat characteristics. The condition for extra sampling was based on occupancy by the animals, so that the adaptive procedure increased the number of units in the sample with which the relationship could be studied.

4.2.2 An Estimator Using Numbers of Initial Intersections

The estimator $\hat{\mu}$ of (4.5) can be expressed in the form

$$\hat{\mu} = \frac{1}{N} \sum_{i=1}^{N} y_i \frac{I_i'}{E[I_i']}. \tag{4.12}$$

From (1.35) there is another estimator that is also unbiased, namely

$$\tilde{\mu} = \frac{1}{N} \sum_{i=1}^{N} y_i \frac{f_i}{E[f_i]}, \tag{4.13}$$

where f_i is the number of units in the initial sample which fall in (intersect) the network A_i that includes unit i. Ignoring the edge units of clusters in the estimation process, f_i is the number of times that the ith unit in the final sample appears in the estimator. We note that $f_i = 0$ if no units in the initial sample intersect A_i.

Since f_i units are selected from the m_i units in A_i, f_i has the hypergeometric distribution Hyper(N, m_i, n_1). From E$[f_i] = n_1 m_i/N$, we have

$$\tilde{\mu} = \frac{1}{n_1} \sum_{i=1}^{N} \frac{y_i f_i}{m_i}. \tag{4.14}$$

To find the variance of $\tilde{\mu}$, it is helpful to rewrite this estimate in terms of the n_1 (not necessarily distinct) networks intersected by the initial sample. Since m_i is the same for all the units in A_i, we see that

$$\tilde{\mu} = \frac{1}{n_1} \sum_{i=1}^{n_1} \frac{1}{m_i} \sum_{j \in A_i} y_j$$

$$= \frac{1}{n_1} \sum_{i=1}^{n_1} w_i,$$

$$= \bar{w}, \quad \text{say}, \tag{4.15}$$

where w_i is the mean of the m_i observations in A_i (Thompson [1990]).

We recognize $\tilde{\mu}$ as the sample mean obtained by taking a simple random sample of size n_1 from a population of w_i values rather than y_i values. Since w_i $(= \bar{v}_k$, say) is the same for each unit in the kth network, and there are x_k units in the kth network, we have

$$\text{E}[\tilde{\mu}] = \frac{1}{N} \sum_{i=1}^{N} w_i$$

$$= \frac{1}{N} \sum_{k=1}^{K} x_k \bar{v}_k$$

$$= \frac{1}{N} \sum_{k=1}^{K} \sum_{i \in B_k} y_i$$

$$= \mu,$$

where B_k is the set of units in the kth network.

Although the unbiasedness of $\tilde{\mu}$ follows immediately from (4.13), the above derivation is instructive. Using the theory of simple random sampling, we have from (1.27) and (1.28) that

$$\text{var}[\tilde{\mu}] = \frac{N - n_1}{Nn_1(N-1)} \sum_{i=1}^{N} (w_i - \mu)^2, \tag{4.16}$$

with unbiased estimate

$$\widehat{\text{var}}[\tilde{\mu}] = \frac{N - n_1}{N n_1 (n_1 - 1)} \sum_{i=1}^{n_1} (w_i - \tilde{\mu})^2. \tag{4.17}$$

We note from (4.15) that $\tilde{\mu}$ is the sample mean of the w-values of n_1 networks, of which κ, say, will be distinct. Hence

$$\tilde{\mu} = \frac{1}{n_1} \sum_{k=1}^{\kappa} b_k \bar{v}_k, \tag{4.18}$$

where b_k is the number of times the kth network appears in (4.15) and κ is the number of distinct networks intersected by the initial sample. Since $b_k = 0$ for those networks not intersected, we can also write

$$\tilde{\mu} = \frac{1}{n_1} \sum_{k=1}^{K} b_k \bar{v}_k$$

$$= \frac{1}{n_1} \sum_{k=1}^{K} \frac{b_k y_k^*}{x_k}, \tag{4.19}$$

where y_k^* is the sum of the y-values in the kth network. As b_k has the hypergeometric distribution Hyper(N, x_k, n_1), E$[b_k] = n_1 x_k / N$ and

$$\tilde{\mu} = \frac{1}{N} \sum_{k=1}^{K} y_k^* \frac{b_k}{\text{E}[b_k]}. \tag{4.20}$$

We thus find that $\tilde{\mu}$ can be expressed in terms of either units, as in (4.13), or networks, as in (4.20). Both representations, which use expected numbers of initial intersections, show that $\tilde{\mu}$ is unbiased. However, we have seen that for finding variances, expressing $\tilde{\mu}$ as a mean as in (4.15) is a more useful representation. All three forms of $\tilde{\mu}$ are used in later extensions of this theory.

In conclusion, we note that we have two types of estimator. The first, $\hat{\mu}$, as in (4.20), is based on the initial intersection probabilities. The second, $\tilde{\mu}$, is based on the numbers of initial intersections. Examples of the computations are given in Section 4.6.1. While $\hat{\mu}$ is related to the Horvitz–Thompson estimator, $\tilde{\mu}$ is related to the Hansen–Hurwitz estimator. Just as the true Horvitz–Thompson estimator would not be possible to use with adaptive cluster sampling because the inclusion probabilities are not known for every unit in the sample, the Hansen-Hurwitz estimator could not be used because the draw-by-draw selection probabilities are not known for every unit in the sample.

4.3   INITIAL RANDOM SAMPLE WITH REPLACEMENT

The estimator $\hat{\mu}$ of (4.5) can also be used with initial random sampling with replacement, *rswr*. From the representation (4.12), it is unbiased irrespective of whether the sampling is with or without replacement. Expressions for var$[\hat{\mu}]$ and $\widehat{\text{var}}[\hat{\mu}]$ are still given by (4.10) and (4.11), except that now

$$\alpha_k = 1 - \left(1 - \frac{x_k}{N}\right)^{n_1} \tag{4.21}$$

and

$$\alpha_{jk} = 1 - \left\{\left(1 - \frac{x_j}{N}\right)^{n_1} + \left(1 - \frac{x_k}{N}\right)^{n_1} - \left(1 - \frac{x_j + x_k}{N}\right)^{n_1}\right\}. \tag{4.22}$$

We also find that $\tilde{\mu}$ can be used here. By (4.13) the estimator will be unbiased irrespective of whether the sampling is with or without replacement. When sampling is with replacement, $f_i \sim \text{Bin}(n_1, m_{i/N})$ and we still have E$[f_i] = n_1 m_i/N$, so that (4.14) and therefore (4.15) are unchanged. Hence $\tilde{\mu}$ is still a sample mean, but for random sampling with replacement. To find the variance of $\tilde{\mu}$, we note that if W is the outcome of choosing a unit at random then $P(W = w_i) = 1/N, (i = 1, 2, \ldots, N)$, and

$$\sigma_W^2 = \text{var}[W]$$

$$= \frac{1}{N}\sum_{i=1}^{N}(w_i - \mu)^2. \tag{4.23}$$

Hence

$$\text{var}[\tilde{\mu}] = \text{var}[\bar{w}] = \frac{\sigma_W^2}{n_1}$$

with unbiased estimator

$$\widehat{\text{var}}[\tilde{\mu}] = \frac{1}{n_1(n_1 - 1)}\sum_{i=1}^{n_1}(w_i - \tilde{\mu})^2. \tag{4.24}$$

The preceding estimators were proposed by Thompson [1990].

4.4   INITIAL UNEQUAL PROBABILITY SAMPLING

In the previous section, the initial sample was selected by random sampling, with or without replacement, so that each unit had the same probability of inclusion in the initial sample. We now consider adaptive cluster sampling designs in which the units in the initial sample are selected with unequal probabilities. Adaptive cluster sampling with initial unequal probability designs has been investigated by Roesch [1993] for a survey of forest trees and by Smith et al. [1995] for a survey of water-fowl.

In a simulation study using data from a census of several species of ducks in a refuge, Smith et al. [1995] compare the efficiency of sampling with several types of designs and different unit sizes. The designs used included simple random sampling, conventional unequal probability sampling, adaptive cluster sampling with an initial simple random sample, and adaptive cluster sampling with an initial unequal probability sample. The unequal probability design used selection with probability proportional to an auxiliary variability (ppx) with replacement. The auxiliary variable used was available habitat based on the amount of the unit that was covered by water during the survey period.

With the unequal probability initial sampling, the estimator $\hat{\mu}$ of (4.6) and variance estimator (4.11) are used, but in the expressions (4.21) and (4.22) for intersection probabilities α_k and α_{jk}, one uses $x_k/N = a_k/A$, where a_k is the total area of available habitat in the kth network and A is the total area of available habitat in the study region. The efficiency comparisons among the different designs depended on the species of interest, the plot size chosen, and the condition used for adaptive addition of units to the sample.

In forestry surveys, several methods of selecting trees reduce to the following procedure. Suppose that at a certain (basal) height (usually 4.5 feet in the USA), trees have a circular trunk, and imagine each tree to be surrounded by a circle with center at the center of the tree and radius proportional to the radius of the tree at basal height. Let t_i be the area of this circle associated with the ith tree ($i = 1, 2, \ldots, N$). Then one sampling procedure is to drop a point at random onto the population area so that the probability of selecting tree i is t_i/A, where A is the area of the study region. Naturally we would want the t_i large enough so that the probability of not landing in a circle is not too big. This would mean that these circles would overlap so that a single sample point would generally select more than one tree. Thus, if n_0 random points were chosen, n_1 (not necessarily distinct) trees would be selected, where n_1 is now random and $n_1 \geq n_0$. It should be emphasized at this stage that the tree is the unit, not the selection area, the latter being a conceptual way of describing the probability of selection. Also the n_0 random points are chosen with replacement; it is unlikely that two such points will coincide.

We now imagine a second circle of fixed radius R around each tree defining the neighborhood (search area) of that tree. Suppose tree i is selected by an initial sample point. Then, if a specified criterion C on an associated variable y_i is satisfied, the neighborhood of that tree is searched to see if the center of any other tree is in the neighborhood. If it is, and the criterion is again satisfied, then its neighborhood is also searched, and so on.

The use of adaptive sampling in forestry was proposed by Roesch [1993], and the theory of this section is based on his paper. By way of illustration he defined y_i to be 1 if the ith tree suffered pollution damage, and 0 otherwise. His criterion was $y_i > 0$, which means search the neighborhood of a tree if it was damaged. Then $\tau_i = \sum_{i=1}^{N} y_i$ is the population total number of damaged trees, and $\mu = \tau/N$ is the proportion.

Referring to the previous section, we find that we can still use the two estimators $\hat{\mu}$ and $\tilde{\mu}$ described in Section 4.2. Networks of trees and edge units (now edge trees) are defined in a similar fashion, and once again the population is partitioned into K networks of trees (not networks of selection areas). Trees are included in the sample estimators only if they are part of the initial sample of n_1 trees or satisfy the criterion C. Thus, referring to (4.6),

$$\hat{\mu} = \frac{1}{N} \sum_{k=1}^{K} \frac{y_k^* J_k}{\alpha_k},$$

where J_k takes a value of 1 (with probability α_k) if network k is selected by a sample point and y_k^* is the sum of the y-values (the number of polluted trees) for the kth network. Arguing as in Section 4.3, we now find that

$$\alpha_k = 1 - \left(1 - \frac{U_k^*}{A}\right)^{n_0},$$

where U_k^* is the union of the selection areas t_i for all trees i in network k. Here U_k^*/A represents the probability that a randomly selected point falls into the selection area of network k. Defining U_{jk}^* to be the union of the selection areas U_j^* and U_k^*, the probability that networks j and k are both intersected by the initial point is, from Section 4.3,

$$\alpha_{jk} = 1 - \left[\left(1 - \frac{U_j^*}{A}\right)^{n_0} + \left(1 - \frac{U_k^*}{A}\right)^{n_0} - \left(1 - \frac{U_{jk}^*}{A}\right)^{n_0}\right].$$

Thus equations (4.10) and (4.11) still apply but with the above definitions of α_k and α_{jk}. It should be noted that although the tree networks do not overlap, the corresponding selection areas can overlap. This will happen when the selection areas of two trees in different networks overlap but the center of the selection area of one tree does not lie in the neighborhood of the other. In this case a single sample point falling in the overlap will select both networks.

The second estimator $\tilde{\mu}$ takes the form (4.13), namely

$$\tilde{\mu} = \frac{1}{N} \sum_{i=1}^{N} \frac{y_i f_i}{\mathrm{E}[f_i]}, \tag{4.25}$$

where f_i is the number of random points that intersect the network A_i defined by tree i. Here each network is counted just once for each time it is selected by a randomly placed point, irrespective of how many of its component trees are selected at that point (since selection areas can overlap). Let U_i be the union of selection areas for this network. Since we have n_0 independent points (as sampling is with replacement) and U_i/A is the probability of a point falling in the network A_i, then $f_i \sim \mathrm{Bin}(n_0, U_i/A)$.

Hence

$$E[f_i] = \frac{n_0 U_i}{A}$$

and, substituting in (4.25),

$$\tilde{\mu} = \frac{A}{Nn_0} \sum_{i=1}^{N} \frac{y_i f_i}{U_i}. \tag{4.26}$$

Following our previous development leading to (4.15), we would partition the values of $y_i f_i / U_i$ into networks corresponding to the initially selected trees. However, we cannot do this as we now have a random number of n_1 initially selected trees. Each one of the initial n_0 points can give rise to the selection of more than one tree. Suppose then that K_h is the random number of networks sampled by point h ($h = 1, 2, \ldots, n_0$): previously $K_h = 1$. Since U_i is the same for all the trees i in the same network k (and equal to U_k^*), we can sum the y_i over network k to get y_k^* and thus write

$$\tilde{\mu} = \frac{A}{Nn_0} \sum_{h=1}^{n_0} \sum_{k=1}^{K_h} \frac{y_k^*}{U_k^*}, \tag{4.27}$$

which is computationally more useful than (4.26).

To find the variance of $\tilde{\mu}$, we can use the same method as that used in Section 4.3, namely express $\tilde{\mu}$ as a sample mean, as follows:

$$\tilde{\mu} = \frac{A}{N} \frac{1}{n_0} \sum_{h=1}^{n_0} w_h, \tag{4.28}$$

where

$$w_h = \sum_{k=1}^{K_h} \frac{y_k^*}{U_k^*}.$$

Since the n_0 points are selected with replacement, the w_h are a random sample from a distribution with mean μ_w and variance σ_w^2, say. Therefore re-expressing w_h in the form

$$w_h = \sum_{k=1}^{K} \frac{y_k^*}{U_k^*} J_k,$$

where J_k takes the value 1 (with probability U_k^*/A) if network k is selected by point h, and 0 otherwise, we have

$$\mu_w = \mathrm{E}[w_h]$$

$$= \sum_{k=1}^{K} \frac{y_k^*}{U_k^*} \mathrm{E}[J_k]$$

$$= \sum_{k=1}^{K} \frac{y_k^*}{U_k^*} P(J_k = 1)$$

$$= \frac{1}{A} \sum_{k=1}^{K} y_k^*$$

$$= \frac{1}{A} \sum_{i=1}^{N} y_i$$

$$= \frac{\tau}{A},$$

and

$$\sigma_w^2 = \mathrm{var}[w_h]$$

$$= \sum_{j=1}^{K} \sum_{k=1}^{K} \frac{y_j^* y_k^*}{U_j^* U_k^*} \mathrm{cov}[J_j, J_k].$$

Since

$$\mathrm{E}[J_j J_k] = P\left([J_j = 1] \cap [J_k = 1]\right)$$

$$= \frac{\Lambda_{jk}}{A},$$

where Λ_{jk} is the intersection of the selection areas of the jth and kth networks, we have from (1.3)

$$\mathrm{cov}[J_j, J_k] = \mathrm{E}[J_j J_k] - \mathrm{E}[J_j]\mathrm{E}[J_k]$$

and hence

$$\sigma_w^2 = \sum_{j=1}^{K} \sum_{k=1}^{K} \frac{y_j^* y_k^*}{U_j^* U_k^*} \left(\frac{\Lambda_{jk}}{A} - \frac{U_j^* U_k^*}{A^2} \right).$$

We finally have from (4.28)

$$\tilde{\mu} = \frac{A}{N} \bar{w},$$

$$\mathrm{E}[\tilde{\mu}] = \frac{A}{N} \mu_w = \mu,$$

which we already knew from Equation (4.25) above, and

$$\mathrm{var}[\tilde{\mu}] = \frac{A^2}{N^2} \frac{\sigma_w^2}{n_0}$$

$$= \frac{1}{N^2 n_0} \sum_{j=1}^{K} \sum_{k=1}^{K} \frac{y_j^* y_k^*}{U_j^* U_k^*} \left(A\Lambda_{jk} - U_j^* U_k^* \right). \tag{4.29}$$

An unbiased estimate of this variance is

$$\widehat{\mathrm{var}}[\tilde{\mu}] = \frac{A^2}{N^2} \widehat{\mathrm{var}}[\bar{w}]$$

$$= \frac{A^2}{N^2} \frac{s_w^2}{n_0}$$

$$= \frac{A^2}{N^2 n_0 (n_0 - 1)} \sum_{h=1}^{n_0} (w_h - \bar{w})^2. \tag{4.30}$$

The formulas (4.27), (4.29), and (4.30) were obtained by Roesch [1993: Equations (4), (5) and (6)]. He focused on the total $\tau = N\mu$ because in forestry N would not generally be known. It is readily seen that for the corresponding estimates $\hat{\tau}$ and $\tilde{\tau}$, N disappears from all the above formulas.

Roesch also proposed a modification of $\tilde{\mu}$. In constructing $\tilde{\mu}$, a network is counted only once for each time it is selected by a randomly selected point, no matter how many of its component trees are selected at that point. Another estimator $\tilde{\mu}_m$ is obtained if we count a network once for each of its component trees selected by the initial sample point. Thus, if three of the network's trees are selected at a point, the network is counted three times. The probability of selecting the network is now proportional to the sum of the selection areas of the component trees rather than their union. Let S_i be the sum of the selection areas for all trees in the network defined by tree i and let y_i' be the sum of the y-values associated with that network. Working with trees rather than networks, we therefore consider

$$\tilde{\mu}_m = \frac{A}{Nn_0} \sum_{h=1}^{n_0} \sum_{i=1}^{n_h} \frac{y_i'}{S_i}$$

$$= \frac{A}{N} \bar{w},$$

where n_h is the number of trees selected by point h, and w_h is redefined as

$$w_h = \sum_{i=1}^{n_h} \frac{y_i'}{S_i}$$

$$= \sum_{i=1}^{N} \frac{y_i'}{S_i} J_i'.$$

Here J_i' takes the value 1 (with probability t_i/A) if the ith tree is selected by the hth sample point. Then

$$
\begin{aligned}
\mu_w &= \sum_{i=1}^{N} \frac{y_i'}{S_i} \mathrm{E}[J_i'] \\
&= \sum_{i=1}^{N} \frac{y_i' t_i}{S_i A} \\
&= \sum_{k=1}^{K} \sum_{i=1}^{k} \frac{y_i' t_i}{S_i A},
\end{aligned}
$$

where the second summation is over all trees in the kth network. Now S_i ($= T_k$, say), and y_i' ($= y_k^*$) will be the same for all trees in the kth network so that

$$
\begin{aligned}
\mu_w &= \sum_{k=1}^{K} \frac{y_k^*}{T_k} \sum_{i=1}^{k} \frac{t_i}{A} \\
&= \frac{1}{A} \sum_{k=1}^{K} \frac{y_k^*}{T_k} T_k \\
&= \frac{1}{A} \sum_{i=1}^{N} y_i \\
&= \frac{\tau}{A},
\end{aligned}
$$

as before. Finally,

$$
\begin{aligned}
\sigma_w^2 &= \sum_{i=1}^{N} \sum_{j=1}^{N} \frac{y_i' y_j'}{S_i S_j} \mathrm{cov}[J_i', J_j'] \\
&= \sum_{i=1}^{N} \sum_{j=1}^{N} \frac{y_i' y_j'}{S_i S_j} \left(\frac{W_{ij}}{A} - \frac{t_i t_j}{A^2} \right),
\end{aligned}
$$

where W_{ij} is the intersection of the selection areas for trees i and j (symbolically $W_{ij} = t_i \cap t_j$). An unbiased estimate of $\mathrm{var}[\tilde{\mu}_m]$ is given by (4.30). These formulas, but related to τ rather than μ, were given by Roesch [1993: Equations (8), (9), and (10)].

Roesch carried out a simulation experiment to compare the performances of the three estimators $\hat{\tau}$, $\tilde{\tau}$, and $\tilde{\tau}_m$ ($\tau = \mu N$). He also included the standard (nonadaptive), unbiased estimator based on the "average" Hansen–Hurwitz estimator, namely

$$
\tau' = \frac{1}{n_0} \sum_{h=1}^{n_0} \sum_{i=1}^{n_h} \frac{y_i}{p_i},
$$

where p_i, the probability of selecting tree i by a randomly chosen point is t_i/A, that is

$$\tau' = \frac{A}{n_0} \sum_{h=1}^{n_0} \sum_{i=1}^{n_h} \frac{y_i}{t_i}.$$

His simulations demonstrated that when adaptive sampling is adequate,

$$\text{MSE}[\hat{\tau}] < \text{MSE}[\tilde{\tau}] < \text{MSE}[\tilde{\tau}_m],$$

where MSE denotes Mean-Square Error. The differences between $\text{MSE}[\hat{\tau}]$ and $\text{MSE}[\tilde{\tau}]$ were mostly fairly small, and $\tilde{\tau}_m$ was decidedly inferior. Reductions in the MSE of over 50% were demonstrated for $\hat{\tau}$ over τ' in some cases.

Finally we mention one or two practical points relating specifically to forestry. The method usually used for the initial sampling process is the so-called horizontal point sampling method. This is described as follows (Husch et al. [1982, pp. 220–227]). A sampling point is chosen at random. The observer then occupies the sample point and views every visible tree at breast height through an angle gauge set at some angle θ. A tree i of radius r_i is selected if it is close enough to cover the field of view, that is, it is selected if its center lies within a circle of radius R_i, where (Figure 4.1)

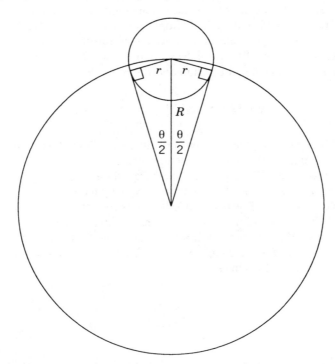

Figure 4.1. Horizontal point sampling method: $r = R \sin \theta/2$, where r is the radius of the tree and R is the radius of the selection area.

$R_i = r_i/\sin(\theta/2)$. This means that the probability of selection is $\pi R_i^2/A$, which is proportional to πr_i^2. Now the point and the tree are, in a sense, interchangeable in that the tree is selected if the point lies inside a circle of radius R_i with center at the center of the tree. If another tree j is also selected at the same time by the same point, then the point must also lie in a circle centered on the tree with radius R_j, say. This scheme is equivalent to the one used in the above theory. If the tree is not circular at breast height, one could generally choose an appropriate measure such as an "average" radius r_i with the property that the cross-sectional area is approximately πr_i^2. Various measures are described by Husch at al. [1982].

4.5 APPLYING THE RAO–BLACKWELL METHOD

By Theorem 2.3, the reduced data D_R are minimal sufficient for the parameter vector θ ($= \mathbf{y}$) of population y-values. This means that given any unbiased estimator T of $\phi = \phi(\theta)$ that is not a function of D_R, we can apply the Rao–Blackwell theorem (Theorem 2.4, Corollary 2) and obtain another unbiased estimator $T_{RB} = \mathrm{E}[T \mid D_R = d_R]$ with smaller variance.

Now the unbiased estimators $\bar{y}_1$, $\hat{\mu}$, and $\tilde{\mu}$ of μ all depend on the order of selection, as they depend on which n_1 units are in the initial sample. Estimator $\tilde{\mu}$ also depends on repeat selections; and when the initial sample is selected with replacement, all three estimators depend on repeat selections. Since all three estimators are not functions of D_R, we can apply the Rao–Blackwell theorem. If T is any one of the three estimators, we shall find T_{RB} and an unbiased estimator for [cf. (2.15)]

$$\mathrm{var}[T_{RB}] = \mathrm{var}[T] - \mathrm{E}\{\mathrm{var}[T \mid d_R]\}. \tag{4.31}$$

Let ν denote, as usual, the number of distinct units in the final adaptive sample. If the initial sample is without replacement, define $G = \binom{\nu}{n_1}$, the number of possible combinations or "groups" of n_1 distinct units from the ν in the sample. Suppose that these combinations are indexed in an arbitrary way by the label g ($g = 1, 2, \ldots, G$). Let t_g denote the value of T when the initial sample consists of combination g and let $\widehat{\mathrm{var}}_g[T]$ denote the value of the unbiased estimator $\widehat{\mathrm{var}}[T]$ when computed using the gth combination. We define the indicator variable I_g to be 1 if the gth combination could give rise to d_R (i.e., is compatible with d_R), and 0 otherwise. The number of compatible combinations is then

$$\xi = \sum_{g=1}^{G} I_g \tag{4.32}$$

and, conditional on d_R, each of these is equally likely. Hence, given d_R, $T = t_g$ with probability $1/\xi$ for all compatible g so that

$$T_{RB} = \mathrm{E}[T \mid d_R]$$

$$= \frac{1}{\xi} \sum_{g=1}^{\xi} t_g$$

$$= \frac{1}{\xi} \sum_{g=1}^{G} t_g I_g. \tag{4.33}$$

To find an unbiased estimator of $\mathrm{var}[T_{RB}]$, we need unbiased estimators of the two terms in (4.31). Since $\mathrm{var}[T]$ is also a function of $\boldsymbol{\theta}$ and $\widehat{\mathrm{var}}[T]$ is an unbiased estimator, we can apply the Rao–Blackwell theorem once again to obtain another unbiased estimator with smaller variance, namely

$$\widehat{\mathrm{var}}_{RB}[T] = \mathrm{E}\{\widehat{\mathrm{var}}[T] \mid d_R\}$$

$$= \frac{1}{\xi} \sum_{g=1}^{\xi} \widehat{\mathrm{var}}_g[T]. \tag{4.34}$$

Also

$$\mathrm{var}[T \mid d_R] = \mathrm{E}[(T - T_{RB})^2 \mid d_R]$$

$$= \frac{1}{\xi} \sum_{g=1}^{\xi} (t_g - T_{RB})^2 \tag{4.35}$$

is an unbiased estimator of its expected value. Combining (4.34) and (4.35), we obtain from (4.31) the unbiased estimator (cf. Thompson [1990])

$$\widehat{\mathrm{var}}[T_{RB}] = \frac{1}{\xi} \sum_{g=1}^{\xi} \{\widehat{\mathrm{var}}_g[T] - (t_g - T_{RB})^2\}$$

$$= \frac{1}{\xi} \sum_{g=1}^{G} \{\widehat{\mathrm{var}}_g[T] - (t_g - T_{RB})^2\} I_g. \tag{4.36}$$

We note that the above estimator could be negative with some sets of data. If the initial sample is selected with replacement, then the above formulas hold with $G = \nu^{n_1}$, the number of sequences of n_1 units (distinguishing order and allowing repeats) from the ν distinct units.

We now consider what happens when T is $\bar{y}_1$, $\hat{\mu}$, or $\tilde{\mu}$. To do this we introduce another statistic D_f, which consists of D_R together with the frequencies f_i. We recall that f_i is the frequency of intersection of the initial sample with the network A_i, containing i, for each unit i in the final sample. Thus

$$D_f = \{(i, y_i, f_i) : i \in s_R\}.$$

We note that f_i is also the number of units of the initial sample in network A_i. It will be zero for those edge units not in the initial sample.

Theorem 4.1. Suppose that sampling is with or without replacement. If $\tilde{\mu}$ is defined by (4.18), then:

(i) D_f is sufficient for $\boldsymbol{\theta}$.
(ii) $E[\bar{y}_1 \mid D_f = d_f] = \tilde{\mu}$.
(iii) $E[\bar{y}_1 \mid D_R = d_R] = E[\tilde{\mu} \mid D_R = d_R] \quad (= \tilde{\mu}_{RB})$.

Proof:

(i) We recall that $D_R = \{(i, y_i) : i \in s_R\}$ is sufficient (in fact minimal sufficient) for $\boldsymbol{\theta}$. Since D_f determines D_R, D_R is a function of D_f, say, $D_R = g(D_f)$, and D_f is sufficient for $\boldsymbol{\theta}$ by Theorem 2.2.

(ii) Let the distinct networks be $B_1, B_2, \ldots, B_K$. Then we have the representation

$$\bar{y}_1 = \frac{1}{n_1} \sum_{k=1}^{\kappa} \sum_{j=1}^{b_k} y_j, \tag{4.37}$$

where κ is the number of distinct networks selected in the initial sample and b_k is the number of units from the initial sample which fall in B_k, that is, the number of times network B_k is selected in the initial sample. Since $f_i = b_k$ for every $i \in B_k$, the b_k's are determined by D_f. Also the b_k units selected from B_k may or may not be distinct depending on whether sampling is with or without replacement. If there are x_k units in B_k, then, with fixed D_f, the only part of $\bar{y}_1$ that can vary is the choice of which b_k units are chosen with or without replacement from the x_k units. Since we have a random sample of b_k units from x_k, and the sample mean is an unbiased estimator of the population mean for sampling with or without replacement,

$$E\left[\frac{1}{b_k} \sum_{j=1}^{b_k} y_j \mid d_f\right] = \bar{v}_k,$$

where $\bar{v}_k$ is the mean of the y-values in B_k. Hence, from (4.37),

$$E[\bar{y}_1 \mid d_f] = \frac{1}{n_1} \sum_{k=1}^{\kappa} b_k \bar{v}_k$$

$$= \tilde{\mu}.$$

(iii) Since $D_R = g(D_f)$,

$$E[\bar{y}_1 \mid d_f] = E[\bar{y}_1 \mid d_f, d_R]. \tag{4.38}$$

We have from **(ii)** that $\tilde{\mu}$ is a function of d_f. Therefore letting $h(\cdot)$ denote any appropriate probability function (a convenient misuse of notation), we have

$$\mathrm{E}[\tilde{\mu} \mid d_R] = \sum_{d_f} \tilde{\mu} h(d_f \mid d_R),$$

where d_f is summed over all d_f such that $g(d_f) = d_R$. Then, from (4.38) and **(ii)**,

$$\begin{aligned}
\mathrm{E}[\tilde{\mu} \mid d_R] &= \sum_{d_f} \mathrm{E}[\bar{y}_1 \mid d_f, d_R] \, h(d_f \mid d_R) \\
&= \sum_{d_f} \sum_{\bar{y}_1} \bar{y}_1 h(\bar{y}_1 \mid d_f, d_R) \frac{h(d_f, d_R)}{h(d_R)} \\
&= \sum_{d_f} \sum_{\bar{y}_1} \bar{y}_1 h(\bar{y}_1, d_f \mid d_R) \\
&= \sum_{\bar{y}_1} \bar{y}_1 h(\bar{y}_1 \mid d_R) \\
&= E[\bar{y}_1 \mid d_R],
\end{aligned}$$

which completes the proof.   □

The preceding theorem, due to Thompson [1990], shows that when the Rao–Blackwell method is applied to both $\bar{y}_1$ and $\tilde{\mu}$, we end up with the same estimator $\tilde{\mu}_{RB}$. We shall see in a small example given later in Section 4.6.1 that $\hat{\mu}$ leads to a different estimator $\hat{\mu}_{RB} = \mathrm{E}[\hat{\mu} \mid d_R]$ and that neither $\tilde{\mu}_{RB}$ nor $\hat{\mu}_{RB}$ is uniformly better (i.e., for all $\theta \in \Theta$) than the other. Finally we note from Theorem 4.1 that the sufficiency of D_f implies that $\mathrm{var}[\tilde{\mu}] \leq \mathrm{var}[\bar{y}_1]$.

4.6   EXPECTED SAMPLE SIZE AND COST

Let I_i take the value 1 when unit i is included in the sample, and 0 otherwise. Then the number of distinct units ν in an adaptive sample is a random variable, namely

$$\nu = \sum_{i=1}^{N} I_i. \tag{4.39}$$

If π_i is the probability that unit i is included in the sample, then $\pi_i = P(I_i = 1) = \mathrm{E}[I_i]$ and

$$\mathrm{E}[\nu] = \sum_{i=1}^{N} \pi_i. \tag{4.40}$$

If sampling is without replacement, then π_i is given by (4.2), namely

$$\pi_i = 1 - \left[\binom{N - m_i - a_i}{n_1} \bigg/ \binom{N}{n_1} \right].$$

If sampling is without replacement, then

$$\pi_i = 1 - (1 - p_i)^{n_1},$$

where

$$p_i = \frac{m_i + a_i}{N}.$$

We recall that m_i is the number of units in the network A_i which contains unit i, and a_i is the total number of units in networks of which unit i is an edge unit. If unit i satisfies C, then $a_i = 0$, otherwise a_i is unknown. Equation (4.39) is therefore not very helpful in designing an experiment as the π_i are not all known. The result, however, is useful for theoretical comparisons and simulation experiments.

If comparisons are to be made on the basis of cost, it will often be cheaper in time or money to sample units within a cluster than to select a new cluster. Given C_T, the total cost, a possible cost equation would be

$$C_T = c_0 + c_1 n_1 + c_2 (\nu - n_1),$$

where c_0 is a fixed cost (e.g., to go to the location), and c_1 and c_2 are the marginal costs per unit in the initial and subsequent samples, respectively. In addition, we may find in some situations that it costs more to observe a unit that satisfies the condition C than one that doesn't. For example, plots with information can take longer to investigate than empty plots. This would be the case if the y-variable is the biomass of a plant species on a plot. When the above conditions apply, the relative advantage of the adaptive to the nonadaptive strategy would tend to be greater than in comparisons based solely on sample size.

Since C_T is a random variable, we would use the equation

$$E[C_T] = c_0 + c_1 n_1 + c_2 \big(E[\nu] - n_1 \big). \tag{4.41}$$

Cost issues are considered in more detail in Chapter 5.

4.6.1 Examples

To illustrate the computations associated with adaptive cluster sampling strategies, a number of examples (from Thompson [1990]) are worked, with some detail added. In each of the examples the neighborhood of a unit is defined to consist of the unit itself and the four adjacent units sharing a common boundary line (Figure 4.2).

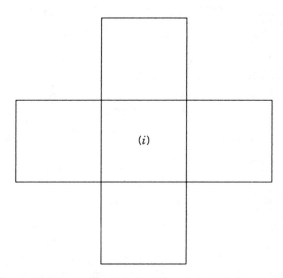

(i)

Figure 4.2. The type of neighborhood for unit i used in the examples.

Example 4.1   Computation of Estimates.   This example is introduced to demonstrate the computations for $\hat{\mu}$ and $\tilde{\mu}$. In Figure 4.3a the initial sample consists of $n_1 = 10$ units selected by simple random sampling from $N = 400$ units. Adding adaptively to those initial units that contain at least one individual leads to the final sample shown in Figure 4.3b. One of the initial sample units, near the top of the study region, intersected a network of $m_1 = 6$ units containing a total of $y_1^* = 36$ point objects. Another intersected a network of $m_2 = 11$ units containing $y_2^* = 107$ objects. For the other eight units of the initial sample, $y_i = 0$ and $m_i = 1$. There were also 20 edge units which are not used in the calculation of the estimates. In Figure 4.3c the networks within the two adaptively added clusters are outlined in heavy bold. The remaining units within each of these clusters are edge units.

First Estimate
The intersection probability for the first network is

$$\alpha_1 = 1 - \left[\binom{400 - 6}{10} \Big/ \binom{400}{10} \right]$$
$$= 1 - 0.8582 = 0.1418.$$

For the other large network intersected by the initial sample,

$$\alpha_2 = 1 - \left[\binom{400 - 11}{10} \Big/ \binom{400}{10} \right] = 1 - 0.7542 = 0.2458.$$

For the networks of size 1, the probability is $\alpha_k = 10/400 = 0.025$. Hence the estimate using intersection probabilities is, from (4.6),

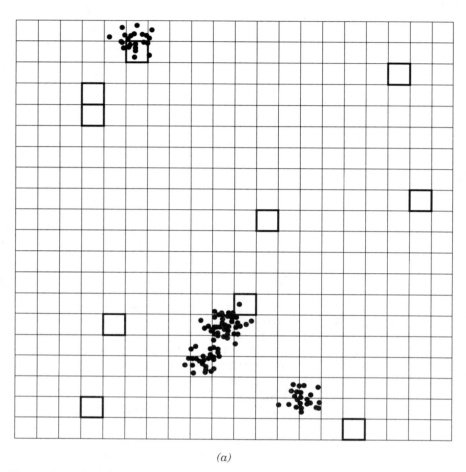

(a)

Figure 4.3a. An initial simple random sample of 10 units. (From S.K. Thompson, Adaptive cluster sampling, 1990, *Journal of the American Statistical Association* **85**, 1050–1059. With permission from the American Statistical Asssociation.)

$$\hat{\mu} = \frac{1}{400} \left(\frac{36}{0.1418} + \frac{107}{0.2458} + \frac{0}{0.025} + \cdots + \frac{0}{0.025} \right) = 1.723$$

objects per unit or 400(1.723) = 689 total objects in the population.

To compute $\widehat{\mathrm{var}}[\hat{\mu}]$ we need [cf. (4.8)]

$$\alpha_{12} = \alpha_1 + \alpha_2 - \left(1 - \left[\binom{400-17}{10} \bigg/ \binom{400}{10} \right] \right)$$

$$= 0.1418 + 0.2458 - 1 + \frac{374 \cdot 375 \cdots 383}{391 \cdot 392 \cdots 400}$$

$$= 0.1418 + 0.2458 - 1 + 0.6444$$

$$= 0.0320.$$

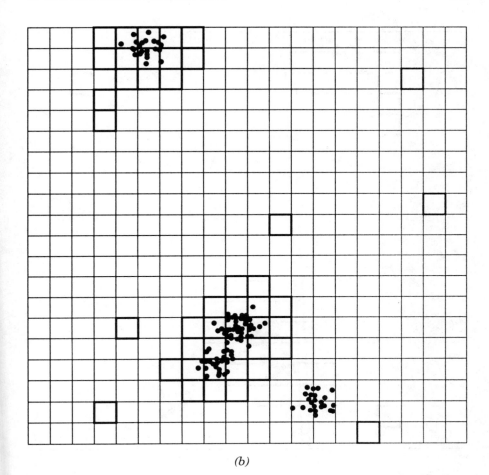

(b)

Figure 4.3b. The resulting adaptive cluster sample. (From S.K. Thompson, Adaptive cluster sampling, 1990, *Journal of the American Statistical Association* **85**, 1050–1059. With permission from the American Statistical Asssociation.)

Then, from (4.11),

$$\widehat{\text{var}}[\hat{\mu}] = \frac{1}{N^2}\left[\frac{y_1^{*2}}{\alpha_1}\left(\frac{1}{\alpha_1}-1\right) + \frac{y_2^{*2}}{\alpha_2}\left(\frac{1}{\alpha_2}-1\right) + \frac{2y_1^*y_2^*}{\alpha_{12}}\left(\frac{\alpha_{12}}{\alpha_1\alpha_2}-1\right)\right]$$

$$= \frac{1}{400^2}\left[\frac{36^2}{.1418}\left(\frac{1}{.1418}-1\right) + \frac{107^2}{.2458}\left(\frac{1}{.2458}-1\right)\right.$$

$$\left. + \frac{2(36)(107)}{.0320}\left(\frac{.0320}{(.1418)(.2458)}-1\right)\right]$$

$$= 1.1157,$$

giving an estimated standard error of $\sqrt{1.1157} = 1.056$.

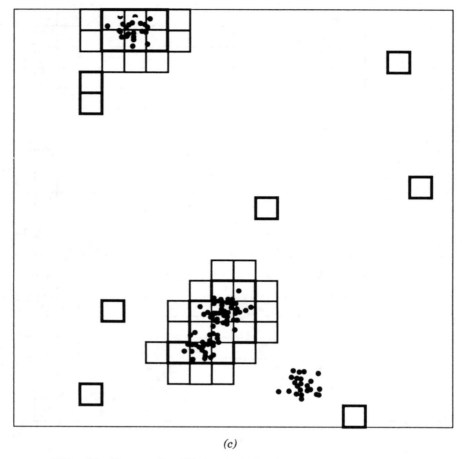

(c)

Figure 4.3c. The networks within the adaptively added clusters are outlined in bold.

Second Estimate

For the first network there was an average of $w_1 = 36/6 = 6$ objects per unit; for the second, $w_2 = 107/11 = 69.727$; and for the rest, $w_i = 0$. Hence from (4.15),

$$\tilde{\mu} = \frac{1}{10}\left(\frac{36}{6} + \frac{107}{11} + \frac{0}{1} + \cdots + \frac{0}{1}\right)$$

$$= 0.1(6 + 9.727 + 0 + \cdots + 0) = 1.573$$

objects per unit, or $N\tilde{\mu} = 400(1.573) = 629$ objects in the population.

From (4.17), the estimated variance is

$$\widehat{var}[\hat{\mu}] = \frac{(400 - 10)}{400(10)(10 - 1)} \left[(6 - 1.573)^2 + (9.727 - 1.573)^2 + \cdots \right.$$

$$\left. +(0 - 1.573)^2 + (0 - 1.573)^2 \right]$$

$$= 1.147.$$

For the estimate of the total, the estimated variance is $400^2(1.147) = 183,520$.

Note that the ordinary sample mean of the 45 units (which includes the 25 edge units) in the final sample would have given $\bar{y} = 143/45 = 3.178$ for an estimate of $N\bar{y} = 1271$ objects in the study region. The fact that the adaptive selection procedure produces a high yield of observed objects gives the ordinary sample mean a tendency to overestimate.

Example 4.2   A Small Population.   In this example, the sampling strategies are applied to a very small population in order to shed light on the computations and properties of the adaptive estimates in relation to each other and to conventional estimates. The population consists of just five units, the y-values of which are

$$\mathbf{y} = (1, 0, 2, 10, 1000)'.$$

The neighborhood of each unit includes all adjacent units (of which there are either one or two). The condition is defined by $C = \{y : y \geq 5\}$. The initial sample size is $n_1 = 2$.

With the adaptive design in which the initial sample is selected by simple random sampling without replacement, there are $\binom{5}{2} = 10$ possible samples, each having probability $1/10$. The resulting observations and the values of each estimator are listed in Table 4.1.

In this population, the 3rd, 4th, and 5th units, with y-values 2, 10, and 1000, form a cluster consisting of a network (4th and 5th units) and an edge unit (3rd unit). In the fourth row of the table, the 1st and 5th units, with y-values 1 and 1000, were selected initially. Since $1000 \geq 5$, the single neighbor of the fifth unit, having y-value 10, was then added to the sample. Since 10 also exceeds 5, the neighboring unit with y-value 2 was also added to the sample. Ignoring the edge unit, the computations for the estimators are $\alpha_1 = 1 - \binom{4}{2}/\binom{5}{2} = 0.4$ and $\alpha_2 = \alpha_3 = 1 - \binom{3}{2}/\binom{5}{2} = 0.7$, leading to

$$\hat{\mu} = \left[(1/.4) + (10/.7) + (1000/.7)\right]/5 = 289.07$$

and $\tilde{\mu} = [1 + (10 + 1000)/2]/2 = 253$. The classical estimator $\bar{y} = 253.25$ is obtained by averaging all four observations in the sample. Also we can compute $\bar{\bar{y}} = [1 + (10 + 2 + 1000)/3]/2 = 169.67$ which, like $\tilde{\mu}$, is an average of the cluster means but using the edge units as well.

The six distinct values of the minimal sufficient statistic D_R are indicated by the distinct values of the Rao–Blackwell estimators $\hat{\mu}_{RB}$ and $\tilde{\mu}_{RB}$. These are obtained by

Table 4.1. All Possible Outcomes of Adaptive Cluster Sampling for a Population of Five Units With y-Values 1, 0, 2, 10, 1000, in Which the Neighborhood of Each Unit Consists of Itself Plus Adjacent Units.*

Observations	$\bar{y}_1$	$\hat{\mu}$	$\hat{\mu}_{RB}$	$\tilde{\mu}$	$\tilde{\mu}_{RB}$	$\bar{y}$	$\bar{\bar{y}}$
1,0	0.50	0.50	0.50	0.50	0.50	0.50	0.50
1,2	1.50	1.50	1.50	1.50	1.50	1.50	1.50
1,10; 2,1000	5.50	289.07	289.07	253.00	253.00	253.25	169.67
1,1000;10,2	500.50	289.07	289.07	253.00	253.00	253.25	169.67
0,2	1.00	1.00	1.00	1.00	1.00	1.00	1.00
0,10; 2,1000	5.00	288.57	288.57	252.50	252.50	253.00	168.67
0,1000;10,2	500.00	288.57	288.57	252.50	252.50	253.00	168.67
2,10;1000	6.00	289.57	289.24	253.50	337.33	337.33	337.33
2,1000;10	501.00	289.57	289.24	253.50	337.33	337.33	337.33
10,1000; 2	505.00	288.57	289.24	505.00	337.33	337.33	337.33
Mean:	202.6	202.6	202.6	202.60	202.60	202.75	169.17
Bias:	0	0	0	0	0	0.15	−33.43
MSE:	59,615	17,418.4	17,418.3	22,862	18,645	18,660	18,086

NOTE: The initial sample of two units is selected by simple random sampling without replacement. Whenever an observed y-value exceeds 5, the neighboring units are added to the sample. Initial observations are separated from subsequent observations in the table by a semicolon. For each possible sample, the value of each estimator is given. The bottom line of the table gives the mean square error for each estimator. The sample mean of a simple random sample of equivalent sample size has variance 24,359.

*Adapted from S.K. Thompson, Adaptive cluster sampling, 1990, *Journal of the American Statistical Association* **85**, 1050–1059. With permission from the American Statistical Association.

averaging $\hat{\mu}$ and $\tilde{\mu}$, respectively, over all samples with the same value of D_R. We note that in the last row of Table 4.1, initial selection of the units with y-values 10 and 1000 leads to addition of the adjacent unit with value 2, which receives no weight in the estimators $\hat{\mu}$ and $\tilde{\mu}$ since it does not satisfy the condition and was not included in the initial sample. But the Rao–Blackwell estimates based on this sample do utilize the value 2 by averaging over the last three rows of the table (which all lead to the same reduced set $\{2, 10, 1000\}$).

The population mean is 202.6, and the population variance (defined with $N - 1$ in the denominator) is 198,718. One sees from the table that the unbiased adaptive strategies indeed have expectation 202.6, while the estimators $\bar{y}$ and $\bar{\bar{y}}$, used with the adaptive design, are biased.

With the adaptive design, the effective sample size ν varies from sample to sample, with expected sample size 3.1. For comparison, the sample mean from a simple random sampling design (without replacement) based on a sample size of 3.1 has variance (using the usual formula)

$$\frac{\sigma^2}{E[\nu]}\left(1 - \frac{E[\nu]}{N}\right) = \frac{198,718(5 - 3.1)}{5(3.1)} = 24,359. \quad (4.42)$$

From the variances and mean-square errors given in the last row of the table, one sees that, for this population, the adaptive design with the estimator $\hat{\mu}_{RB}$ has the lowest variance among the unbiased strategies (noting, however, that the extra digit of reporting precision is necessary in the table to show that var$[\hat{\mu}_{RB}]$ is slightly less than var$[\hat{\mu}]$). Also all of the adaptive strategies are more efficient than simple random sampling. Among the five unbiased adaptive strategies, the four that make use of labels in the data have lower variance than $\bar{y}_1$, the mean of the initial sample, which does not.   □

4.6.2 Comparative Efficiencies

In this section, the efficiency of adaptive cluster sampling for a rare, clustered population is illustrated using, first, the clustered population of Figure 4.3 and, second, the same population with each y-value converted to a zero–one variable indicating the presence or absence of objects in the unit. General results on factors influencing the efficiency of adaptive cluster sampling are given in Chapter 5. The population of each example is contained in a square region partitioned into $N = 20 \times 20 = 400$ units. As in Example 4.1 of the previous section, the neighborhood of each unit consists, in addition to itself, of all adjacent units (i.e., those that share a common boundary line). A unit satisfies the condition for additional sampling if the y-value associated with the unit is greater than or equal to 1.

For each example, variances are computed for the estimators $\hat{\mu}$ and $\tilde{\mu}$ under the design of adaptive cluster sampling with the initial sample of n_1 units selected by simple random sampling without replacement. Results are listed in the tables for a selection of initial sample sizes from $n_1 = 1$ to $n_1 = 200$.

For comparison, the variance is also computed for the sample mean μ_0^* of a simple random sample (without replacement) with sample size equal to the expected (effective) sample size $E[\nu]$ under the adaptive design [cf. (4.42)]. For each adaptive strategy, the relative efficiency—the variance of the simple random sampling strategy divided by the variance of the adaptive strategy—is also listed. The population y-values for the two examples are listed at the end of the chapter.

Example 4.3   Population 1.   The population of point objects illustrated in the figures was produced as a realization in the unit square of a Poisson cluster process (see Diggle [1983]) with five "parent locations" from a uniform distribution and random [Poisson (mean = 40)] numbers of point objects dispersed in relation to the parent locations with a symmetric normal distribution having standard deviation 0.02. The population mean is 190/400=0.475.

Table 4.2 lists the expected sample sizes, variances, and relative efficiencies for the different sampling strategies for a selection of initial sample sizes. With an initial sample size of 1, the variances of the adaptive strategies are about equal to that obtained with simple random sampling. The relative advantage of the adaptive strategies increases with increasing n_1. An initial sample of size 10, as illustrated in Figure 4.3a, leads to an average final sample of size of about 18 units and, with the estimator $\tilde{\mu}$, an efficiency gain of 12% over simple random sampling with equivalent sample size. With an initial sample size of 200, the adaptive strategy leads on average to observing a total of about 224 units and is 15.36 times as efficient as simple random sampling.

Table 4.2. Population 1: Variances of $\hat{\mu}$ and $\tilde{\mu}$ With Adaptive Cluster Sampling and Initial Sample Size n_1 for the Population Illustrated in Figure 4.3.

n_1	E[ν]	var[$\hat{\mu}$]	var[$\tilde{\mu}$]	var[μ_0^*]	eff($\hat{\mu}$)	eff($\tilde{\mu}$)
1	1.92	4.29705	4.29705	4.28364	1.00	1.00
2	3.82	2.12386	2.14314	2.14420	1.01	1.00
10	18.26	0.38655	0.42001	0.43240	1.12	1.03
20	34.66	0.17097	0.20462	0.21805	1.28	1.07
30	49.56	0.10030	0.13282	0.14627	1.46	1.10
40	63.26	0.06587	0.09693	0.11012	1.67	1.14
50	76.00	0.04593	0.07539	0.08819	1.92	1.17
60	87.97	0.03322	0.06103	0.07338	2.21	1.20
100	130.80	0.01096	0.03231	0.04258	3.89	1.32
200	223.86	0.00106	0.01077	0.01628	15.36	1.51

NOTE: The variance of $\bar{y}$ (var[μ_0^*]) with simple random sampling is calculated for sample size E[ν], the expected sample size with the adaptive design. Relative efficiencies in the last two columns are eff($\hat{\mu}$) = var[μ_0^*]/var[$\hat{\mu}$] and eff($\tilde{\mu}$) = var[μ_0^*]/var[$\tilde{\mu}$].
*Adapted from S.K. Thompson, Adaptive cluster sampling, 1990, *Journal of the American Statistical Association* **85**, 1050–1059. With permission from the American Statistical Association.

Example 4.4   Population 2.   The y-values for this population are either zero or 1. The population was obtained from that of Example 4.3 by letting the y-value of each unit indicate the presence or absence of point objects in that unit. Thus, the pattern of the "1's" in the population is identical to the pattern of nonzero units in Figure 4.3.

For such a population, the within-network variance is zero, since every network in the population consists either of a single unit with $y_i = 0$ or a group of one or more units each with $y_i = 1$. Therefore, as we shall see later in Chapter 5, the adaptive strategy with the estimator $\tilde{\mu}$ cannot do better than simple random sampling in this situation. The variance computations in Table 4.3 reveal, however, that the estimator $\hat{\mu}$ used with the adaptive design does turn out to be more efficient than simple random sampling for initial sample sizes of 50 or larger.   □

Table 4.3. Population 2: Variance Comparisons With the y-variable Indicating Presence or Absence of Objects in the Population of Figure 4.3.*

n_1	E[ν]	var[$\hat{\mu}$]	var[$\tilde{\mu}$]	var[μ_0^*]	eff($\hat{\mu}$)	eff($\tilde{\mu}$)
1	1.92	0.04974	0.04974	0.02581	.52	.52
2	3.82	0.02459	0.02481	0.01292	.53	.52
10	18.26	0.00448	0.00486	0.00261	.58	.54
20	34.66	0.00198	0.00237	0.00131	.66	.55
30	49.56	0.00116	0.00154	0.00088	.76	.57
40	63.26	0.00076	0.00112	0.00066	.87	.59
50	76.00	0.00053	0.00087	0.00053	1.00	.61
60	87.97	0.00038	0.00071	0.00044	1.15	.63
100	130.80	0.00012	0.00037	0.00026	2.06	.69
200	223.86	0.00001	0.00012	0.00010	9.52	.79

*Adapted from S.K. Thompson, Adaptive cluster sampling, 1990, *Journal of the American Statistical Association* **85**, 1050–1059. With permission from the American Statistical Association.

4.7 PRIMARY AND SECONDARY UNITS

4.7.1 Introduction

In surveys of natural and human populations, sample units larger than single plots or elements are often used. For example a common sampling unit is the strip transect, which we call the primary unit. In its adaptive modification, the strip would be divided up into smaller secondary units; and if we find animals in a secondary unit, we would sample the units on either side of that unit with still further searching if additional animals are sighted while on this search. Strips are widely used in both aerial and ship surveys of animals and marine mammals, where the aircraft or vessel travels down a line (transect) and the area is surveyed on either side out to a given distance. An example of such an adaptive scheme is given in Figure 4.4 taken from Thompson [1991a] in which the initial sample consists of five randomly selected strips (primary

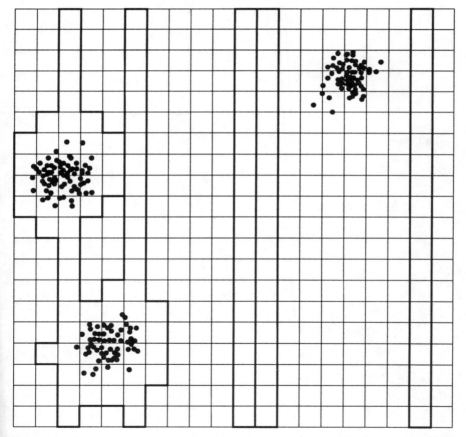

Figure 4.4. An adaptive cluster sample with an initial simple random sample of five strips. (From S.K. Thompson, Adaptive cluster sampling: designs with primary and secondary units, 1991, *Biometrics* **47**, 1103–1115. With permission from the Biometric Society.)

units). The secondary units are small square plots, and the neighborhood of such a plot consists of the adjacent plots to the left, right, top and bottom as well as the plot itself. Whenever a plot in the sample contains one or more animals, the rest of its neighborhood is also added to the sample. If, in turn, any of the additional plots contain animals, then their neighborhoods are also added. The final sample is shown in the figure.

In some wildlife surveys, the selection of sites chosen for observation is done systematically and a single systematic selection then forms the primary unit. This primary unit contains secondary units that are no longer contiguous. For example, in Figure 4.5, taken from Thompson [1991a], the initial sample is a spatial systematic sample with two randomly selected starting points thus giving a sample of two primary units. Whenever animals are observed in any plot of the sample, the neighborhood of that plot is added to the sample, and so on. The final sample is illustrated in the figure.

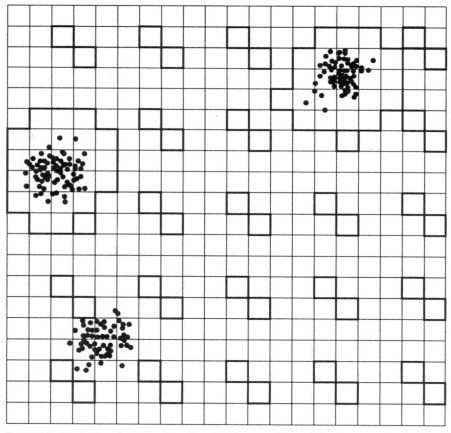

Figure 4.5. An adaptive cluster sample with an initial systematic sample having two randomly selected starting points. (From S.K. Thompson, Adaptive cluster sampling: designs with primary and secondary units, 1991, *Biometrics* **47**, 1103–1115. With permission from the Biometric Society.)

For both these sample schemes, the initial sample of primary units would normally be chosen without replacement. In terms of transect sampling, the ship or aircraft would not travel the same strip twice. In systematic sampling, a different starting point would be used each time so that the primary units would be selected without replacement. For these reasons we shall consider only simple random sampling when choosing the primary units. With the designs described below, once a primary unit is selected, every secondary unit within it is included in the sample. Thus, the initial designs of the adaptive cluster sampling are conventional cluster or systematic samples. Adaptive cluster sampling with initial two-stage designs are described in Salehi and Seber [1995]. With those designs, only a sample of secondary units is selected within each selected primary unit of the initial sample. A pilot survey can be used to determine samples sizes for a given cost or precision of estimation.

To estimate rare household characteristics, adaptive cluster sampling with an initial systematic sample of households was investigated by Danaher and King [1994]. The design was initially motivated by a market research survey wanting to estimate potential purchases of child-care products. The conventional strategy was inefficient because the characteristic of having children aged 2 or less was rare and distributed in an uneven, clustered pattern. The study to investigate the efficiency of the adaptive strategy was based on a census of a rare, clustered characteristic that was easier to observe—the presence or absence of "no circular" (no junk mail) notices. The adaptive strategy with the estimator based on initial intersection probabilities was assessed to be more efficient than the the the conventional strategy.

4.7.2   Simple Random Sampling of Primary Units

Suppose we have MN secondary units in the population and these are divided into N primary units, each consisting of M secondary units. Instead of labeling the units by a single integer, we now use a double integer notation. Thus, unit (i, j) represents the jth secondary unit in the ith primary unit, and y_{ij} is the associated y-value. Unit (i, j) is said to satisfy the condition of interest C if y_{ij} is in a specified set, defined for example by $y_{ij} > c$. In Figures 4.4 and 4.5, we have $c = 0$.

The initial sample consists of a selection of n_1 primary units and we add further secondary units using the adaptive method described above. Our aim is to estimate the population total

$$\tau = \sum_{i=1}^{N} \sum_{j=1}^{M} y_{ij}, \qquad (4.43)$$

or the mean $\mu = \tau/MN$. This can be done for μ using the mean of all the y-values in just the initial sample, namely

$$\bar{y}_1 = \frac{1}{Mn_1} \sum_{i=1}^{n_1} y_{i\cdot} \quad \left(= \frac{\bar{y}.}{M}, \text{ say}\right),$$

where $y_{i\cdot} = \sum_{j=1}^{M} y_{ij}$ is the sum of the y-values in the ith primary unit. Since $\bar{y}_{\cdot}$ is the mean of a simple random sample of the primary units, we have

$$E[\bar{y}_{\cdot}] = \frac{1}{N} \sum_{i=1}^{N} y_{i\cdot} = \frac{\tau}{N} = M\mu$$

and

$$E[\bar{y}_1] = \mu.$$

Also, from (1.27) with Y being a random variable taking values $y_{i\cdot}$,

$$\text{var}[\bar{y}_1] = \frac{1}{M^2} \text{var}[\bar{y}_{\cdot}]$$

$$= \frac{1}{M^2 n_1} \sigma_Y^2 \left(1 - \frac{n_1}{N} \right), \qquad (4.44)$$

where

$$\sigma_Y^2 = \frac{1}{N-1} \sum_{i=1}^{N} (y_{i\cdot} - M\mu)^2. \qquad (4.45)$$

A unbiased estimator of (4.44) is

$$\widehat{\text{var}}[\bar{y}_1] = \frac{1}{M^2 n_1} s_Y^2 \left(1 - \frac{n_1}{N} \right),$$

where

$$s_Y^2 = \frac{1}{n_1 - 1} \sum_{i=1}^{n_1} (y_{i\cdot} - \bar{y}_{\cdot})^2.$$

Another estimate of μ we can consider is (4.6), namely

$$\hat{\mu} = \frac{1}{MN} \sum_{k=1}^{K} \frac{y_k^* J_k}{\alpha_k} = \frac{1}{MN} \sum_{k=1}^{\kappa} \frac{y_k^*}{\alpha_k}, \qquad (4.46)$$

with the symbols appropriately redefined below. As before, we can divide up the population of MN secondary units into K distinct networks using clusters without their edge units. Let κ denote the number of networks in the sample. Let y_k^* denote the sum of the y-values in the kth network ($k = 1, 2, \ldots, K$) and let x_k now denote the number of *primary* units in the population that intersect the kth network [for the special case when the primary unit is just a single secondary unit, x_k then becomes the number of secondary units in the kth network, as in the development following Equation (4.6)]. We define J_k to take the value of 1 with (intersection) probability α_k

if the initial sample of primary units intersects the kth network, and 0 otherwise. As in (4.7), but with primary units, we have

$$\alpha_k = 1 - \left[\binom{N - x_k}{n_1} \Big/ \binom{N}{n_1} \right]$$

and, from (4.8),

$$\alpha_{kk'} = 1 - \left[\binom{N - x_k}{n_1} + \binom{N - x_{k'}}{n_1} - \binom{N - x_k - x_{k'} + x_{kk'}}{n_1} \right] \Big/ \binom{N}{n_1},$$

where $\alpha_{kk'}$ is the probability that the initial sample intersects both networks k and k', and $x_{kk'}$ is the number of primary units that intersect both networks.

Since $E[J_k] = P(J_k = 1) = \alpha_k$,

$$E[\hat{\mu}] = \frac{1}{MN} \sum_{k=1}^{K} y_k^* = \mu,$$

and $\hat{\mu}$ is unbiased. From (1.5) with $z_i = y_i^*/\alpha_i$, $\pi_i = \alpha_i$ and $\pi_{ij} = \alpha_{ij}$, we have

$$\mathrm{var}[\hat{\mu}] = \frac{1}{M^2 N^2} \sum_{k=1}^{K} \sum_{k'=1}^{K} y_k^* y_{k'}^* \left(\frac{\alpha_{kk'} - \alpha_k \alpha_{k'}}{\alpha_k \alpha_{k'}} \right), \qquad (4.47)$$

with the convention that $\alpha_{kk} = \alpha_k$. An unbiased estimator of this variance is given by (1.6), namely

$$\widehat{\mathrm{var}}[\hat{\mu}] = \frac{1}{M^2 N^2} \sum_{k=1}^{K} \sum_{k'=1}^{K} y_k^* y_{k'}^* I_k I_{k'} \left(\frac{\alpha_{kk'} - \alpha_k \alpha_{k'}}{\alpha_{kk'} \alpha_k \alpha_{k'}} \right)$$

$$= \frac{1}{M^2 N^2} \sum_{k=1}^{K} \sum_{k'=1}^{K} \frac{y_k^* y_{k'}^*}{\alpha_{kk'}} \left(\frac{\alpha_{kk'}}{\alpha_k \alpha_{k'}} - 1 \right). \qquad (4.48)$$

The other estimator we can generalize from the beginning of this chapter is (4.20), namely

$$\tilde{\mu} = \frac{1}{MN} \sum_{k=1}^{K} \frac{b_k y_k^*}{E[b_k]},$$

which is also unbiased. Here b_k is the number of times network k is intersected by the initial sample. Since b_k has the hypergeometric distribution $\mathrm{Hyper}(N, x_k, n_1)$, $E[b_k] = n_1 x_k / N$ and

$$\tilde{\mu} = \frac{1}{M n_1} \sum_{k=1}^{K} \frac{b_k y_k^*}{x_k}. \qquad (4.49)$$

To find the variance of $\tilde{\mu}$, it is more helpful to use a representation such as (4.15) and write $\tilde{\mu}$ as a sample mean. To do this, we note that

$$b_k = \sum_{i=1}^{n_1} I_{ik},$$

(4.50)

where $I_{ik} = 1$ if the ith primary unit intersects the kth network, and 0 otherwise. Using (4.50) in (4.49) leads to

$$\tilde{\mu} = \frac{1}{n_1 M} \sum_{i=1}^{n_1} \sum_{k=1}^{K} \frac{I_{ik} y_k^*}{x_k}$$

$$= \frac{1}{n_1} \sum_{i=1}^{n_1} w_i$$

$$= \bar{w},$$

(4.51)

where

$$w_i = \frac{1}{M} \sum_{k=1}^{K} \frac{I_{ik} y_k^*}{x_k} = \frac{1}{M} \sum_{k=1}^{\kappa_i} \frac{y_k^*}{x_k}$$

(4.52)

and κ_i is the number of networks that intersect the ith primary unit. Since $\tilde{\mu}$ is the mean of a simple random sample of w_i values, and $E[\bar{w}] = \mu$, we have

$$\mathrm{var}[\tilde{\mu}] = \frac{\sigma_w^2}{n_1} \left(1 - \frac{n_1}{N}\right),$$

(4.53)

where

$$\sigma_w^2 = \frac{1}{N-1} \sum_{i=1}^{N} (w_i - \mu)^2.$$

An unbiased estimate of the variance of $\tilde{\mu}$ is

$$\widehat{\mathrm{var}}[\tilde{\mu}] = \frac{s_w^2}{n_1} \left(1 - \frac{n_1}{N}\right),$$

(4.54)

where

$$s_w^2 = \frac{1}{n_1 - 1} \sum_{i=1}^{n_1} (w_i - \tilde{\mu})^2.$$

Most of the results of this section were given in Thompson [1991a].

4.7.3 Examples With Primary and Secondary Units

Examples of the types of designs described in this section are illustrated in Figures 4.4 and 4.5 (pp. 121–122), in which the aim is to estimate the mean number of point objects in the study region. These point objects can, for example, represent the locations of animals, plants, mineral deposits, and so on, in a clumped population. In Figure 4.4, the initial sample consists of five randomly selected strips (primary units). The secondary units are small, square plots. Whenever a plot in the sample contains one or more of the animals, adjacent plots are added adaptively to the sample as in the examples of Section 4.6.1. The final sample resulting from this procedure is shown in the figure.

In Figure 4.5, the initial sample is a spatial systematic sample with two starting points chosen at random in a four-by-four square, and the positions are repeated throughout the study area. Again, whenever objects are observed in any plot of the sample, adjacent plots are added adaptively to the sample. The final sample is illustrated in the figure.

In each case there are 400 secondary units. The locations of individuals or objects in the study region were produced with a realization of a Poisson cluster process (cf. Diggle [1983]), with three parent locations selected at random and Poisson (mean = 100) numbers of offspring distributed about each parent with a bivariate normal distribution (with standard deviation 0.03 in the unit square). The object of sampling is to estimate the number of objects in the study region (the correct answer: 326) or, equivalently, the mean number per unit (0.815).

Example 4.5 Initial Strip Plots. In this sample, the first (leftmost) of the primary five units in the sample intersects two collections of units which satisfy the condition, leading to the additional units added to the sample. The network of units satisfying the condition within the uppermost of these clusters has a total y-value of $y_1^* = 106$; it can be determined from the sample that this network intersects four of the primary units. The lower cluster has total y-value of $y_2^* = 105$ and also intersects four primary units. All other observations in the sample give zero y-values.

Since $n_1 = 5$, the intersection probability for each of the two "nonzero" networks in the sample is $1 - \binom{20-4}{5}/\binom{20}{5} = .7183$. Hence from (4.46)

$$\hat{\mu} = \frac{1}{400}\left(\frac{106}{.7183} + \frac{105}{.7183}\right) = .7344.$$

Since two primary units intersect both networks, the joint intersection probability for the two sample networks is

$$\alpha_{12} = 1 - \left\{\binom{20-4}{5} + \binom{20-4}{5} - \binom{20-4-4+2}{5}\right\}/\binom{20}{5} = .5657.$$

The sample estimate of variance is (cf. (4.48))

$$\widehat{\text{var}}[\hat{\mu}] = \frac{1}{400^2} \left\{ \frac{106^2}{.7183} \left(\frac{1}{.7183} - 1 \right) + \frac{105^2}{.7183} \left(\frac{1}{.7183} - 1 \right) \right.$$
$$\left. + 2(106)(105)\frac{1}{.5676} \left(\frac{.5676}{.7183^2} - 1 \right) \right\}$$
$$= 0.09963.$$

The variable w_i associated with the first sample primary unit is

$$w_1 = (1/20)[(106/4) + (105/4)] = 2.6375.$$

For the second sample primary unit, the term is $w_2 = (1/20)(105/4) = 1.3125$, and, for the other three primary units in the sample, $w_i = 0$. Then (cf. (4.51))

$$\tilde{\mu} = \frac{1}{5}(2.6375 + 1.3125 + 0 + 0 + 0) = 0.79.$$

The variance estimate is, by (4.54),

$$\widehat{\text{var}}[\tilde{\mu}] = \frac{(20 - 5)}{20(5)}(1.38966) = 0.2084,$$

in which 1.38966 is the sample variance of the w-values 2.6375, 1.3125, 0, 0, and 0.

Example 4.6   *Systematic Initial Sample.*   In Figure 4.5, the first primary unit (based on the starting position in the third column of the second row) intersects the central left network, with a total y-value of $y_2^* = 106$, whereas the second primary unit (starting position in fourth column of third row) intersects both that network and the top right network, which has total y-value of $y_1^* = 115$. There are $N = 16$ possible systematic samples (primary units) and $M = 25$ secondary units in each primary unit. By dividing the study area into four-by-four squares, it can be determined that 10 ($= x_1$) primary units intersect the top right network, 13 ($= x_2$) intersect the central left network, and 9 ($= x_{12}$) intersect both. The respective intersection probabilities are therefore found to be $\alpha_1 = .875$, $\alpha_2 = .975$, and $\alpha_{12} = .8583$. Hence

$$\hat{\mu} = \frac{1}{400} \left(\frac{115}{.875} + \frac{106}{.975} \right) = 0.6004,$$

and

$$\widehat{\text{var}}[\hat{\mu}] = \frac{1}{400^2} \left[\frac{115^2}{.875} \left(\frac{1}{.875} - 1 \right) + \frac{106^2}{.975} \left(\frac{1}{.975} - 1 \right) \right.$$
$$\left. + 2\frac{(115)(106)}{.8583} \left(\frac{.8583}{(.875)(.975)} - 1 \right) \right]$$
$$= 0.01684.$$

We also have $\bar{\mu} = (1/2)(.32615 + .78615) = 0.5562$ and

$$\widehat{\mathrm{var}}[\bar{\mu}] = \frac{(16 - 2)}{(16)(2)}(.1058) = 0.0463.$$ □

4.7.4 Comparative Efficiencies

The actual variances for each of the unbiased estimators are given in Table 4.4 for the design with the initial strips and in Table 4.5 for the initial systematic design. In addition to the estimators $\bar{y}_1$, $\hat{\mu}$, and $\bar{\mu}$, the variance has once again been computed for the sample mean μ_0^* of a simple random sample of primary units with sample size equal to the expected sample size under the adaptive designs. (Thus, the sample size used for comparison need not be a whole number, though in practice of course an integer sample size would need to be used.) The variance of μ_0^* is computed using the formula (4.42). Thus, μ_0^* offers one way to compare the adaptive strategies with

Table 4.4. Variances With Initial Long, Thin Strip Plots.*

n_1	$E[\nu]$	$\mathrm{var}[\bar{y}_1]$	$\mathrm{var}[\mu_0^*]$	$\mathrm{var}[\hat{\mu}]$	$\mathrm{var}[\bar{\mu}]$
1	1.57	1.30628	0.80706	0.79253	0.79253
2	3.01	0.61876	0.38751	0.34713	0.37541
3	4.35	0.38959	0.24758	0.19944	0.23637
4	5.58	0.27501	0.17749	0.12651	0.16685
5	6.74	0.20625	0.13530	0.08378	0.12514
6	7.82	0.16042	0.10702	0.05636	0.09733
7	8.85	0.12768	0.08666	0.03788	0.07746
8	9.82	0.10313	0.07123	0.02510	0.06257
9	10.76	0.08403	0.05907	0.01621	0.05098
10	11.66	0.06875	0.04917	0.01008	0.04171

NOTE: n_1 is the initial sample size and $E[\nu]$ is the expected sample size with the adaptive design. Nonadaptive strategies are represented by $\bar{y}_1$ and μ_0^*, with sample sizes n_1 and $E[\nu]$, respectively. Adaptive strategies are represented by $\hat{\mu}$ and $\bar{\mu}$.
*From S. K. Thompson, Adaptive cluster sampling: designs with primary and secondary units, 1991, *Biometrics* **47**, 1103–1115. With permission from the Biometric Society.

Table 4.5. Variances With Initial Systematic Samples.*

n_1	$E[\nu]$	$\mathrm{var}[\bar{y}_1]$	$\mathrm{var}[\mu_0^*]$	$\mathrm{var}[\hat{\mu}]$	$\mathrm{var}[\bar{\mu}]$
1	2.98	0.44078	0.12825	0.08441	0.08441
2	4.36	0.20570	0.07846	0.01684	0.03939
3	5.31	0.12734	0.05919	0.00363	0.02439
4	6.15	0.08816	0.04701	0.00072	0.01688
5	6.98	0.06465	0.03798	0.00011	0.01238
6	7.80	0.04898	0.03089	0.00001	0.00938
7	8.62	0.03778	0.02516	0.00000	0.00724
8	9.44	0.02939	0.02042	0.00000	0.00563

*From S.K. Thompson, Adaptive cluster sampling: designs with primary and secondary units, 1991, *Biometrics* **47**, 1103–1115. With permission from the Biometric Society.

nonadaptive counterparts of equivalent sample size. Sample sizes in the table are expressed in terms of primary units. One primary unit consists of 20 secondary units in the strip design and 25 secondary units in the systematic design. The tables give variances obtained with initial sample sizes ranging from 1 up to a sampling fraction of 1/2.

With the initial strip design, the adaptive strategies with an initial sample size of 1 primary unit are slightly more efficient than the comparable nonadaptive strategy for the population used. The relative advantage of the adaptive strategies increases with increasing initial sample size, and also the efficiency of $\hat{\mu}$ relative to $\tilde{\mu}$ increases. With an initial sample size of 10 (initial sampling fraction of 1/2), the adaptive cluster sampling strategy increases the expected sample size by 16.6% ($E[\nu] = 11.66$), but is almost five times as efficient as the equivalent nonadaptive strategy ($\text{var}[\mu_0^*]/\text{var}[\hat{\mu}] = 0.04917/0.01008 = 4.88$).

With the initial systematic sampling design, the adaptive strategies are dramatically more efficient than their nonadaptive counterparts for the population used in the examples. Also, comparing Tables 4.4 and 4.5, one sees that even with conventional systematic strategies ($\bar{y}_1$ and μ_0^*), variances are considerably lower than with the conventional strategies using strips. (This result would be expected owing to the positive, monotonically decreasing covariance density function of the Poisson cluster process—see for example Matérn [1986] and Thompson and Ramsey [1987].) The efficiency of the adaptive strategy with $\hat{\mu}$ relative to the comparable nonadaptive systematic strategy with μ_0^* ranges from 152% ($= 100 \times 0.12852/0.08441$) for a single initial systematic sample to infinity. The adaptive strategy has zero variance for initial selections of more than six, as the intersection probability for each of the three networks in the population becomes 1 with such a design.

4.8   SIMPLE LATIN SQUARE SAMPLING +1 DESIGNS

In the previous section, we introduced primary and secondary units and described one method of choosing a primary unit using systematic sampling and a random starting point. This primary unit contains secondary units that are no longer contiguous. It has the advantage that this initial sample will often give a more uniform coverage of the population area than a simple random sample, the latter sometimes giving a very uneven coverage with large areas not being represented. Unfortunately, a systematic sample with a single randomly selected starting point does not allow design-unbiased estimation of the variance of estimators.

We consider below a scheme due to Munholland and Borkowski [1993a, b], which they call the simple Latin square sample +1 (or SLSS +1), that combines features of a systematic sample with additional random sampling and allows for unbiased estimation of variance. This can be described using Figure 4.6. Here we have a population of 16 units with crosses down the diagonal (Figure 4.6a). To get an SLSS of size 4 we randomly permute the rows (or columns) to get a single cross in every row and every column. One further unit (described by a circle), which is the +1 unit (Figure 4.6b), is then selected at random from the remaining 12 units, thus

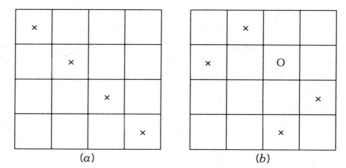

Figure 4.6. Constructing a 4 × 4 simple Latin square +1 design: (a) starting position, (b) final design with five units selected.

giving a final sample of size 5. Once such a sample has been selected, any unit in the sample can then be added to adaptively. We now examine the theory for such adaptive designs.

Suppose we have a $p \times p$ grid of population units (i.e., $N = p^2$). We can select an SLSS of size p by randomly selecting one unit from the p units in the first row. Then, deleting the corresponding column, a second unit is selected at random from the remaining $(p-1)$ columns in the second row. Further, omitting this column a unit is randomly drawn from the $(p-2)$ columns in the third row, and so on. The "+1" unit is then selected at random from the remaining $(p^2 - p)$ units. Thus in all there are $p!(p^2 - p)$ possible samples and they are all equally likely. The estimator Munholland and Borkowski [1993b] use is the modified Horvitz–Thompson estimator (4.4),

$$\hat{\mu} = \frac{1}{N} \sum_{k=1}^{K} \frac{y_k^* J_k}{\alpha_k},$$

where J_k takes the value 1 with (intersection) probability α_k if the initial sample intersects the kth network, and 0 otherwise, and y_k^* is the sum of the y-values for the kth network. The variance and variance estimators for $\hat{\mu}$ are given by (4.10) and (4.11). We now have to find α_k and α_{jk}, the latter being the probability that the initial sample intersects the jth and kth networks. The following theory comes from Borkowski and Munholland [1993b].

We begin by relating α_k and α_{jk} to α_k' and α_{jk}', the corresponding intersection probabilities for the initial SLSS (without the +1 unit). If x_k is the number of units in network k, then $(1 - \alpha_k)$ is the probability $(1 - \alpha_k')$ that the SLSS does not intersect the kth network multiplied by the conditional probability $(p^2 - p - x_k)/(p^2 - p)$ that the +1 unit does not intersect it either, that is,

$$1 - \alpha_k = \frac{p^2 - p - x_k}{p^2 - p}(1 - \alpha_k').$$

As the networks do not overlap, there are just three possible ways in which networks j and k can be included in the final (adaptive) sample, namely (1) the SLSS intersects both (with probability α'_{jk}); (2) the SLSS intersects the jth only and the $+1$ unit intersects the kth (with probability $(\alpha'_j - \alpha'_{jk})[x_k/(p^2 - p)]$; and (3) the SLSS intersects the kth only and the $+1$ unit intersects the jth (with probability $(\alpha'_k - \alpha'_{jk})[x_j/(p^2 - p)]$. Hence

$$\alpha_{jk} = \alpha'_{jk} + (\alpha'_j - \alpha'_{jk})\frac{x_k}{p^2 - p} + (\alpha'_k - \alpha'_{jk})\frac{x_j}{p^2 - p}$$

$$= \alpha'_{jk}\left(1 - \frac{x_j + x_k}{p^2 - p}\right) + \frac{\alpha'_j x_k + \alpha'_k x_j}{p^2 - p}.$$

The final stage in the argument is to find α'_k and α'_{jk}. To do this we refer again to Figure 4.6b and ignore the circle. If we delete any row and any column, then the remaining 3×3 block of units may be an SLSS of size 3 or else contain an embedded SLSS of size 2 or just size 1. Similarly any 2×2 block may be an SLSS of size 2 or contain an embedded SLSS of size 1. In general, if we have an SLSS of size p, then any $p - q \times p - q$ block of units can contain SLSS's of size r for $r = 1, 2, \ldots, p - q$. Also the biggest SLSS that can be embedded in the kth network (of x_k points) is $x_k^* \times x_k^*$, where $x_k^* = [1/2(x_k + 1)]$, the integral part of $(1/2)(x_k + 1)$. Hence an SLSS of size p cannot intercept the kth network at more than x_k^* units. Now let β_{kr} and γ_{jkr} denote the number of SLSS's of order r which are respectively contained in network k and contained in the union of networks j and k with both networks intersected. If $x_{jk}^* = [(1/2)(x_j + x_k + 1)]$, the biggest SLSS that can be embedded in the union of networks j and k, then we have the following lemma.

Lemma

(i) $\displaystyle \alpha'_k = \sum_{r=1}^{x_k^*}(-1)^{r+1}\beta_{kr}\frac{(p - r)!}{p!}$.

(ii) $\displaystyle \alpha'_{jk} = \sum_{r=2}^{x_{jk}^*}(-1)^r\gamma_{jkr}\frac{(p - r)!}{p!}$.

Proof:

(i) Let E_i be the event that the SLSS of size p intersects the kth network in at least the ith unit ($i = 1, 2, \ldots, x_k$) of the network. Then using the inclusion/exclusion principle,

$\alpha'_k = P$(probability of intersecting network k)

$= P(E_1 \cup E_2 \cup \cdots \cup E_{x_k})$

$= \sum P(E_i) - \sum P(E_i \cap E_j) + \cdots + (-1)^{x_k+1}P(E_1 \cap E_2 \cap \cdots \cap E_{x_k}).$

However, $P(E_{i_1} \cap E_{i_2} \cap \cdots \cap E_{i_r})$ will be zero for $r > x_k^*$ as x_k^* is the maximum number of points that can be intercepted by the SLSS. Also, for $r \leq x_k^*$,

$$P(E_{i_1} \cap E_{i_2} \cap \cdots \cap E_{i_r}) = \frac{1}{p} \cdot \frac{1}{p-1} \cdots \frac{1}{p-r+1}$$

$$= \frac{(p-r)!}{p!},$$

and the number of terms is β_{kr}, that is,

$$\sum P(E_{i_1} \cap E_{i_2} \cap \cdots \cap E_{i_r}) = \beta_{kr} \frac{(p-r)!}{p!}.$$

This completes the proof.

(ii) The proof for this part follows in a similar fashion. Here r begins at 2 as at least one point must be intersected in each of the jth and kth networks. Also E_i is now defined to be the event that the SLSS of size p intersects the union of the jth and kth networks (consisting of $x_{jk} = x_j + x_k$ units) in at least the ith unit $(i = 1, 2, \ldots, x_{jk})$. Let F be the event that at least one unit from each of the jth and kth networks is selected; then

$$\alpha'_{jk} = P\left([E_1 \cup E_2 \cup \cdots \cup E_{x_{jk}}] \cap F \right)$$

$$= \sum P(E_g \cap E_h \cap F) - \sum P(E_g \cap E_h \cap E_i \cap F) + \cdots$$

$$+ (-1)^{x_{jk}} P(E_1 \cap E_2 \cap \cdots \cap E_{x_{jk}} \cap F).$$

The argument now follows as in (i).   □

To evaluate the selection probabilities, we need to be able to compute the β_{kr} and the γ_{jkr}. Algorithms for doing this are provided by Munholland and Borkowski [1993b].

Having developed the above theory, a natural question to ask is, why the +1 unit? Looking at the unbiased variance estimator (4.11), we see that it will only exist if all the α_{jk} are nonzero. Without the +1 unit certain combinations of units are not possible in the initial SLSS, for example two units in the same row or the same column. If these units were networks of size 1, then α_{jk} would be zero for that pair of networks. However, by adding the extra +1 unit, every pair of units can be in the initial sample and each $\alpha_{jk} > 0$.

When will an SLSS +1 do better than a simple random sample of size $p + 1$? For nonadaptive sampling, Munholland and Borkowski [1993a] showed that the SLSS +1 performs better, that is, has smaller variance if and only if

$$E[y_i y_j \mid I_{ij} = 1] \leq E[y_i y_j \mid I_{ij} = 0],$$

where $I_{ij} = 1$ when units i and j are located in different rows and columns, and $I_{ij} = 0$ otherwise. Clearly the strict inequality will hold when similar y-values tend to cluster together. However, there will be no difference if the expectations are independent of the relative location of the units. Similar properties can be expected to carry over into the adaptive case. Further research is needed to determine the efficiency of the unbiased variance estimator and to compare this procedure with Latin square $+k$ designs with k units added by simple random sampling.

4.9   STRATIFIED ADAPTIVE CLUSTER SAMPLING

In some populations there may be prior information about where aggregations are likely to be so that stratification can be used to reduce some of the variability in the estimators. We can therefore apply the adaptive sampling schemes of Section 4.2 to each stratum. After the initial samples are taken in each stratum (e.g., Figure 4.7), clusters begin to grow as appropriate neighborhoods are added according to condition C. If these growing clusters are truncated at the stratum boundaries, then the stratum estimators are independent and can be combined in the usual fashion to provide an overall (weighted) estimator of the population mean in the form of (4.55) below. Truncating clusters, however, leads to a loss of efficiency, and a more efficient scheme would be to allow clusters to overlap stratum boundaries as for example in Figure 4.8. Stratified adaptive cluster sampling was introduced in Thompson [1991b], on which the following is based. We shall assume simple random sampling in each stratum.

Suppose that the total population of N units is partitioned into H strata, with n_h units in the hth stratum ($h = 1, 2, \ldots, H$). Define unit (h, i) to be the ith unit in the hth stratum with associated y-value y_{hi}. A simple random sample of n_h units is taken from stratum h, and we now define $n_0 = \sum_{h=1}^{H} n_h$ to be the initial total sample size. If we ignored the adaptive additions and used just the initial samples, then we could compute the usual stratified mean

$$\bar{y}_0 = \sum_{h=1}^{H} \frac{N_h}{N} \bar{y}_h, \tag{4.55}$$

where $\bar{y}_h = \sum_{i=1}^{n_h} y_{hi}/n_h$ is the sample mean for stratum h. Also

$$\text{var}[\bar{y}_0] = \frac{1}{N^2} \sum_{h=1}^{H} N_h(N_h - n_h)\frac{\sigma_h^2}{n_h}, \tag{4.56}$$

where

$$\sigma_h^2 = \frac{1}{N_h - 1} \sum_{i=1}^{N_h} (y_{hi} - \mu_h)^2$$

and μ_h is the population mean for stratum h.

Using the full adaptive sample, the first estimator we can consider is given by (4.6) based on the initial intersection probabilities, namely

$$\hat{\mu}_{st} = \frac{1}{N} \sum_{k=1}^{K} \frac{y_k^* J_k}{\alpha_k}, \tag{4.57}$$

where the K distinct networks are labeled $1, 2, \ldots, K$ *without* regard to stratum boundaries, J_k equals 1 (with probability α_k) if the initial sample of size n_0 intersects network k, and 0 otherwise, and y_k^* is the sum of the y-values for network k. To derive α_k we need to consider probabilities of intersecting network k with initial samples in each of the strata. An initial unit in a given stratum can only fall in (intersect) that part of each network in the stratum. Therefore we define x_{hk} to be the number of units in stratum h that lie in network k. This will be zero if network k lies totally outside stratum h. If the network straddles a boundary, then we ignore the network units that lie outside stratum h in the definition of x_{hk}. With this definition of x_{hk} we see that

$$\alpha_k = 1 - \left[\prod_{h=1}^{H} \binom{N_h - x_{hk}}{n_h} \middle/ \binom{N_h}{n_h} \right].$$

Since $E[J_k] = \alpha_k$, $\hat{\mu}_{st}$ is unbiased. To find its variance we need $\alpha_{kk'}$, the probability that the initial sample intersects both networks k and k', namely [cf. (4.8)]

$$\alpha_{kk'} = 1 - (1 - \alpha_k) - (1 - \alpha_{k'}) + \left[\prod_{h=1}^{H} \binom{N_h - x_{hk} - x_{hk'}}{n_h} \middle/ \binom{N_h}{n_h} \right].$$

Then, from (4.10) with $\alpha_{kk} = \alpha_k$, we have

$$\text{var}[\hat{\mu}_{st}] = \frac{1}{N^2} \sum_{k=1}^{K} \sum_{k'=1}^{K} y_k^* y_{k'}^* \left(\frac{\alpha_{kk'} - \alpha_k \alpha_{k'}}{\alpha_k \alpha_{k'}} \right)$$

with unbiased estimate

$$\widehat{\text{var}}[\hat{\mu}_{st}] = \frac{1}{N^2} \sum_{k=1}^{K} \sum_{k'=1}^{K} y_k^* y_{k'}^* \left(\frac{\alpha_{kk'} - \alpha_k \alpha_{k'}}{\alpha_{kk'} \alpha_k \alpha_{k'}} \right) I_k I_{k'}.$$

The other estimate we can use is based on the numbers of initial intersections [cf. (4.13)]. To do this we introduce A_{hi}, the network containing unit (h, i), and A_{ghi}, that part of A_{hi} in stratum g. Let f_{ghi} be the number of units from the initial sample in stratum g that fall in A_{ghi}, and let m_{ghi} be the number of units in A_{ghi}. Then

$$f_{\cdot hi} = \sum_{g=1}^{H} f_{ghi}$$

is the number of units from the initial sample of n_0 units that fall in A_{hi}. Equation (4.13) now translates into

$$\tilde{\mu}_{st} = \frac{1}{N} \sum_{h=1}^{H} \sum_{i=1}^{N_h} y_{hi} \frac{f_{\cdot hi}}{E[f_{\cdot hi}]}, \tag{4.58}$$

which is unbiased. Since f_{ghi} has a hypergeometric distribution Hyper(N_g, m_{ghi}, n_g),

$$E[f_{ghi}] = \frac{n_g m_{ghi}}{N_g}$$

and

$$E[f_{\cdot hi}] = \sum_{g=1}^{H} \frac{n_g}{N_g} m_{ghi}. \tag{4.59}$$

Substituting (4.59) into (4.58) leads to the estimate given by Thompson [1991b, Equation (1)].

To find var[$\tilde{\mu}_{st}$] we adapt the approach of (4.15) and write $\tilde{\mu}_{st}$ in terms of weighted sample means. To do this we need to relate the observations to the networks intersected by the initial sample. The term $y_{hi} f_{\cdot hi}$ tells us that A_{hi} is intersected $f_{\cdot hi}$ times by the initial sample, so that $\tilde{\mu}_{st}$ is just a weighted sum of all units in all the networks corresponding to the initial sample, with some networks being repeated. Since the weight $E[f_{\cdot hi}]$ is the same for each unit in A_{hi}, we have

$$\tilde{\mu}_{st} = \frac{1}{N} \sum_{h=1}^{H} \sum_{i=1}^{n_h} \frac{1}{E[f_{\cdot hi}]} \sum_{(h',i') \in A_{hi}} y_{h'i'}$$

$$= \frac{1}{N} \sum_{h=1}^{H} \sum_{i=1}^{n_h} \frac{Y_{hi}}{E[f_{\cdot hi}]}, \tag{4.60}$$

where Y_{hi} is the sum of the y-observations in A_{hi}.

We can also express (4.60) in the form

$$\tilde{\mu}_{st} = \sum_{h=1}^{H} \frac{N_h}{N} \bar{w}_h, \tag{4.61}$$

where $\bar{w}_h = \sum_{i=1}^{n_h} w_{hi}/n_h$, and

$$w_{hi} = \frac{n_h}{N_h} \cdot \frac{Y_{hi}}{E[f_{\cdot hi}]}.$$

When n_h/N_h is the same for all strata, it follows from (4.59) that $w_{hi} = Y_{hi}/\sum_g m_{ghi}$. The representation (4.61) takes the form of a stratified sample mean of a stratified random sample (without replacement) in which the variable of interest is now w_{hi}. Hence

$$\text{var}[\tilde{\mu}_{st}] = \frac{1}{N^2} \sum_{h=1}^{H} N_h(N_h - n_h)\frac{\sigma_h^2}{n_h}, \tag{4.62}$$

where the stratum population variance term is now

$$\sigma_h^2 = \frac{1}{N_h - 1} \sum_{i=1}^{N_h} (w_{hi} - \overline{W}_h)^2, \tag{4.63}$$

and the stratum population mean is $\overline{W}_h = \sum_{i=1}^{N_h} w_{hi}/N_h$. An unbiased estimate $\widehat{\text{var}}[\tilde{\mu}_{st}]$ is obtained by replacing σ_h^2 in (4.62) with the sample variance

$$s_h^2 = \frac{1}{n_h - 1} \sum_{i=1}^{n_h} (w_{hi} - \bar{w}_h)^2. \tag{4.64}$$

A variation $\tilde{\mu}'_{st}$ on the estimator $\tilde{\mu}_{st}$ may be constructed which is related to the stratified "multiplicity" estimator of network sampling (Birnbaum and Sirken [1965], Sirken [1970, 1972a, b], Sirken and Levy [1974], and Levy [1977]). We now replace f_{ghi} by $N_g f_{ghi}/n_g$ and use

$$f'_{\cdot hi} = \sum_{g=1}^{H} \frac{N_g}{n_g} f_{ghi}$$

with

$$\text{E}[f'_{\cdot hi}] = \sum_{g=1}^{H} \frac{N_g}{n_g} \cdot \frac{n_g}{N_g} m_{ghi} = \sum_{g=1}^{H} m_{ghi}$$

in (4.58) to get the unbiased estimator

$$\tilde{\mu}'_{st} = \frac{1}{N} \sum_{h=1}^{H} \sum_{i=1}^{N_h} y_{hi} \frac{f'_{\cdot hi}}{\text{E}[f'_{\cdot hi}]}. \tag{4.65}$$

In attempting to express (4.65) in the form of (4.61), we note that $f'_{\cdot hi}$ differs from $f_{\cdot hi}$ in that all observations sampled from stratum g are weighted by the factor N_g/n_g. Hence

$$\tilde{\mu}'_{st} = \frac{1}{N} \sum_{h=1}^{H} \frac{N_h}{n_h} \sum_{i=1}^{n_h} \frac{Y_{hi}}{\text{E}[f'_{\cdot hi}]} \tag{4.66}$$

$$= \sum_{h=1}^{H} \frac{N_h}{N} \bar{w}'_h,$$

where $\bar{w}'_h = \sum_{i=1}^{x_h} w'_{hi}/n_h$, and $w'_{hi} = Y_{hi}/\sum_{g=1}^{H} m_{ghi}$ is the mean of the observations in network A_{hi}. Replacing w_{hi} by w'_{hi} in (4.62), (4.63), and (4.64) leads to $\text{var}[\tilde{\mu}'_{st}]$ and its unbiased estimator $\widehat{\text{var}}[\tilde{\mu}'_{st}]$.

Thompson [1991b] mentions one other estimator which ignores any units added through crossing stratum boundaries. This estimator is given by

$$\tilde{\mu}''_{st} = \sum_{h=1}^{H} \frac{N_h}{N} \tilde{\mu}_h,$$

where $\tilde{\mu}_h = \sum_{i=1}^{n_h} w''_{hi}/n_h$ and w''_{hi} is the total of the y-values in the intersection of stratum h with A_{hi} divided by the number of units in the intersection, that is, w''_{hi} is the network mean for that part of the network A_{hi} in stratum h. Since $\tilde{\mu}_h$ is the same as $\tilde{\mu}$ of (4.15) for the unstratified case, $\tilde{\mu}_h$ is an unbiased estimator of μ_h and

$$E[\tilde{\mu}''_{st}] = \sum_{h=1}^{H} \frac{N_h}{N} \mu_h$$

$$= \frac{1}{N} \sum_{h=1}^{H} \sum_{i=1}^{N_h} y_{hi}$$

$$= \mu.$$

Thus, $\tilde{\mu}''_{st}$ is unbiased for μ. Since the $\tilde{\mu}_h$ are independent, expressions for $\text{var}[\tilde{\mu}''_{st}]$ and its unbiased estimator $\widehat{\text{var}}[\mu''_{st}]$ are again given by (4.62), (4.63), and (4.64), with w_{hi} replaced by w''_{hi}.

The Rao–Blackwell method can be used in the stratified situation for the same reasons as those given in Section 4.5. The unbiased estimators above are not functions of the minimal sufficient statistic D_R since they depend on order of selection or frequencies of intersections. Thus, starting with any one of the unbiased estimators such as $\hat{\mu}_{st}$, for example, we can obtain the Rao–Blackwell version $\hat{\mu}_{RB}$ by taking the conditional expectation given the reduced data D_R.

The Rao–Blackwell version of the initial stratified sample mean $\bar{y}_0$ is obtained by seeing which initial samples give rise to the the final value of D_R and then averaging $\bar{y}_0$ over these initial samples. If we assume that networks do not cross boundaries, then the above averaging is equivalent to the following two-step procedure (as can be demonstrated by considering a simple example). The first step consists of finding the Rao–Blackwell estimate for each stratum, that is, by finding the initial samples that give rise to the final reduced data for that stratum and then averaging over those initial sample means. The second step consists of then simply combining the stratum Rao–Blackwell estimates using the usual (weighted average) formula for a stratified mean. By Theorem 4.1 in Section 4.5, the first step applied to either $\bar{y}_0$ or $\hat{\mu}''$ leads to the same estimator in each stratum so that the overall Rao–Blackwell version of $\bar{y}_0$ is the same as the Rao–Blackwell version $\hat{\mu}''_{RB}$ of $\hat{\mu}''$. The Rao–Blackwell versions $\hat{\mu}_{RB}$, $\tilde{\mu}_{RB}$, $\hat{\mu}'_{RB}$ and $\tilde{\mu}''_{RB}$ are, however, distinct estimators, as demonstrated later in Example 4.8.

4.9.1 Examples for Stratified Populations

Example 4.7   Clumped Population.   The spatially clumped population of Figure 4.7 was produced as a realization of a Poisson cluster process (cf. Diggle [1983]). Four "parent" locations were randomly located in the study region, and "offspring" locations were distributed about each parent location according to a symmetric normal distribution with standard deviation $\sigma = 0.02$. The numbers of offspring were Poisson random variables, each with mean 100. The y-values for each of the 400 units (plots) in the population are listed in Thompson [1990]. The actual number of point objects in the region is 397, so that the true population mean is $\mu = 397/400 = 0.9925$.

For the first design, the study region is divided into two strata, and initial samples are selected by stratified random sampling with equal sample sizes in each stratum. A unit satisfies the condition if it contains one or more of the point objects. The

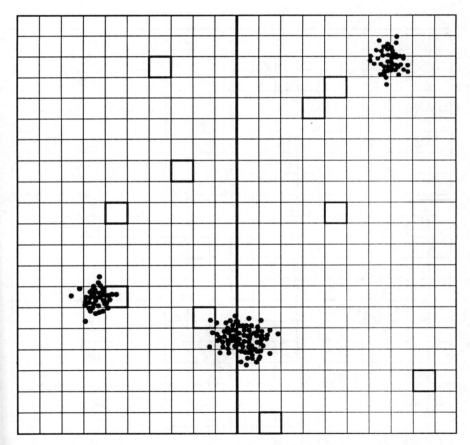

Figure 4.7. An initial stratified random sample, with five units in each of the two strata. (From S.K. Thompson, Stratified adaptive cluster sampling, 1991, *Biometrika* **78**, 389–397. With permission from the Biometrika Trustees.)

neighborhood of a unit includes all adjacent units, so that a typical neighborhood away from the boundary consists of five plots in a cross shape.

Consider the design with initial sample sizes of five units in each stratum. An outcome of the initial sample selection is shown in Figure 4.7, and Figure 4.8 shows the final sample which results. Sample computations will be illustrated using Figure 4.9, which emphasizes the network structure. In stratum 1 (on the left), the initial sample has intersected two networks of larger than single-unit size. The first network (on the left) consists of six units, the total y-value of which is 96. The second network has five units within the first stratum and six units within the second stratum. The total of the y-values in the intersection of this network with the first stratum is 78, while the total of the y-values in the intersection of the network with the second stratum is 114. Thus, the second network has a total of 11 units and a total y-value of 192. In the second stratum, none of the five units of the initial sample satisfied the condition.

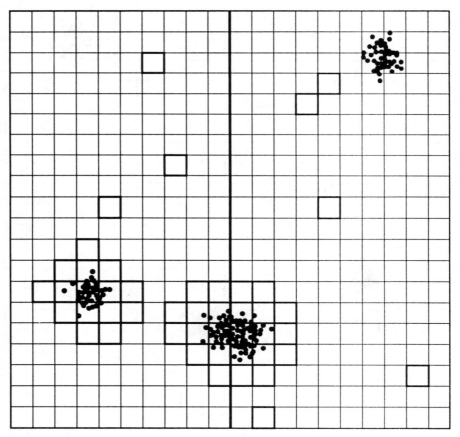

Figure 4.8. The resulting stratified adaptive cluster sample. (From S.K. Thompson, Stratified adaptive cluster sampling, 1991, *Biometrika* **78**, 389–397. With permission from the Biometrika Trustees.)

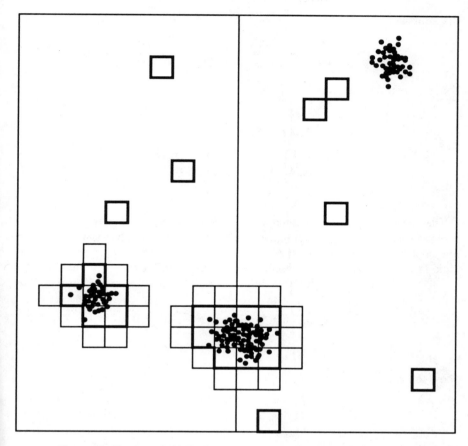

Figure 4.9. The networks within the adaptively added clusters are outlined in bold.

For the estimator $\hat{\mu}$, the intersection probabilities must first be calculated. For every unit in the initial sample not satisfying the condition, the intersection probability is $n_h/N_h = 5/200 = .025$, and is the same in each stratum because of the equal sample and stratum sizes. For the first of the large networks intersected (the one on the left), the inclusion probability is $\alpha_1 = 1 - \left[\binom{200-6}{5} / \binom{200}{5} \right] = .14261$. For the second network, since it intersects both strata, the intersection probability is $\alpha_2 = 1 - \left[\binom{200-5}{5} \binom{200-6}{5} / \binom{200}{5} \binom{200}{5} \right] = .24554$. The joint inclusion probability for both networks is

$$\alpha_{12} = 1 - (1 - .14261) - (1 - .24554)$$
$$+ \left[\binom{200-6-5}{5} \binom{200-0-6}{5} / \binom{200}{5} \binom{200}{5} \right]$$
$$= .03240.$$

The estimate is

$$\hat{\mu} = \frac{1}{400}\left(\frac{96}{.14261} + \frac{192}{.24554} + \frac{0}{.025} + \cdots + \frac{0}{.025}\right) = 3.64.$$

The estimated variance is

$$\widehat{\mathrm{var}}[\hat{\mu}] = \frac{1}{400^2}\left\{\frac{96^2}{.14261}\left[\frac{1}{.14261} - 1\right] + \frac{192^2}{.24554}\left[\frac{1}{.24554} - 1\right]\right.$$

$$\left. +2\frac{(96)(192)}{.0324}\left[\frac{.0324}{(.14261)(.24554)} - 1\right] + 0 + \cdots + 0\right\}$$

$$= 4.78.$$

The value of w''_{hi} for the estimator $\tilde{\mu}''$, which ignores crossover between strata, is zero for all units not satisfying the condition. In the first network intersected in stratum 1, the value is $w''_{11} = 96/6 = 16$. For the second network intersected, the value is $w'_{12} = 78/5 = 15.6$, based only on units within stratum 1. There are no intersections in the second stratum. The estimate of the population mean is therefore

$$\tilde{\mu}'' = \frac{1}{400}\left[\frac{200}{5}(16 + 15.6 + 0 + 0 + 0) + \frac{200}{5}(0 + 0 + 0 + 0 + 0)\right]$$

$$= 3.16.$$

The estimated variance is

$$\widehat{\mathrm{var}}[\tilde{\mu}''] = \frac{1}{400^2}\left[\frac{200(200 - 5)(74.9)}{5} + 0\right] = 3.65,$$

in which 74.9 is the sample variance of the five numbers 16, 15.6, 0, 0, and 0.

To calculate the estimator $\tilde{\mu}$, we note that n_h/N_h is the same for both strata. Therefore the variables w_{hi} are $w_{11} = 96/6 = 16$ for the first network and $w_{12} = 192/11 = 17.45$ for the second. The estimate is then

$$\tilde{\mu} = \frac{1}{400}\left[\frac{200}{5}(16 + 17.45 + 0 + 0 + 0) + 0\right] = 3.35.$$

The variance estimate is

$$\widehat{\mathrm{var}}[\tilde{\mu}] = \frac{1}{400^2}\left[\frac{200(200 - 5)(84.2)}{5} + 0\right] = 4.10,$$

in which 84.2 is the sample variance of the five sample values of w_{1i} in the first stratum. The estimators $\tilde{\mu}'$ and $\tilde{\mu}$ are identical because of the equal stratum and sample sizes.

To illustrate the differences between the computations of $\tilde{\mu}$ and $\tilde{\mu}'$ when the initial sample sizes are not equal, consider an initial sample of five units in the first stratum, as illustrated in Figure 4.7, but with an initial sample of only three units in the second stratum. We suppose again that none of the sample units in the second stratum contain any point objects. For the estimator $\tilde{\mu}$, the values are $w_{11} = [(5/200)(96)]/[(5/200)(6)] = 16$ and $w_{12} = [(5/200)(192)]/[(5/200)(5) + (3/200)(6)] = 22.33$. The estimate is

$$\tilde{\mu} = \frac{1}{400}\left[\frac{200}{5}(16 + 22.33 + 0 + 0 + 0) + 0\right] = 3.83.$$

For the estimator $\tilde{\mu}'$, $w'_{11} = 96/6 = 16$ and $w'_{12} = 192/11 = 17.45$ (as in the previous example), so that $\tilde{\mu}' = 3.35$ with $\widehat{\text{var}}[\tilde{\mu}'] = 4.10$, as before.

Comparative Efficiencies

The actual variances of the estimators for this population with the design having equal initial sample sizes in each of the two strata are given in Table 4.6 for a selection of initial sample sizes, ranging from samples of size 1 in each stratum to initial samples of 100 units in each stratum (an initial sampling fraction of 1/2). The first column of the table gives the total initial sample size, while the second column gives $E[\nu]$, the total expected number of distinct units in the final sample. The third column gives the variance of the initial stratified sample mean $\bar{y}_0$. For comparisons with adaptive strategies, the fourth column gives the variance of the stratified sample mean μ_0^* of the stratified mean of a conventional stratified random sample with total sample size equal to $E[\nu]$, the expected sample size under the adaptive design. The last four columns of the table give the variances of the basic four types of unbiased estimators used with stratified adaptive cluster sampling.

For this population, efficiency increases from left to right across the table. In addition, there is a tendency for the relative advantage of the adaptive strategies, compared to the nonadaptive strategies $\bar{y}_0$ and μ_0^*, to increase with increasing sam-

Table 4.6. Two Strata, Equal Sample Sizes.*

n_0	$E[\nu]$	var[$\bar{y}_0$]	var[μ_0^*]	var[$\tilde{\mu}''$]	var[$\tilde{\mu}'$]	var[$\tilde{\mu}$]	var[$\hat{\mu}$]
2	3.99	19.28371	9.61712	8.63371	8.58974	8.58974	8.53095
4	7.89	9.59340	4.81846	4.29516	4.27329	4.27329	4.17070
6	11.69	6.36330	3.21874	2.84898	2.83447	2.83447	2.71809
8	15.41	4.74825	2.41875	2.12589	2.11506	2.11506	1.99240
10	19.04	3.77922	1.93866	1.69203	1.68342	1.68342	1.55748
20	36.08	1.84116	0.97749	0.82432	0.82013	0.82013	0.69143
40	65.59	0.87213	0.49402	0.39047	0.38848	0.38848	0.26809
50	78.64	0.67832	0.39598	0.30370	0.30215	0.30215	0.18734
60	90.84	0.54912	0.32979	0.24585	0.24460	0.24460	0.13564
80	113.31	0.38761	0.24517	0.17354	0.17266	0.17266	0.07564
100	133.99	0.29071	0.19239	0.13016	0.12949	0.12949	0.04413
200	226.06	0.09690	0.07456	0.04339	0.04316	0.04316	0.00322

ple size. For example, consider the design depicted in Figure 4.9, having initial
sample sizes of 5 in each stratum and total initial sample size $n_0 = 10$. As the
effective sample size is 19.04, we allocate $19.04/2 = 9.52$ units to each stratum in
the variance formula for the variance of the conventional stratified random sample
mean μ_0^*. This gives a relative efficiency of the adaptive design with the estimator $\hat{\mu}$ to
the nonadaptive strategy μ_0^* with equivalent sample size of $1.93866/1.55748 = 1.24$.
The adaptive strategy is therefore 24% more efficient than the comparable nonadaptive
one. However, with initial stratum sample sizes of 100 (total sample size 200), the
effective sample size is 226 (113 in each stratum) and the relative efficiency is now
$.07456/.00322 = 23.16$. The adaptive strategy is about 23 times more efficient.

Table 4.7 summarizes variances for the strategies with initial sample sizes in the
ratio 2:1 (but rounded to whole numbers) in the two strata, so that the initial sample
size in the first stratum is as near as possible to twice the sample size in the second.
For example, when the total initial sample size n_0 is 6, the two stratum initial sample
sizes are $n_1 = 4$ and $n_2 = 2$. When $n = 8$, rounding gives initial sample sizes of
$n_1 = 5$ and $n_2 = 3$, as in the example just worked. When $n_0 = 200$, $n_1 = 133$ and
$n_2 = 67$. With the unequal sample sizes, the estimator $\tilde{\mu}$ performs slightly better than
$\tilde{\mu}'$ for this population.

With more than two strata, networks may intersect more than two strata—an
initial sample unit in any of these strata may then result in units of the others being
added to the sample. Figure 4.10 shows the study area with the clumped population
divided into 16 strata, with 4 of these strata intersecting the middle network. Table 4.8
summarizes the variances obtained for the various strategies with equal sample sizes
in each of the 16 strata. With this stratification, the efficiency of the strategies again
shows a tendency to increase from left to right in the table. The relative efficiency
of the most efficient adaptive strategy $\hat{\mu}$ to the comparable nonadaptive strategy μ_0^*
ranges from $1.25567/.86451 = 1.45$ with initial samples of size 1 in each stratum (a
total sample size of 16) to $.07360/.00147 = 50.07$ with sample sizes of 13 in each
stratum (total sample size 208). Thus, with an initial sampling fraction of just over
1/2, the adaptive strategy has 50 times the precision of conventional stratified random
sampling for estimating the mean of the clumped population illustrated.

NOTE: For each of Tables 4.6 through 4.8, the first column gives the total initial
sample size. The second column gives $E[\nu]$, the total expected number of distinct
units in the final sample. The third column gives the variance of the initial stratified
sample mean $\bar{y}_0$. For comparisons with adaptive strategies, the fourth column gives
the variance of the stratified sample mean μ_0^* of the stratified mean of a conventional
stratified random sample with total sample size equal to $E[\nu]$, the expected sample
size under the adaptive design. The last four columns of the table give the variances
of the basic four types of unbiased estimators used with stratified adaptive cluster
sampling.

Example 4.8   A Small Population.   Consider a population of 5 units with y-
values

$$\mathbf{y} = (1, 2, 10; 1000, 3)'$$

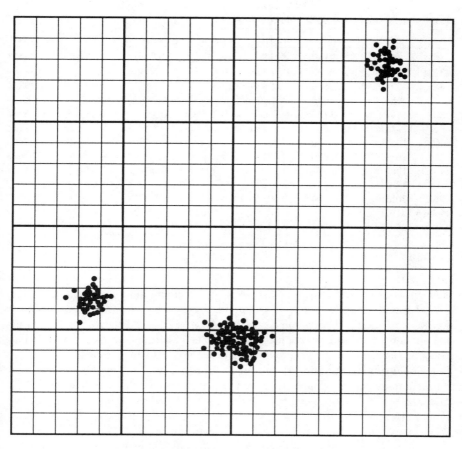

Figure 4.10. A study region partitioned into 16 strata.

Table 4.7. Two Strata, Sample Sizes in Ratio 2:1.*

n_0	$E[\nu]$	$\mathrm{var}[\bar{y}_0]$	$\mathrm{var}[\mu_0^*]$	$\mathrm{var}[\tilde{\mu}'']$	$\mathrm{var}\,\tilde{\mu}'$	$\mathrm{var}[\tilde{\mu}]$	$\mathrm{var}[\hat{\mu}]$
4	7.86	12.90630	6.52072	6.44492	6.16427	5.51233	5.41609
6	11.67	7.20188	3.65693	3.47436	3.36374	3.20237	3.09003
8	15.39	5.08782	2.59904	2.41122	2.34938	2.28465	2.16485
10	18.98	4.54119	2.34637	2.22361	2.14148	1.99178	1.86911
20	35.98	2.04102	1.09135	0.97973	0.95023	0.91060	0.78354
40	65.25	1.01236	0.58305	0.49313	0.47579	0.44640	0.32721
50	78.24	0.77030	0.45726	0.37339	0.36089	0.34123	0.22723
60	90.28	0.63297	0.38816	0.30839	0.29753	0.27916	0.17094
80	112.55	0.44709	0.28977	0.21834	0.21047	0.19687	0.10031
100	132.95	0.34328	0.23418	0.16907	0.16249	0.14994	0.06498
200	225.27	0.12151	0.09701	0.06181	0.05874	0.05166	0.01092

*From S.K. Thompson, Stratified adaptive cluster sampling, 1991, *Biometrika* **78**, 389–397. With permission from the Biometrika Trustees.

Table 4.8. Sixteen Strata, Equal Sample Sizes.*

n_0	$E[\nu]$	var$[\bar{y}_0]$	var$[\mu_0^*]$	var$[\tilde{\mu}'']$	var$[\tilde{\mu}']$	var$[\tilde{\mu}]$	var$[\hat{\mu}]$
16	30.31	2.47109	1.25567	1.25161	1.13492	1.13492	0.86451
32	55.65	1.18407	0.63716	0.59973	0.54381	0.54381	0.34784
48	77.48	0.75506	0.42859	0.38244	0.34678	0.34678	0.18356
64	96.87	0.54055	0.32220	0.27379	0.24826	0.24826	0.10721
80	114.55	0.41185	0.25658	0.20860	0.18915	0.18915	0.06573
96	131.04	0.32605	0.21133	0.16514	0.14975	0.14975	0.04131
112	146.70	0.26476	0.17777	0.13410	0.12160	0.12160	0.02628
128	161.80	0.21879	0.15158	0.11082	0.10049	0.10049	0.01678
144	176.49	0.18304	0.13039	0.09271	0.08407	0.08407	0.01068
160	190.91	0.15444	0.11277	0.07823	0.07093	0.07093	0.00673
176	205.14	0.13104	0.09780	0.06637	0.06019	0.06019	0.00417
192	219.25	0.11154	0.08488	0.05650	0.05123	0.05123	0.00252
208	233.27	0.09504	0.07360	0.04814	0.04365	0.04365	0.00147

*From S.K. Thompson, Stratified adaptive cluster sampling, 1991, *Biometrika* **78**, 389–397. With permission from the Biometrika Trustees.

divided into two strata, with the first stratum containing the 3 units with the values $\{1, 2, 10\}$ and the second stratum containing the 2 units with the values $\{1000, 3\}$. Let the condition of interest be $C = \{y : y \geq 5\}$, so that, whenever a value greater than or equal to 5 is observed, the units in the neighborhood of that observation are added to the sample. The neighborhood of each unit is defined to be the unit and its immediately adjacent units. For example, if the unit with value 10 is observed, the adjacent units, having values 2 and 1000, are added to the sample; then, since 1000 also exceeds 5, the adjacent unit with value 3 is also added to the sample. The two units with values $\{10, 1000\}$, the only units in the population that satisfy the condition, form a network which crosses the boundary between strata.

Consider a stratified adaptive cluster sampling design with an initial sample size of 1 in each stratum, namely, $n_1 = n_2 = 1$. The six possible samples obtainable under this design, each with equal probability, are listed in Table 4.9. The initial observations for each of the six possible samples is followed, after the semicolon,

Table 4.9. Values of Estimators for the Six Possible Samples in the Example*

Observations	$\tilde{\mu}''$	$\tilde{\mu}'$	$\tilde{\mu}$	$\hat{\mu}$
1, 1000; 10, 2, 3	400.6	202.6	243.0	303.6
1, 3	1.8	1.8	1.8	1.8
2, 1000;10, 3	401.2	203.2	243.6	304.2
2, 3	2.4	2.4	2.4	2.4
10, 1000; 2, 3	406.0	505.0	484.8	303.0
10, 3; 2, 1000	7.2	304.2	243.6	304.2
Mean:	203.2	203.2	203.2	203.2
Variance:	39,766.2	30,361.2	27,504.9	20,220.8

*From S.K. Thompson, Stratified adaptive cluster sampling, 1991, *Biometrika* **78**, 389–397. With permission from the Biometrika Trustees.

by the observations subsequently added to the sample with the adaptive procedure. For each possible sample, the value of each of the unbiased estimators, other than the initial stratified sample mean, is computed. At the bottom of the table are given the means (equal in each case to 203.2, the population mean) and variances of the estimators under the adaptive design.

To illustrate the calculations we consider the third row of the table. Here the initial sample selected the unit with value 2 from the first stratum and the unit with value 1000 from the second stratum, resulting in the addition to the sample of the units with values 10 and 3. We therefore have two networks intercepted by the initial sample, namely $\{2\}$ and $\{10, 1000\}$, and an edge unit of $\{3\}$. The computations for each of the estimators are as follows: The intersections of the second network with each stratum have only one unit each, so that $\tilde{\mu}'' = (1/5)[3(2) + 2(1000)] = 401.2$. The sample unit with value 3, being an edge unit (i.e., not satisfying the condition and not intersected by the initial sample), is not included in the calculations, so that $\tilde{\mu}' = (1/5)[2 + (10 + 1000)/2] = 203.2$. The expected number of times the unit with value 2 is intersected by an initial sample is 1/3. The expected number for the unit with value 10, as well as for the unit with value 1000, is $(1/3) + (1/2) = 5/6$. Thus, $\tilde{\mu} = (1/5)[2/(1/3) + 10/(5/6) + 1000/(5/6)] = 243.6$. The intersection probability α_i for the unit with value 2 is 1/3. For the units with values 10 and 1000, the intersection probability is $1 - (2/3)(1/2) = 2/3$. Thus, $\hat{\mu} = (1/5)[2/(1/3) + 10/(2/3) + 1000/(2/3)] = 304.2$.

The conventional stratified sample mean for the sample of the third row would be $(1/5)[3(2 + 10)/2 + 3(1000 + 3)/2] = 204.2$. The mean of these estimates over the six possible samples is 136.67. Hence, the conventional stratified sample mean is biased when used with the adaptive design.

Three of the possible samples in the table—those in the third, fifth, and sixth rows—have the same set of four distinct observations, and hence have the same value of the minimal sufficient statistic. The Rao–Blackwell version of any of the estimators for each of these samples is obtained by averaging the value of the corresponding estimator over the three samples. The values of the Rao–Blackwell versions of each of the estimators are listed in Table 4.10 for each of the six possible samples, and the variances of these improved unbiased estimators are given at the bottom of the table. □

Table 4.10. Values of Estimators Improved by the Rao–Blackwell Method*

Observations	$\tilde{\mu}''_{RB}$	$\tilde{\mu}'_{RB}$	$\tilde{\mu}_{RB}$	$\hat{\mu}_{RB}$
1, 1000; 10, 2, 3	400.6	202.6	243.0	303.6
1, 3	1.8	1.8	1.8	1.8
2, 1000; 10, 3	271.47	337.47	324.0	303.8
2, 3	2.4	2.4	2.4	2.4
10, 1000; 2, 3	271.47	337.47	324.0	303.8
10, 3; 2, 1000	271.47	337.47	324.0	303.8
Mean:	203.2	203.2	203.2	203.2
Variance:	22,305.1	22,494.3	21,040.8	20,220.6

*From S.K. Thompson, Stratified adaptive cluster sampling, 1991, *Biometrika* **78**, 389–397. With permission from the Biometrika Trustees.

NOTE: Variances of estimators with the small population of five units. Any unit with $y \geq 5$ satisfies the condition, and the neighborhood of a unit contains the adjacent units. The observations obtained in the initial sample are followed, after the semicolon, by observations obtained subsequently.

CHAPTER 5

Efficiency and Sample Size Issues

5.1 INTRODUCTION

In the previous chapter, we saw, from efficiency comparisons in examples with spatially clustered and rare populations, that adaptive cluster sampling has the potential of producing estimates of the population mean or total with high efficiency relative to comparable conventional designs. In general, however, neither type of strategy is uniformly better than the other over every possible population configuration. The relative efficiency of the two strategies depends on a number of factors involving the characteristics of the population and the sample sizes or costs involved in the survey methods. In this chapter, factors are identified and evaluated that influence the relative efficiency of simple random sampling to adaptive cluster sampling with an initial simple random sample. We use a design-based approach so that properties such as design unbiasedness hold, no matter what the population itself is like. Many of the results in this chapter are based on Thompson [1994]. Model-based methods with adaptive cluster sampling are promising also, as indicated at the end of the chapter in the comments about maximum likelihood methods with flexible designs.

5.2 ADAPTIVE VERSUS SIMPLE RANDOM SAMPLING

Before comparing the relative efficiencies of adaptive cluster sampling and simple random sampling in the nonstratified situation, it is helpful to summarize the setup described in Section 4.2. Consider a finite population of N units with variable of interest y_i associated with the ith unit. The object of sampling is to obtain a design-unbiased estimator of the population mean $\mu = (1/N) \sum_{i=1}^{N} y_i$ or population total $\tau = N\mu$ having low variance under the design used.

For the adaptive cluster sampling design, an initial random sample without replacement of n_1 units is selected from the N units in the population. For each unit in the population, a neighborhood is defined. The neighborhood of the i unit may be defined to consist of that unit plus all adjacent units, but many other neighborhood configurations are possible; for example the neighborhood could consist of a systematic grid of units centered on unit i. Whenever the y-value of a unit i in the sample satisfies a specified condition C (for instance, $y_i > c$ where c is a specified constant), then all units in the neighborhood of unit i are added to the sample. If in turn any of the added units satisfies the condition, still more units are added until the sample contains the neighborhood of any sample unit satisfying the condition. The condition and neighborhoods specified by the design uniquely partition the population into networks with the property that if any unit of a network is included in the initial sample, then all units in the network are included in the final sample. A unit not satisfying the condition is in a network consisting only of itself. A unit not satisfying the condition will be included in the sample whenever either (1) it is selected in the initial sample, or (2) a unit in its neighborhood satisfying the condition belongs to a network from which a unit was included in the initial sample. Units added to the sample only by (2) are called "edge units."

Let K denote the number of networks in the population and let $k(i)$ label the network that includes unit i, so that k takes values in $\{1, 2, \ldots, K\}$. Let B_k denote the set of units that comprise the kth network. Let $m_{k(i)}$ denote the number of units in the network $B_{k(i)}$ $(= A_i$ in Section 4.2) that includes unit i. For any unit i, the average $w_{k(i)}$ of the y-values of the units in the network that includes unit i is

$$w_{k(i)} = \frac{1}{m_{k(i)}} \sum_{j \in B_{k(i)}} y_j.$$

With the adaptive cluster sampling design described above, a design-unbiased estimator of the population mean μ is given by (4.15), namely

$$\tilde{\mu} = \frac{1}{n_1} \sum_{i=1}^{n_1} w_{k(i)},$$

which is the sample mean of the w-values associated with each unit in the initial sample. From (4.16), the variance of $\tilde{\mu}$ is

$$\mathrm{var}[\tilde{\mu}, acs, n_1] = \frac{(N - n_1)}{n_1 N(N - 1)} \sum_{i=1}^{N} (w_{k(i)} - \mu)^2.$$

With a simple random sample of m units, an unbiased estimator of the population mean μ is the sample mean $\bar{y}$, having variance

$$\mathrm{var}[\bar{y}, srs, m] = \frac{(N - m)}{m N(N - 1)} \sum_{i=1}^{N} (y_i - \mu)^2.$$

The relative efficiency of the sample mean $\bar{y}$ of a simple random sample of m units compared to an adaptive cluster sample with an initial simple random sample of n_1 units with the estimator $\tilde{\mu}$ is defined as the ratio $\mathrm{var}[\tilde{\mu}; acs, n_1]/\mathrm{var}[\bar{y}; srs, m]$. If the ratio is less than 1, the adaptive cluster sampling strategy is more efficient, while if the ratio is greater than 1, simple random sampling with sample size m is more efficient. It is often of interest to report the reciprocal of the above ratio, that is, the relative efficiency of adaptive cluster sampling to simple random sampling. However, the ratio as given above will be somewhat simpler to interpret.

Given a partition $B_1, \ldots, B_K$ of the population into networks, the total sum of squares for the variable of interest can be partitioned into between-network and within-network components as in a one-way analysis of variance (with the observations in each network forming a "sample" and $w_{k(i)}$ being the sample mean), namely,

$$
\sum_{i=1}^{N}(y_i - \mu)^2 = \sum_{k=1}^{K}\sum_{i \in B_k}(y_i - w_{k(i)})^2 + \sum_{k=1}^{K}\sum_{i \in B_k}(w_{k(i)} - \mu)^2
$$

$$
= \sum_{k=1}^{K}\sum_{i \in B_k}(y_i - w_{k(i)})^2 + \sum_{i=1}^{N}(w_{k(i)} - \mu)^2.
$$

Thus, the variance of the estimator $\tilde{\mu}$ with adaptive cluster sampling may be written

$$
\mathrm{var}[\tilde{\mu}] = \frac{N - n_1}{n_1 N(N - 1)}\left[\sum_{i=1}^{N}(y_i - \mu)^2 - \sum_{k=1}^{K}\sum_{i \in B_k}(y_i - w_{k(i)})^2\right].
$$

The relative efficiency of a conventional simple random sampling design of sample size m, with the unbiased estimator $\bar{y}$, to the adaptive cluster sampling strategy of initial sample size n_1, with the estimator $\tilde{\mu}$, is

$$
\frac{\mathrm{var}[\tilde{\mu}; acs, n_1]}{\mathrm{var}[\bar{y}; srs, m]} = \frac{m}{n_1}\left(\frac{N - n_1}{N - m}\right)\left[1 - \frac{\sum_{k=1}^{K}\sum_{i \in B_k}(y_i - w_{k(i)})^2}{\sum_{i=1}^{N}(y_i - \mu)^2}\right]. \qquad (5.1)
$$

5.3 WITHIN-NETWORK VARIATION

The last factor (in square brackets) in the relative efficiency (5.1) contains the proportion of total population variance that is contained within networks. This factor thus depends on the structure of the population and is not influenced by the sample sizes used in the comparison. For given sample sizes m and n_1, the adaptive cluster sampling strategy will be efficient relative to the conventional strategy if the within-network variation represents a relatively large proportion of the overall variation. The larger the proportion, the smaller will be the factor in the square brackets. The relative efficiency depends on the partitioning of the population into networks, which depends on the condition C and on the definition of neighborhoods.

When the condition for extra sampling is of the form $y_i > c$ for some constant c, so that the extra sampling is associated with high abundance, then the networks of units satisfying the condition correspond to the natural aggregations or clusters of the variable of interest. The adaptive cluster sampling strategy will have relatively high efficiency if there is high variability in abundance within these aggregations or clusters.

5.4 SAMPLE SIZE AND COST ISSUES

A question of practical interest is how precise are the estimates obtained with adaptive cluster sampling compared to what could be obtained with a conventional design at a comparable cost. With conventional cluster sampling, it is usually found that a simple random sample of units would produce a more precise estimate for a fixed total sample size. However, field costs are less in obtaining units in clusters, so that for a given cost a more precise estimate may be obtained with cluster sampling (cf. Cochran [1977, p. 233] and Matérn [1986]). With adaptive cluster sampling, some of the same principles may apply, such as lower cost in adding a cluster or network of nearby units than in traveling to additional units selected at random in the study region.

Denote the first factor in the relative efficiency (5.1) as

$$b(n_1, m, N) = (m/n_1)[(N - n_1)/(N - m)].$$

For a fixed size m of the simple random sample, this factor depends only on the sample sizes to be compared and on the number of units in the population. Small values of b give more efficiency to the adaptive strategy whereas large values give more efficiency to simple random sampling. For fixed n_1, b is an increasing function of m. When $m = n_1$, $b = 1$. If m and n_1 both increase at a constant ratio, that is, $m = a_1 k$ and $n_1 = a_2 k$, then b is an increasing function of k. Finally, with n_1 and m fixed, b is a decreasing function of N if $m > n_1$ and increasing if $m < n_1$. For most comparisons of interest, m will be greater than or equal to n_1. In summary, adaptive cluster sampling will be more efficient if m is not much bigger than n_1. If m is less than or equal to n_1, adaptive cluster sampling will always be more efficient than simple random sampling.

Ideally, the values of m and n_1 with which to compute the relative efficiency are based on respective cost functions, with m and n_1 each the largest sample sizes possible for a given expected cost measured in time or money. Because of travel time and other factors, it is generally the case that, with a given amount of time or money, more units can be obtained in clusters—conventionally or adaptively—than can be obtained as a simple random sample selected throughout the study region.

Let ν denote the final sample size of the adaptive cluster sample with initial sample size n_1 so that the expected final sample size is $E[\nu]$. Consider a linear cost function with initial cost c_0, a cost c_1 for each unit obtained by simple random sampling, and a cost c_2 for each adaptively added unit. Then, from (4.41) in Section 4.6, the total expected cost c_T with adaptive cluster sampling is

$$\begin{aligned} c_T &= c_0 + c_1 n_1 + c_2(E[\nu] - n_1) \\ &= c_0 + (c_1 - c_2)n_1 + c_2 E[\nu]. \end{aligned} \tag{5.2}$$

For the same total cost, the conventional simple random sampling design could use a sample size m_c satisfying

$$c_T = c_0 + c_1 m_c. \tag{5.3}$$

Equating the two cost functions (5.2) and (5.3) and solving for m_c gives

$$m_c = \left(1 - \frac{c_2}{c_1}\right) n_1 + \frac{c_2}{c_1} E[\nu]. \tag{5.4}$$

In many surveys of natural populations, the cost per added unit c_2 will be less than the cost per initial unit c_1 because of the reduced travel time and other logistic savings associated with obtaining sample units in clusters. In such situations, $0 \le c_2/c_1 \le 1$ so that m_c is a weighted average of n_1, and $E[\nu]$ and satisfies $n_1 \le m_c \le E[\nu]$. Since the variance for the simple random sample decreases as the sample size m increases, bounds for the relative efficiency of simple random sampling to adaptive cluster sampling for fixed expected cost are given by

$$\frac{\text{var}[\tilde{\mu}; acs, n_1]}{\text{var}[\bar{y}; srs, n_1]} \le \frac{\text{var}[\tilde{\mu}; acs, n_1]}{\text{var}[\bar{y}; srs, m_c]} \le \frac{\text{var}[\tilde{\mu}; acs, n_1]}{\text{var}[\bar{y}; srs, E(\nu)]}. \tag{5.5}$$

At the end of Section 4.5, it was noted that the variance of $\tilde{\mu}$ with adaptive cluster sampling is always less than or equal to the variance of $\bar{y}$ with a simple random sample of n_1 units. Hence the left side of the efficiency bound (5.5) is always less than or equal to 1, favoring adaptive cluster sampling. In the absence of cost information, the right side of the efficiency bound, with $m = E[\nu]$, can serve as a conservative assessment of efficiency giving the benefit of the doubt to the conventional strategy. This was the basis of the comparisons in the examples discussed in Chapter 4. The relative efficiency based on expected final sample size $m = E[\nu]$ can be either less than or greater than 1, depending on the proportion of variance within networks and other factors to be analyzed below.

For the right-hand side of the efficiency bound (5.5), it is of interest to note that the expected final sample size $E[\nu]$ depends for a given population on the initial sample size n_1. The quantity

$$g(n_1) = \frac{E[\nu]}{n_1} \cdot \frac{(N - n_1)}{(N - E[\nu])},$$

and hence the upper bound on the relative efficiency, are similarly influenced by the choice of initial sample size n_1. Although $E[\nu]$ is an increasing function of n_1, the ratio $E[\nu]/n_1$ is decreasing in n_1. The derivation of these results follows.

For adaptive cluster sampling with an initial simple random sample, the probability π_i that unit i is included in the sample is, from (4.2),

$$\pi_i = 1 - \left[\binom{N - m_i - a_i}{n_1} \Big/ \binom{N}{n_1}\right],$$

where m_i is the number of units in the network that includes unit i and a_i is the number of units in networks of which i is an edge unit. From (4.40) in Section 4.6, the expected sample size is

$$
\begin{aligned}
\mathrm{E}[\nu] &= \sum_{i=1}^{N} \pi_i \\
&= \sum_{i=1}^{N} \left[1 - \binom{N - m_i - a_i}{n_1} \Big/ \binom{N}{n_1} \right] \\
&= N - \sum_{i=1}^{N} \frac{(N - m_i - a_i)!\, n_1!\, (N - n_1)!}{n_1!\,(N - m_i - a_i - n_1)!\, N!}.
\end{aligned}
$$

Dividing $(N - n_1)!$ by $(N - m_i - a_i - n_1)!$ gives

$$
\mathrm{E}[\nu] = N - \frac{1}{N!} \sum_{i=1}^{N} (N - m_i - a_i)!\,(N - n_1 - m_i - a_i + 1) \times \cdots \times (N - n_1).
$$

$$(5.6)$$

Each term of the sum contains the product of $m_i + a_i$ factors that depend on n_1, but the number of factors does not depend on n_1. Writing $f(n_1) = \mathrm{E}[\nu]$, the first derivative with respect to n_1 is

$$
\begin{aligned}
f'(n_1) &= -\frac{1}{N!} \sum_{i=1}^{N} (N - m_i - a_i)! \sum_{j=1}^{m_i+a_i} (-1) \prod_{\substack{k=1 \\ k \neq j}}^{m_i+a_i} (N - n_1 + k) \\
&= \frac{1}{N!} \sum_{i=1}^{N} (N - m_i - a_i)! \sum_{j=1}^{m_i+a_i} \prod_{\substack{k=1 \\ k \neq j}}^{m_i+a_i} (N - n_1 + k).
\end{aligned}
$$

Every factor is positive provided $n_1 < N - m_i - a_i$. Hence $f'(n_1)$ is positive and $f(n_1)$ is an increasing function of n_1.

The second derivative is

$$
f''(n_1) = \frac{1}{N!} \sum_{i=1}^{N} (N - m_i - a_i)! \sum_{j=1}^{m_i+a_i} \sum_{\substack{r=1 \\ r \neq j}}^{m_i+a_i} (-1) \prod_{\substack{k=1 \\ k \neq j \\ k \neq r}}^{m_i+a_i} (N - n_1 + k).
$$

Since $f''(n_1)$ is negative, the function $f(n_1)$, which is increasing and equals zero at zero, is concave downward. Therefore the ratio $f(n_1)/n_1$ of expected to initial sample size is decreasing.

Finally, by taking logarithms, the derivative of the function $g(n_1) = [f(n_1)/n_1] \{(N - n_1)/[N - f(n_1)]\}$ is

$$g'(n_1) = \frac{f'(n_1)(N - n_1)}{n_1[N - f(n_1)]} - \frac{Nf(n_1)}{n_1^2[N - f(n_1)]} - \frac{f(n_1)(N - n_1)[-f'(n_1)]}{n_1[N - f(n_1)]^2}$$

$$= \{n_1 f'(n_1)(N - n_1) - f(n_1)[N - f(n_1)]\} \left\{ \frac{N}{n_1^2[N - f(n_1)]^2} \right\}.$$

Thus, $g(n_1)$ decreases in n_1 only if $n_1 f'(n_1)(N - n_1) - f(n_1)[N - f(n_1)] < 0$, that is, if

$$f'(n_1) < \frac{f(n_1)[N - f(n_1)]}{n_1(N - n_1)},$$

so that g may be increasing or decreasing in n_1 depending on the population configuration and on the sample sizes. In the examples considered in Sections 4.6.1 and 4.6.2, the function $g(n_1)$, and hence the relative efficiency of simple random sampling, decreased with increasing n_1 for the values computed, ranging from $n_1 = 1$ to $n_1 = N/2$.

In adaptive cluster sampling with the simple estimators $\bar{\mu}$ and $\hat{\mu}$, edge units add to the sample size while not contributing to the efficiency of the estimator. A population in which all of the units satisfying the condition were scattered far apart from each other would best be sampled using a conventional design. Such units represent networks with zero variance. With adaptive cluster sampling, such units trigger additional observations that add only edge units to the sample, increasing the cost while not increasing precision.

In many real situations it will be true not only that observing units in clusters or networks is less costly than observing the same number by simple random sampling, but also that the cost per unit of observing units not satisfying the condition will be lower than the cost of observing units that do satisfy the condition. For example, in a survey of lichen biomass with condition $y_i > 0$ for additional sampling, zero biomass can be determined at a glance, whereas biomass above zero involves careful collection and laboratory analysis.

5.5 RARITY OF THE POPULATION

Another factor that influences the relative efficiency of simple random sampling to adaptive cluster sampling is the degree of geographic rarity of the population. A population with a given clustering or aggregation pattern has more geographic rarity if the given clusters or aggregations are scattered over a large study region than one of the same population size and number of clusters with the clusters crowded into a smaller study region. For example, a population of whales with a tendency to aggregate into groups and spread unpredictably over an entire ocean has more geographic rarity—and is harder to sample—than a population of the same number of whales in the same number of groups known to be confined to a specific bay during the survey. Similarly, an environmental pollutant with a patchy distribution is rare if high concentrations are found at very few sites.

To analyze the effect of rarity on the relative efficiency, we consider a population of a fixed number of animals, plants, or other variable with a fixed pattern of spatial aggregation, and examine the effect on relative efficiency of the total number of units in the population. For simplicity, suppose that the condition for extra sampling is $y_i > 0$. Increasing N, with the population size and spatial aggregation characteristics fixed, corresponds to increasing the study region to contain more units with zero y-values. With the population size $\tau = \sum_{i=1}^{N} y_i$, the within-network sum of squares $A = \sum_{k=1}^{K} \sum_{i \in B_k} (y_i - w_{k(i)})^2$, and the total (uncorrected) sum of squares $B = \sum_{i=1}^{N} y_i^2$ all held fixed, the effect on efficiency of a larger value of N is examined. Let $C = \tau^2$, which is also fixed. Let η denote the total number of units satisfying the condition plus edge units in the population. Because of the fixed spatial aggregation pattern, η is assumed to remain fixed. With the initial sample size n_1 also held fixed, the expected final sample size $E[\nu]$ is a function of N, for which we again write $f(N) = E[\nu]$. Define $g(N) = [f(N)/n_1]\{(N - n_1)/[N - f(N)]\}$ and $k(N) = 1 - \{A/[B - (C/N)]\}$.

With this notation, the relative efficiency of simple random sampling with sample size $E[\nu]$ to adaptive cluster sampling with initial sample size n_1 is, by (5.1), $h(N) = g(N)k(N)$. As a function of N, the expected sample size can be written

$$f(N) = E[\nu] = N - \sum_{i=1}^{N} \frac{(N - n_1 - m_i - a_i + 1)}{(N - m_i - a_i + 1)} \cdots \frac{(N - n_1)}{N}.$$

which was obtained by dividing $(N - m_i - a_i)!$ by $N!$ in (5.6). The number of units satisfying the condition and edge units is fixed at η (we can assume without loss of generality that these are the first η units). For all other units, the network size is $m_i = 1$ and the number of units satisfying the condition in adjacent networks is $a_i = 0$, giving only the one factor $(N - n_1)/N$ in the ith term. Thus,

$$f(N) = N - (N - \eta)\frac{(N - n_1)}{N}$$

$$- \sum_{i=1}^{\eta} \frac{(N - n_1 - m_i - a_i + 1)}{(N - m_i - a_i + 1)} \cdots \frac{(N - n_1)}{N}$$

$$= n_1 + \eta - \frac{n_1 \eta}{N} - \sum_{i=1}^{\eta} \frac{(N - n_1 - m_i - a_i + 1)}{(N - m_i - a_i + 1)} \cdots \frac{(N - n_1)}{N}.$$

The sum contains η terms each consisting of a product of $(m_i + a_i)$ ratios that approach 1 as N approaches infinity, so that

$$\lim_{N \to \infty} f(N) = n_1 + \eta - \eta = n_1.$$

That is, the expected sample size approaches the initial sample size with increasing geographic rarity. Further, $\lim_{N \to \infty} g(N) = 1$ and $\lim_{N \to \infty} k(N) = 1 - (A/B)$, so that

$$\lim_{N \to \infty} h(N) = 1 - \frac{A}{B}.$$

Thus, with increasing geographic rarity, for a population of fixed size and spatial aggregation pattern, the relative efficiency based on $E[\nu]$ of simple random sampling to adaptive cluster sampling approaches the limit $1 - (A/B)$. This is strictly less than 1 if the within-network variance is greater than zero, which suggests that many aggregated populations, if geographically rare, will be more efficiently sampled with adaptive than with conventional designs. Additional considerations in the sampling of rare populations are discussed by Kalton and Anderson [1986], Czaja et al. [1986], and Sudman et al. [1988].

Relative efficiency based on equal expected total cost, with m_c given by (5.4), also increases with increasing rarity, since

$$\lim_{N \to \infty} m_c = \lim_{N \to \infty} \left[\left(1 - \frac{c_2}{c_1} \right) n_1 + \frac{c_2}{c_1} f(N) \right] = \left(1 - \frac{c_2}{c_1} \right) n_1 + \frac{c_2}{c_1} n_1 = n_1,$$

so that the previous limits are unchanged, that is, the limiting relative efficiency as N approaches infinity is still $1 - A/B$.

5.6 ALTERNATIVE ESTIMATORS AND DESIGNS

The efficiency formula (5.1) applies only to the estimator $\tilde{\mu}$ based on expected numbers of intersections. In the examples with clustered populations computed in Thompson [1990, 1991a, b], and considered in the previous chapter, the estimator $\hat{\mu}$, based on probabilities of intersection, consistently gave a lower variance than $\tilde{\mu}$. With the estimator $\hat{\mu}$, the efficiency $\text{var}[\hat{\mu}; acs, n_1]/\text{var}[\bar{y}; srs, m]$ of simple random sampling to adaptive cluster sampling is straightforward to compute but involves the individual and joint intersection probabilities and does not lend itself to easy interpretation.

Each of the estimators $\tilde{\mu}$ and $\hat{\mu}$ can be improved through the Rao–Blackwell method. The improved estimators do use some information from edge units in the sample. Let $\tilde{\mu}_{RB} = E[\tilde{\mu} \,|\, D_R = d_R]$, the Rao–Blackwell estimator based on $\tilde{\mu}$. Let $\bar{y}_1$ denote the sample mean of a simple random sample of n_1 units. Since $\tilde{\mu}_{RB} = E[\bar{y}_1 \,|\, d_R]$ (Theorem 4.1 in Section 4.5), the variance of $\tilde{\mu}_{RB}$ can be written (cf.(4.31))

$$\text{var}[\tilde{\mu}_{RB}] = \text{var}[\bar{y}_1] - E\{\text{var}[\bar{y}_1 \,|\, d_R]\}.$$

The efficiency of a simple random sample of size m to adaptive cluster sampling with $\tilde{\mu}_{RB}$ is

$$\frac{\text{var}[\tilde{\mu}_{RB}; acs, n_1]}{\text{var}[\bar{y}_1; srs, m]} = \frac{(N - n_1)\sigma^2/(Nn_1) - E\{\text{var}[\bar{y}_1 \,|\, d_R]\}}{(N - m)\sigma^2/(Nm)}$$

$$= \frac{m}{n_1} \left(\frac{N - n_1}{N - m} \right) \left[1 - \frac{(Nn_1)E\{\text{var}[\bar{y}_1 | D_R]\}}{(N - n_1)\sigma^2} \right],$$

where

$$\sigma^2 = \frac{\sum_{i=1}^{N}(y_i - \mu)^2}{N - 1}$$

is the population variance.

Adaptive cluster sampling strategies with initial designs other than simple random sampling have been compared through examples with corresponding conventional designs, but analytic expressions of component factors influencing relative efficiency have not been obtained. For example, in previous sections of this chapter, adaptive cluster sampling with initial cluster samples and initial systematic samples were compared with conventional cluster and systematic strategies, and stratified adaptive cluster sampling was compared with conventional stratified sampling. In the examples, some of the greatest efficiency gains were obtained with initial designs, such as systematic sampling and stratified sampling with many small strata, which are themselves efficient for spatially aggregated populations. That is, the initial design was an efficient conventional designs in itself and then the adaptive procedure substantially increased the efficiency further. Adaptive cluster sampling with an initial unequal probability sampling design is compared with conventional sampling with unequal probabilities for hardwood forest inventories by Roesch [1993]. Adaptive cluster sampling with initial random samples and initial samples selected with probability proportional to available habitat is compared to conventional counterparts for surveys of wintering waterfowl on a refuge by Smith et al. [1995]. Adaptive cluster sampling with initial systematic samples of households in a marketing survey were examined by Danaher and King [1994]. Many other such studies are needed.

The effect of imperfect detectability on variances for both conventional and adaptive designs is described in Chapter 9. The effect of imperfect detectability on the relative efficiency of simple random sampling and adaptive cluster sampling would be to add an identical amount to the variance under each strategy, so that the relative efficiency, whether greater than or less than 1, would move slightly closer to 1.

The cost of observing edge units could be decreased considerably if there exists an auxiliary variable x_i for each unit i in the sample that indicates the abundance of the variable of interest y_i but is much easier to measure. For example, suppose y_i is the abundance of animals determined by photographic images from an aerial survey and x_i is a visual indicator, with $x_i = 0$ if the unit has zero abundance in the visual assessment and $x_i = 1$ otherwise. Neighboring units are added to the sample whenever $x_i = 1$. Then all decisions regarding extra sampling are based on the variable x, which uniquely partitions the population into networks. By the results on multivariate adaptive cluster sampling in Chapter 8, the estimates $\bar{\mu}$ and $\hat{\mu}$ remain unbiased for the population mean of the y-variable, even though the adaptive procedure was based on the x-variable. The advantage in cost is that for edge units (for which $x_i = 0$) the y-variable need not be measured. Thus, with an auxiliary variable that is easy or cheap to measure, the edge units, for which only the auxiliary variable need be observed, add very little to cost.

5.7 SUMMARY OF FACTORS INFLUENCING EFFICIENCY

The relative efficiency of adaptive cluster sampling to simple random sampling depends on several characteristics of the population, the design, and cost. In summary, any of the following characteristics tend to increase the efficiency of adaptive cluster sampling relative to simple random sampling:

1. The within-network variance is a high proportion of the total variance of the population. Such a population would be described as clustered or as having aggregation tendencies.

2. The population is rare—the number of units in the population is big relative to the number of units satisfying the condition. The study region is big relative to the area in which the animals or plants would be encountered.

3. The expected final sample size is not much larger than the initial sample size. Additionally, with a larger initial sample, the ratio of the expected final sample size to the initial sample size tends to be smaller. The situation in which adaptive cluster sampling would be less efficient relative to simple random sampling is a population in which the units satisfying the condition are all separate from each other. Here the within-network variance is zero and adaptive sampling adds only edge units. These increase the sample size but do not contribute to the precision of estimators.

4. The cost of observing units in clusters or networks is less than the cost of observing the same number selected at random throughout the study region.

5. The cost of observing units not satisfying the condition is less than the cost of observing units satisfying the condition.

6. In some cases, cost savings may be achieved by basing the condition for additional sampling on an easy-to-observe auxiliary variable so that the variable of interest does not need to be measured on edge units.

7. The analysis above is based on the estimator $\tilde{\mu}$. Greater efficiency in adaptive cluster sampling may be achieved through the use of $\hat{\mu}$ or the Rao–Blackwell improvement of either $\tilde{\mu}$ or $\hat{\mu}$.

5.8 LIMITING SAMPLE SIZE AND DEALING WITH UNEXPECTED EXIGENCIES

The guidelines of the previous section suggest that adaptive cluster sampling is most efficient when the population is rare and clustered and the final sample size tends to be proportionally not too much larger than the initial sample size. Yet, in reality, it can happen that one has underestimated the abundance and range of the population in setting the condition for extra sampling so that rigorous adherence to the adaptive procedure would lead to the addition of more units than time or cost allow. The exact design unbiasedness of the adaptive sampling estimators depends on inclusion in the sample of the neighborhood of every sample unit satisfying the condition. If sampling is curtailed prematurely, biases may arise. In fact, the same problem arises in conventional sampling when the entire selected sample cannot be obtained, as when inclement weather curtails a fishery survey, and biases arise if the completed part of the sample is not representative of the population as a whole. The following are offered as guidelines and suggestions on limiting the total sampling effort in adaptive cluster sampling.

1. If the study region is stratified ahead of the survey into H strata, then the decision whether to use adaptive cluster sampling and what condition to base it on in any stratum can be based on the results in the strata already completed—time spent, abundance observed, variance estimated, or other criteria. In each stratum, design-unbiased estimators of the stratum population total and of estimation variance are used depending on the design implemented in that stratum. The stratified estimate of the population total is the sum of the individual stratum estimates, and the estimate of its variance is the sum of the individual stratum variance estimates. These estimators can be shown to be design unbiased by a conditioning argument similar to that used in Chapter 7 for the situation in which adaptive allocations are based on data from previous strata.

2. A pragmatic procedure to follow in the implementation of an adaptive cluster sampling design, using poststratification at the estimation stage, is as follows. The initial sample, as a conventional probability sample, would provide an unbiased estimate of the population total if no adaptive sampling were planned at all. In the field, priority should be given to completion of the initial sample. On the other hand, it is usually desirable for cost purposes to do the adaptive sampling in sequence whenever a unit satisfies the condition, with travel commencing generally from one side of the study area to the other. If at some point in the survey it is determined that time remains only to finish the initial sample, then all adaptive sampling should stop at that point and the initial sample completed. For estimation purposes the population could be poststratified so that one stratum represents the part of the population represented by the adaptive part of the survey and the other stratum represents the part of the population sampled conventionally. An adaptive cluster sampling estimator is used in the first stratum and a conventional estimator in the second. The two estimates, along with the two estimates of variance, can then be combined as in traditional stratified sampling. More generally, if the condition for extra sampling is modified one or more times during the survey, then the region should be poststratified, with each stratum representing an area in which a given condition was used. The separate stratum estimates may then be combined as stratified estimates. For example, with the condition $y_i > c$, suppose the first few clusters added adaptively to the sample are so large and time consuming that it is decided to subsequently use a higher value c_2 for the condition in the remainder of the survey. For estimation purposes the study region is then poststratified into two strata corresponding to the two conditions used. This procedure would be exactly design unbiased had the strata been delineated ahead of time as in paragraph 1 above.

3. If prior knowledge about the population is too weak for the selection of a suitable condition on which to base the extra sampling, the condition can be based on order statistics as described in the next chapter. Thus, additional sampling might take place only in the vicinity of the top observations out of 100, rather than in the vicinity of every observation exceeding a prespecified level.

4. As one way of controlling final sample size, Brown [1994] describes adaptive cluster sampling in which selection of units for the initial sample continues sequentially until a specified final sample size is reached. With this plan, each time a unit in the initial sample is selected and observed, the associated adaptive additions of surrounding units are carried out. If the cumulative total number of units in the sample is below the specified sample size limit, the next initial unit is added and associated adaptive additions carried out. If the cumulative total number of units in the sample equals or exceeds the sample size limit, no more initial units are selected. The final sample size can then vary only by the size of the cluster associated with the last selected initial unit. The sequential selection of initial units introduces a bias into the estimates of population total. Although overall sampling effort is quite well controlled by this method, the need to sequentially select initial units at random precludes using the most efficient travel path between selected units in the study region.

5. "Partition neighborhoods" can be used to limit the potential size of networks. The study region is partitioned into regions or blocks and any adaptive additions to the sample stop at these partition boundaries. For an illustrative example, consider a city partitioned into blocks. A marketing survey has the objective of estimating the market for infant care supplies. Some neighborhoods turn out to have few households with children of the appropriate age, while other neighborhoods have many such households, and this information is not available prior to the survey. Whenever a household satisfying the condition (say, spending at least a specified amount on child care products) is encountered, neighboring households are added to the sample, but only from within the same block. Thus, no network can exceed the size of a block, and most neighborhoods are smaller than that. The usual adaptive sampling estimators are design unbiased with such a neighborhood structure. Many spatial configurations of partition neighborhoods are possible.

6. Neighborhoods can consist of systematic grids of units rather than contiguous units, so that the resulting networks will be constrained to be spatially sparse.

7. In Chapter 3, it was shown that with a stochastic model for the population, maximum likelihood predictors of population quantities, unlike design-unbiased predictors, do not depend on the design used for selecting the sample, for adaptive as well as conventional designs. Suppose that in selecting the sample, field workers have adapted in unexpected ways to observed patterns in the population, to limitations in time and other resources, and to unforeseen exigencies. With a good model for the population, maximum likelihood prediction of the population total would not depend on the way the sample was selected, provided the selection did not depend on y-values in units outside the sample or on unknown parameter values or auxiliary variables not available at the estimation stage. As effective maximum likelihood methods are developed for classes of real populations, considerable flexibility in field sampling procedures can thereby be accommodated.

8. An approach utilizing information from a pilot survey is described in Salehi and Seber [1995].

CHAPTER 6

Adaptive Cluster Sampling Based on Order Statistics

6.1   INTRODUCTION

In adaptive cluster sampling designs, neighboring units are added to the sample whenever the value of the variable of interest satisfies a chosen criterion. Commonly, the criterion consists of a fixed, prespecified value, so that additional units are added to the sample whenever a unit is observed exceeding that prespecified value. In some surveys, it may be difficult to pick an appropriate fixed value ahead of time. In such cases, the criterion for additional sampling may be made relative to the observed sample values by basing it on the sample order statistics. For example, in an environmental survey, concentrations of a pollutant are measured at each site in an initial sample of 100 sites. Additional sites are then added to the sample in the neighborhoods of the top 10 sites, that is, those sites with the 10 largest order statistics of the variable of interest. If any of the added sites also have large values, still more sites may be added to the sample (see Example 1.3 in Section 1.5).

Adaptive cluster sampling based on order statistics is particularly appropriate for sampling situations in which the investigators, for reasons inherent to their field of study, wish to search for high values of the variable of interest in addition to estimating the overall mean or total. In ecological, environmental, and epidemiological surveys, investigators often have an interest in observing as many animals or plants of a rare species as possible, finding pollution "hot spots" as in Example 1.3, or observing a large number of cases of a rare disease. In many real populations, high values of the variable of interest tend to be associated with neighboring high values through such factors as spatial pattern or social connections. By adding neighboring units of the highest observed values to the sample, adaptive cluster sampling based on order statistics tends to increase the probability of observing units with high values and to increase the total value of the variable of interest in the sample, while at the same time allowing for unbiased estimation of the population mean or total.

   With the criterion based on sample order statistics, the situation differs mathematically from that of ordinary adaptive cluster sampling with fixed criteria. The usual adaptive cluster sampling estimators are no longer unbiased as the sampling distribution of such an estimator is complicated by the fact that the condition for extra sampling, and hence the size and structure of the networks of units satisfying the condition, depend on the observed sample values. The networks now vary from sample to sample. In this chapter, we give estimators of the population mean (or total) and variance estimators that are design unbiased with adaptive cluster sampling based on order statistics. These estimators can be improved using the Rao–Blackwell method. The main results of this chapter are from Thompson (1996).

6.2   THE DESIGN

In developing the following theory based on order statistics, it is helpful to focus on the pollution problem outlined above and in Example 1.3. We suppose that an initial simple random sample of pollution readings on n_1 sites is taken yielding the ordered readings $y_{(1)} < y_{(2)} < \cdots < y_{(n_1)}$. It is assumed that no ties occur, as is typically the case with a continuous variable. We then decide to sample the neighborhoods of those sites whose y-values are $y_{(r+1)}, y_{(r+2)}, \ldots, y_{(n_1)}$, that is, those sites with the $(n_1 - r)$ largest values. When $r = n_1 - 1$, the additional sampling will be carried out only in the vicinity of the unit with the maximum value. If any of the added sites give readings higher than $y_{(r)}$, then their neighborhoods are added as well. This scheme differs radically from the previous adaptive scheme in which the criterion C for determining cluster membership (e.g., $y_i > c$) is fixed and determined before the experiment is begun. There the clusters and the networks are determined in advance. However, using order statistics, the criterion C is now determined by the initial sample and therefore depends on the data. Such an approach is applicable to the situation where one does not know prior to the survey whether 10 parts per million of the pollutant, 100,000 microroentgens of radiation, one animal of the species, or another fixed criterion is appropriate. If the wrong criterion is used, it may turn out that every unit in the study region, or none of them, satisfies the criterion. A criterion based on the highest values encountered during the survey is then very appealing.

   As in Chapter 4, the edge units of the cluster are those units whose y-values do not exceed the condition level, which in the present case is $y_{(r)}$, and the networks are the clusters without their edge units. An edge unit does not satisfy the condition of exceeding $y_{(r)}$ and was not selected in the initial sample but was picked up in the additional sampling by being in the neighborhood of a unit satisfying the condition. Any unit whose y-value does not exceed $y_{(r)}$ is a network of size 1. The final reduced sample is denoted by the unordered set $s_R = \{s_1, s_{2R}\}$, where s_1 is the set of n_1 unordered labels from the initial sample (which are distinct, as sampling is without replacement), and s_{2R} is the set of distinct unordered labels from the remainder of the sample. In Chapter 2, we defined s to be s_R but with the labels ordered from smallest to largest. For notational convenience, we shall therefore use s to represent s_R in this chapter.

6.3   UNBIASED ESTIMATION

For each unit i in the sample s, let A_{is} denote the network containing unit i and let m_{is} be the number of units in A_{is}. Let f_{is} be the number of units from A_{is} included in the initial sample. In contrast to Section 4.2, we add a subscript s to remind us that network membership is now determined by the sample, in fact the initial sample s_1. Using a predetermined method of augmenting the initial sample so that the final sample can be uniquely ordered, we see that the probability of getting the final ordered sample is the same as the probability of getting the initial sample. Hence Theorem 2.3 of Section 2.4 still applies, and the reduced data D_R are minimal sufficient. However we shall focus on

$$D_f = \{(i, y_i, f_{is}): i \in s\}.$$

Since D_f determines D_R (by dropping the f_{is}), D_R is a function of D_f and D_f is sufficient (by Theorem 2.2 of Section 2.4).

We now ask whether the estimators $\hat{\mu}$ and $\tilde{\mu}$ described in Section 4.2 are still unbiased, given the added level of variability in the criterion C. Since the probability that the initial sample intersects the kth network depends on the other units in the sample, we see from (4.6) that $\hat{\mu}$ is no longer appropriate. However, from Theorem 4.1 in Section 4.5 we have $\tilde{\mu} = E[\bar{y}_1 \,|\, D_f = d_f]$, which may still be true when order statistics are used. Using (4.14), with $f_{is} = 0$ if the initial sample does not intersect A_{is}, we therefore define

$$\tilde{\mu}_{ord} = \frac{1}{n_1} \sum_{i \in s} \frac{y_i f_{is}}{m_{is}}. \tag{6.1}$$

From (4.15) we have the alternative representation

$$\tilde{\mu}_{ord} = \frac{1}{n_1} \sum_{i=1}^{n_1} w_{is}, \tag{6.2}$$

where w_{is} is the mean of the m_{is} observations in A_{is}. The following results come from Thompson [1995].

Theorem 6.1.   For the adaptive scheme based on order statistics, and using simple random sampling for the initial sample,

(i) $E[\bar{y}_1 \,|\, D_f = d_f] = \tilde{\mu}_{ord}$.
(ii) $E[\tilde{\mu}_{ord}] = \mu$.

Proof:

(i) Let $B_{1s}, B_{2s}, \ldots, B_{\kappa_s s}$ represent the κ_s distinct networks intersected by the initial sample. Let x_{ks} be the number of units in B_{ks} and let b_{ks} be the number of times the kth network is sampled by the initial sample. We note that for any unit i in the kth network $x_{ks} = m_{is}$ and $b_{is} = f_{is}$. Let $\mathcal{S}_f$ be the set of all samples s that can give rise to the given value of d_f, and let ξ_f be the number of samples in $\mathcal{S}_f$. Then

$$\xi_f = \binom{x_{1s}}{b_{1s}}\binom{x_{2s}}{b_{2s}}\cdots\binom{x_{\kappa_s s}}{b_{\kappa_s s}}.$$

Similarly, if ξ_{if} is the number of initial samples s_1 containing unit i conditional on the value of d_f, and unit i belongs to B_{ks}, then

$$\xi_{if} = \binom{x_{1s}}{b_{1s}}\binom{x_{2s}}{b_{2s}}\cdots\binom{x_{ks}-1}{b_{ks}-1}\cdots\binom{x_{\kappa_s s}}{b_{\kappa_s s}}$$

and, since $B_{ks} = A_{is}$,

$$\frac{\xi_{if}}{\xi_f} = \binom{x_{ks}-1}{b_{ks}-1}\bigg/\binom{x_{ks}}{b_{ks}} = \frac{b_{ks}}{x_{ks}} = \frac{f_{is}}{m_{is}}.$$

Conditional on $D_f = d_f$, the ξ_f samples will be equally likely, so that

$$E[\bar{y}_1 \mid d_f] = \frac{1}{\xi_f}\sum_{s\in S_f}\frac{1}{n_1}\sum_{i\in s_1}y_i$$

$$= \frac{1}{\xi_f n_1}\sum_{i\in s:f_{is}>0}\xi_{if}y_i$$

$$= \frac{1}{n_1}\sum_{i\in s:f_{is}>0}\frac{f_{is}y_i}{m_{is}}$$

$$= \tilde{\mu}_{ord}, \tag{6.3}$$

 by (6.1)

(ii) The proof follows immediately as

$$\mu = E[\bar{y}_1]$$

$$= EE[\bar{y}_1 \mid d_f]$$

$$= E[\tilde{\mu}_{ord}]. \qquad\square$$

We note from the above theorem that, since $\tilde{\mu}_{ord}$ is the expectation of $\bar{y}_1$ conditional on the sufficient statistic D_f, then $\tilde{\mu}_{ord}$ will be more efficient than $\bar{y}_1$.

Having shown that $\tilde{\mu}_{ord}$ is unbiased, we now focus on variance estimation. From

$$\text{var}[\bar{y}_1] = E\big[\text{var}(\bar{y}_1 \mid d_f)\big] + \text{var}\big[E(\bar{y}_1 \mid d_f)\big]$$

we have

$$\text{var}[\tilde{\mu}_{ord}] = \text{var}\left[E(\bar{y}_1 \mid d_f)\right]$$
$$= \text{var}[\bar{y}_1] - E\left[\text{var}(\bar{y}_1 \mid d_f)\right]. \tag{6.4}$$

Since $\bar{y}_1$ is the mean of a simple random sample of size n_1, $\text{var}[\bar{y}_1]$ may be estimated by the unbiased estimate

$$\widehat{\text{var}}[\bar{y}_1] = \frac{v_1}{n_1}\left(1 - \frac{n_1}{N}\right), \tag{6.5}$$

where

$$v_1 = \frac{1}{n_1 - 1}\sum_{i=1}^{n_1}(y_i - \bar{y}_1)^2. \tag{6.6}$$

The second term in (6.4), namely $E[\text{var}(\bar{y}_1 \mid d_f)]$, can be estimated from the data using

$$\text{var}[\bar{y}_1 \mid d_f] = \frac{1}{\xi_f}\sum_{s \in S_f}(\bar{y}_{1s} - \tilde{\mu}_{ord})^2, \tag{6.7}$$

where, for sample s, $\bar{y}_{1s}$ is the mean of the observations for $i \in s_1$. Thus, an unbiased estimator of $\text{var}[\tilde{\mu}_{ord}]$ is

$$\widehat{\text{var}}[\tilde{\mu}_{ord}] = \widehat{\text{var}}[\bar{y}_1] - \text{var}[\bar{y}_1 \mid d_f]. \tag{6.8}$$

Although unbiased, the variance estimator $\widehat{\text{var}}[\tilde{\mu}_{ord}]$ is computationally complex and can take on negative values with some samples. Because of the computational difficulty and the possibility that $\widehat{\text{var}}[\tilde{\mu}_{ord}]$ can take on negative values, a better approach might be to investigate the use of simpler variance estimators that, although biased, are invariably nonnegative. An obvious candidate is the estimator

$$\widetilde{\text{var}}[\tilde{\mu}_{ord}] = \frac{(N - n_1)}{Nn_1}v_w, \tag{6.9}$$

where v_w is the sample variance of the w_{is}; that is, $v_w = \sum_{i=1}^{n_1}(w_{is} - \tilde{\mu}_{ord})^2/(n_1 - 1)$. The variance estimator $\widetilde{\text{var}}[\hat{\mu}_{ord}]$ is the ordinary variance estimator which is unbiased with adaptive cluster sampling when the condition is fixed in advanced rather than based on order statistics. If $\widetilde{\text{var}}[\tilde{\mu}_{ord}]$ is used except when it turns out negative, in which case $\widetilde{\text{var}}[\hat{\mu}_{ord}]$ is used, we denote the combined estimator $\widetilde{\text{var}}_c[\hat{\mu}_{ord}]$.

6.4 IMPROVED UNBIASED ESTIMATORS

Because $\tilde{\mu}_{ord}$ depends on the order in which the sample is selected, basing its network structure on the units in the initial sample, $\tilde{\mu}_{ord}$ is not a function of the minimal sufficient statistic

$$D_R = \{(i, y_i): i \in s\}.$$

We can therefore use the Rao–Blackwell method to give an improved estimator

$$\tilde{\mu}_{RB} = E[\tilde{\mu}_{ord} \mid d_R] \qquad (6.10)$$

$$= E\left[E(\bar{y}_1 \mid d_f) \mid d_R\right]$$

$$= E[\bar{y}_1 \mid d_R], \qquad (6.11)$$

using the same argument that proved part (iii) of Theorem 4.1 in Section 4.5. From (6.4) we have

$$\text{var}[\tilde{\mu}_{RB}] = \text{var}\left[E(\bar{y}_1 \mid d_R)\right]$$

$$= \text{var}[\bar{y}_1] - E\left[\text{var}(\bar{y}_1 \mid d_R)\right]. \qquad (6.12)$$

The general theory of Section 4.5 can now be applied here with $T = \bar{y}_1$ and $T_{RB} = \tilde{\mu}_{RB}$. As in Section 4.5 we define ξ to be the number of groups (initial samples) of n_1 distinct units chosen from s distinct units in the final sample, indexed by the label g (for $g = 1, 2, \ldots, \xi$), which are compatible with the observed d_R. Let $T_g = \bar{y}_{1g}$, $\widehat{\text{var}}_g[\bar{y}_1]$ and v_{1g} denote the values of $\bar{y}_1$, $\widehat{\text{var}}[\bar{y}_1]$ of (6.5) and v_1 of (6.6), respectively, when computed using the gth combination for the initial sample. Then, from (6.11) and (4.33),

$$\tilde{\mu}_{RB} = \frac{1}{\xi} \sum_{g=1}^{\xi} \bar{y}_{1g}.$$

Also, by the Rao–Blackwell theorem, an improved unbiased estimate of $\text{var}[\bar{y}_1]$ is given by

$$E\left[\widehat{\text{var}}(\bar{y}_1) \mid d_R\right] = \frac{1}{\xi} \sum_{g=1}^{\xi} \widehat{\text{var}}_g[\bar{y}_1]$$

$$= \frac{N - n_1}{n_1 N} E[v_1 \mid d_R],$$

where

$$E[v_1 \mid d_R] = \frac{1}{\xi} \sum_{g=1}^{\xi} v_{1g}. \qquad (6.13)$$

We also need

$$\text{var}[\bar{y}_1 \mid d_R] = \frac{1}{\xi} \sum_{g=1}^{\xi} (\bar{y}_{1g} - \tilde{\mu}_{RB})^2.$$

Putting these together, an unbiased estimator of (6.12) is given by [cf. (4.36)]

$$\widehat{\text{var}}[\tilde{\mu}_{RB}] = \frac{1}{\xi} \sum_{g=1}^{\xi} \left\{ \widehat{\text{var}}_g[\bar{y}_1] - (\bar{y}_{1g} - \tilde{\mu}_{RB})^2 \right\}. \tag{6.14}$$

From (6.10) we can also use

$$\tilde{\mu}_{RB} = \frac{1}{\xi} \sum_{g=1}^{\xi} \tilde{\mu}_{ord,g}, \tag{6.15}$$

where $\tilde{\mu}_{ord,g}$ is $\tilde{\mu}_{ord}$ computed using the gth combination.

We note from the previous section that ξ_f is the number of combinations compatible with d_f so that we can write, from (6.3),

$$\tilde{\mu}_{ord} = \text{E}[\bar{y}_1 \mid d_f]$$

$$= \frac{1}{\xi_f} \sum_{g=1}^{\xi_f} \bar{y}_{1g}. \tag{6.16}$$

Clearly the only difference between $\tilde{\mu}_{RB}$ and $\tilde{\mu}_{ord}$ is that d_R is replaced by d_f and ξ by ξ_f. Hence, using Section 4.5 again but with $T = \bar{y}_1$ and $T_{RB} = \tilde{\mu}_{ord}$, we see that an unbiased estimate of $\text{var}[\tilde{\mu}_{ord}]$ is given by (6.14) but with ξ replaced by ξ_f, namely

$$\widehat{\text{var}}[\tilde{\mu}_{ord}] = \frac{1}{\xi_f} \sum_{g=1}^{\xi_f} \left\{ \widehat{\text{var}}_g[\bar{y}_1] - (\bar{y}_{1g} - \tilde{\mu}_{ord})^2 \right\}. \tag{6.17}$$

The computational efficiency issues involved in computing the estimators improved by the Rao–Blackwell method are related to the "occupancy problem" of urn models (see Johnson and Kotz [1977]).

6.5 EXAMPLES

Two examples illustrating adaptive cluster sampling based on order statistics will be examined. In Example 6.1, additional sampling is carried out only in the vicinity of the largest order statistic of the initial sample. In Example 6.2, it is carried out in the vicinity of the largest two order statistics. For ease of presentation we work with $\tau = N\mu$ rather than μ to avoid a lot of decimals. Though the examples are deliberately small so that every possible sample may be listed, most of the key issues of this chapter are illustrated, including the calculations for the estimator $\tilde{\tau}_{ord}$ $(= N\tilde{\mu}_{ord})$, the possibility of the unbiased variance estimator taking on negative values with some samples, the fact that the network associations vary from sample to sample, the appearance of edge units in the sample, the computation of the unbiased estimator of variance of $\tilde{\tau}_{ord}$, the computation of the Rao–Blackwell estimator, the fact that $\tilde{\tau}_{RB}$ can use information even from the edge units, the unbiasedness of each estimator, and the increased efficiency of $\tilde{\tau}_{ord}$ relative to $N\bar{y}_1$ and of $\tilde{\tau}_{RB}$ relative to $\tilde{\tau}_{ord}$.

Example 6.1   Largest Order Statistic.   We consider first a design in which initially a simple random sample of $n_1 = 3$ units is selected. The neighborhood of each unit consists of itself plus any adjacent units. Following the initial sample, units in the neighborhood of the unit with the largest y-value—that is, the maximum, $y_{(3)}$— are added to the sample. If, in turn, the y-value of any added unit exceeds $y_{(2)}$, the second largest value of the initial sample, then any units in the neighborhood of that added unit are also added, and so on. The design is applied to a population consisting of $N = 6$ units, with y-values

$$\mathbf{y} = (4, 20, 1, 10, 1000, 100)',$$

listed in the order of their spatial arrangement and totaling $\tau = 1135$. The 20 possible samples under the procedure are listed in Table 6.1. For each sample, the y-values for the units in the initial sample are listed, followed after the semicolon by the y-values for units subsequently added. The value of the simple estimator $N\bar{y}_1$ is given for each sample, and the value of the estimator $\tilde{\tau}_{ord}$ is given with the calculations shown. In the last column, the value of $\tilde{\tau}_{RB}$, the improvement of $\tilde{\tau}_{ord}$ using the Rao–Blackwell method, is shown.

In the first row of the table, the initial sample consists of the first, second, and third units of the population, having y-values 4, 20, and 1. The maximum is $y_{(3)} = 20$, but the neighboring units are already in the sample, so no new units are added. Each of the three units in the sample forms a network by itself. From (6.2), $\tilde{\mu}_{ord}$ is just the average of the three units: we multiply by 6 to get $\tilde{\tau}_{ord}$.

In the second row of the table, the initial sample consists of the units with y-values 4, 20, and 10. The maximum is $y_{(3)} = 20$, and the neighboring unit with value 1 is added to the sample. The other neighboring unit is already in the sample. Since the value 1 does not exceed the second largest initial sample value $y_{(2)} = 10$, no other units are added. Each of the four units in the sample forms a network by itself. Since the network of the unit with value 1 was not intersected by the initial sample, that unit receives no weight in the estimator $\tilde{\tau}_{ord}$. That is, $f_{is} = 0$ for that unit with this sample; the unit with value 1 is an edge unit. We now average over the three units and multiply by 6.

In the third row of the table, the initial sample consists of the units with y-values 4, 20, and 1000. The maximum is $y_{(3)} = 1000$, and the neighboring units, with values 10 and 100, are added. The value 10 does not exceed the second largest initial sample value $y_{(2)} = 20$, so no new neighbor is added. The two units with values 1000 and 100 form a network in this sample. The unit with value 10, forming a network by itself and not intersected by the initial sample, receives no weight in $\tilde{\tau}_{ord}$. The three values of w_{is} for this sample are $w_{1s} = 4$, $w_{2s} = 20$, and $w_{3s} = 1/2(1000 + 100) = 550$.

Table 6.1. Adaptive Cluster Sampling Based on the Initial Sample Maximum.*

Sample Values	$N\bar{y}_0$	$\tilde{\tau}_{ord}$	$\tilde{\tau}_{RB}$
4, 20, 1;	50	$\frac{6}{3}(4 + 20 + 1) = 50$	50
4, 20, 10; 1	68	$\frac{6}{3}(4 + 20 + 10) = 68$	65
4, 20, 1000; 10, 100	2048	$\frac{6}{3}\left(4 + 20 + \frac{1000+100}{2}\right) = 1148$	1148
4, 20, 100; 10, 1000	248	$\frac{6}{3}\left(4 + 20 + \frac{1000+100}{2}\right) = 1148$	1148
4, 1, 10; 100, 1000	30	$\frac{6}{3}\left(4 + 1 + \frac{10+1000+100}{3}\right) = 750$	750
4, 1, 1000; 10, 100	2010	$\frac{6}{3}\left(4 + 1 + \frac{10+1000+100}{3}\right) = 750$	750
4, 1, 100; 10, 1000	210	$\frac{6}{3}\left(4 + 1 + \frac{10+1000+100}{3}\right) = 750$	750
4, 10, 1000; 100	2028	$\frac{6}{3}\left(4 + 10 + \frac{1000+100}{2}\right) = 1128$	1488
4, 10, 100; 1000	228	$\frac{6}{3}\left(4 + 10 + \frac{1000+100}{2}\right) = 1128$	1488
4, 1000, 100; 10	2208	$\frac{6}{3}(4 + 1000 + 100) = 2208$	1488
20, 1, 10; 4	62	$\frac{6}{3}(20 + 1 + 10) = 62$	65
20, 1, 1000; 10, 100	2042	$\frac{6}{3}\left(20 + 1 + \frac{1000+100}{2}\right) = 1142$	1142
20, 1, 100; 10, 1000	242	$\frac{6}{3}\left(20 + 1 + \frac{1000+100}{2}\right) = 1142$	1142
20, 10, 1000; 100	2060	$\frac{6}{3}\left(20 + 10 + \frac{1000+100}{2}\right) = 1160$	1520
20, 10, 100; 1000	260	$\frac{6}{3}\left(20 + 10 + \frac{1000+100}{2}\right) = 1160$	1520
20, 1000, 100; 10	2240	$\frac{6}{3}(20 + 1000 + 100) = 2240$	1520
1, 10, 1000; 100	2022	$\frac{6}{3}\left(1 + 10 + \frac{1000+100}{2}\right) = 1122$	1482
1, 10, 100; 1000	222	$\frac{6}{3}\left(1 + 10 + \frac{1000+100}{2}\right) = 1122$	1482
1, 1000, 100; 10	2202	$\frac{6}{3}(1 + 1000 + 100) = 2202$	1482
10, 1000, 100;	2220	$\frac{6}{3}(10 + 1000 + 100) = 2220$	2220
Expectation:	1135	1135	1135
Bias:	0	0	0
Variance:	954,975.4	430,095.4	313,454.5
Standard Error:	977.23	655.82	559.87

*Here $n_1 = 3$, for a population of six units with y-values 4, 20, 1, 10, 1000, 100, in which the neighborhood of each unit consists of itself plus adjacent units. The initial sample of three units is selected by simple random sampling without replacement. Units linked by inclusion in the same network are underlined. The bottom lines of the table give expected values, variances, and standard errors for each estimator.

The value of the estimator is

$$\tilde{\tau}_{ord} = (6/3)(4 + 20 + 550) = 1148.$$

The sample of the fourth row of the table, like the sample of the third row, intersects each of the networks having values (4), (20), and (1000, 100) exactly once. Thus, the value of d_f is the same for each of these two samples, and the value of the estimator is the same for each.

In the fifth row of the table, the initial sample consists of the units with values 4, 1, and 10. The maximum is $y_{(3)} = 10$, so the neighboring unit with value 1000 is added. Since 1000 also exceeds the second largest initial sample value $y_{(2)} = 4$, the neighboring unit with value 100 is in turn added. Now the three units with values (10, 1000, 100) form a network.

Variance estimates are given for each sample in Table 6.2. With the sample of row 1, the initial sample variance is $v_1 = 104.33$. Each of the three units in the sample forms a network, and this is the only sample intersecting each once, that is, the only sample with the given value of d_f. Therefore $E[v_1 \mid d_f] = 104.33$ and $\text{var}[\bar{y}_1 \mid d_f] = 0$, so that, by (6.8) and (6.5),

$$\widehat{\text{var}}[\tilde{\tau}_{ord}] = 6(6 - 3)(104.33)/3 = 626.$$

The alternative estimate (6.9) is also 626.

With the sample of row 3, the initial sample mean is $\bar{y}_1 = 341.33$ and the initial sample variance is $v_1 = 325,445.33$. The sample of row 4 gives the same value of d_f and has $\bar{y}_1 = 41.33$ and $v_1 = 2645.33$. Since these are the only two samples with the given value of d_f, $\xi_f = 2$ gives

$$E[v_1 \mid d_f] = (1/2)(325,445.33 + 2645.33) = 164,045.33,$$

$$E[\bar{y}_1 \mid d_f] = (1/2)(341.33 + 41.33) = 191.33 \quad (= \tilde{\mu}_{ord})$$

Table 6.2. Variance Estimators for the Samples in Table 6.1.

Sample Values	$\widehat{\text{var}}[N\bar{y}_0]$	$\widehat{\text{var}}[\tilde{\tau}_{ord}]$	$\widehat{\text{var}}[\tilde{\tau}_{ord}]$	$\widehat{\text{var}}_c(\hat{\tau})$
4, 20, 1;	626	626	626	626
4, 20, 10; 1	392	392	392	392
4, 20, 1000; 10, 100	1,952,672	174,272	579,272	174,272
4, 20, 100; 10, 1000	15,872	174,272	579,272	174,272
4, 1, 10; 100, 1000	126	−129,474	270,126	270,126
4, 1, 1000; 10, 100	1,990,026	−129,474	270,126	270,126
4, 1, 100; 10, 1000	19,026	−129,474	270,126	270,126
4, 10, 1000; 100	1,972,152	184,752	589,752	184,752
4, 10, 100; 1000	17,352	184,752	589,752	184,752
4, 1000, 100; 10	1,811,232	1,811,232	1,811,232	1,811,232
20, 1, 10; 4	542	542	542	542
20, 1, 1000; 10, 100	1,958,762	177,662	582,662	177,662
20, 1, 100; 10, 1000	16,562	177,662	582,662	177,662
20, 10, 1000; 100	1,940,600	167,600	572,600	167,600
20, 10, 100; 1000	14,600	167,600	572,600	167,600
20, 1000, 100; 10	1,776,800	1,776,800	1,776,800	1,776,800
1, 10, 1000; 100	1,978,182	188,082	593,082	188,082
1, 10, 100; 1000	17,982	188,082	593,082	188,082
1, 1000, 100; 10	1,817,802	1,817,802	1,817,802	1,817,802
10, 1000, 100;	1,798,200	1,798,200	1,798,200	1,798,200
Expectation:	954,975.4	430,095.4	692,535.4	490,035.4
Bias:	0	0	262,440	59,940
$\sqrt{\text{MSE}}$:	946,469	694,450	647,486	662,670

and

$$\text{var}[\bar{y}_1 \mid d_f] = 1/2[(341.33 - 191.33)^2 + (41.33 - 191.33)^2] = 22{,}500.$$

The unbiased variance estimate with either the sample of row 3 or that of row 4 is

$$\widehat{\text{var}}[\tilde{\tau}_{ord}] = [6(6 - 3)(164{,}045.33)/3] - 6^2(22{,}500)$$
$$= 984{,}272 - 810{,}000 = 174{,}242.$$

Also

$$\widehat{\text{var}}[\tilde{\tau}_{ord}] = 6(6 - 3)(96{,}545.33)/3 = 579{,}272$$

where 96,545.33 is the sample variance of the three w-values 4, 20, and 550.

With the sample of row 5 (and the following two samples consisting of the same final set of units), $\widehat{\text{var}}[\tilde{\tau}_{ord}]$ is negative so we use $\widehat{\text{var}}[\tilde{\tau}_{ord}] = 270{,}126$. For the remaining samples, $\widehat{\text{var}}[\tilde{\tau}_{ord}]$ is positive.

To illustrate the Rao–Blackwell estimator, consider the samples of row 2 and row 11. In row 2, the initial sample contained the units with values 4, 20, and 10, to which the unit with value 1 was added. The value 1 receives no weight and $\tilde{\tau}_{ord} = 68$. In row 11, the initial sample contained the units with values 20, 1, and 10, to which the unit with value 4 was added. The value 4 receives no weight and $\tilde{\tau}_{ord} = 62$. Both these samples contain exactly the same set of four distinct units, so that the value of the minimal sufficient statistic D_R is the same for each of the two samples. These are the only two samples with the given value of D_R so that $\xi = 2$. Therefore

$$\tilde{\tau}_{RB} = (1/2)(68 + 62) = 65.$$

Note that the more efficient Rao–Blackwell estimator does give some weight to the units with values 1 and 4. In the table, the distinct values of $\tilde{\tau}_{ord}$ identify distinct values of D_f, while distinct values of $\tilde{\tau}_{RB}$ identify distinct values of D_R.

The expected value, bias, variance, and standard error of each estimator are given at the bottom of Table 6.1. Each of the estimators is unbiased. As expected, the estimator $\tilde{\tau}_{ord}$ has lower variance than the expanded initial sample mean $N\bar{y}_1$ and the improved estimator $\tilde{\tau}_{RB}$ has lower variance still. The expected final sample size with the adaptive design is 4.25, so that another comparison could be made with the expansion estimator for a simple random sample of hypothetical sample size $E[\nu] = 4.25$ using the variance formula $N(N - E[\nu])\sigma^2/E[\nu]$, where $\sigma^2 = \sum_{i=1}^{N}(y_i - \mu)^2/(N - 1)$ is the finite population variance. The comparison for this example gives a variance of $6(6 - 4.25)(159{,}162.6)/4.25 = 393{,}225$, which is less than that of $\tilde{\tau}_{ord}$ but greater than that of $\tilde{\tau}_{RB}$.

We finally note that if the alternative variance estimator $\widehat{\text{var}}[\tilde{\tau}_{ord}]$ is used only when the unbiased estimator is negative, then the bias for this "combined" estimator $\widehat{\text{var}}_c[\hat{\mu}_{ord}]$ is found to be about 14% and the mean-square error is 662,670. However, if the alternative estimator $\widetilde{\text{var}}[\hat{\mu}_{ord}]$ is always used, then the bias is about 25% but the mean-square error is now 647,486. For further details see Thompson [1995].

Example 6.2 Two Largest Order Statistics. In this example, the same population and initial design are used as in Example 6.1, but the additional sampling is carried out in the neighborhoods of both of the top two order statistics of the initial sample. Following the selection and observation of the initial sample, units in the neighborhood of the units with the two largest y-values—that is, the top two order statistics, $y_{(2)}$ and $y_{(3)}$—are added to the sample, so that $r = 1$. If, in turn, the y-value of any added unit exceeds $y_{(1)}$, the third largest value of the initial sample, then any

Table 6.3. Adaptive Cluster Sampling Based on the Top Two Order Statistics.

Sample Values	$N\bar{y}_1$	$\tilde{\tau}_{ord}$	$\tilde{\tau}_{RB}$
4, 20, 1;	50	$\frac{6}{3}\left[2\left(\frac{4+20}{2}\right)+1\right]=50$	50
4, 20, 10; 1000, 100, 1	68	$\frac{6}{3}\left(4+20+\frac{10+1000+100}{3}\right)=788$	843.6
4, 20, 1000; 10, 100, 1	2048	$\frac{6}{3}\left(4+20+\frac{10+1000+100}{3}\right)=788$	843.6
4, 20, 100; 10, 1000, 1	248	$\frac{6}{3}\left(4+20+\frac{10+1000+100}{3}\right)=788$	843.6
4, 1, 10; 20, 100, 1000	30	$\frac{6}{3}\left(\frac{4+20}{2}+1+\frac{10+1000+100}{3}\right)=766$	843.6
4, 1, 1000; 20, 10, 100	2010	$\frac{6}{3}\left(\frac{4+20}{2}+1+\frac{10+1000+100}{3}\right)=766$	843.6
4, 1, 100; 20, 10, 1000	210	$\frac{6}{3}\left(\frac{4+20}{2}+1+\frac{10+1000+100}{3}\right)=766$	843.6
4, 10, 1000; 100, 1	2028	$\frac{6}{3}\left[4+2\left(\frac{10+1000+100}{3}\right)\right]=1488$	1488
4, 10, 100; 1000, 1	228	$\frac{6}{3}\left[4+2\left(\frac{10+1000+100}{3}\right)\right]=1488$	1488
4, 1000, 100; 10, 1	2208	$\frac{6}{3}\left[4+2\left(\frac{10+1000+100}{3}\right)\right]=1488$	1488
20, 1, 10; 4, 1000, 100	62	$\frac{6}{3}\left(\frac{4+20}{2}+1+\frac{10+1000+100}{3}\right)=766$	843.6
20, 1, 1000; 4, 10, 100	2042	$\frac{6}{3}\left(\frac{4+20}{2}+1+\frac{10+1000+100}{3}\right)=766$	843.6
20, 1, 100; 4, 10, 1000	242	$\frac{6}{3}\left(\frac{4+20}{2}+1+\frac{10+1000+100}{3}\right)=766$	843.6
20, 10, 1000; 4, 100, 1	2060	$\frac{6}{3}\left(20+10+\frac{1000+100}{2}\right)=1160$	843.6
20, 10, 100; 4, 1000, 1	260	$\frac{6}{3}\left(20+10+\frac{1000+100}{2}\right)=1160$	843.6
20, 1000, 100; 10	2240	$\frac{6}{3}\left[20+2\left(\frac{1000+100}{2}\right)\right]=2240$	2240
1, 10, 1000; 100	2022	$\frac{6}{3}\left[1+2\left(\frac{10+1000+100}{3}\right)\right]=1482$	1482
1, 10, 100; 1000	222	$\frac{6}{3}\left[1+2\left(\frac{10+1000+100}{3}\right)\right]=1482$	1482
1, 1000, 100; 10	2202	$\frac{6}{3}\left[1+2\left(\frac{10+1000+100}{3}\right)\right]=1482$	1482
10, 1000, 100;	2220	$\frac{6}{3}\left[10+2\left(\frac{1000+100}{2}\right)\right]=2220$	2220
Expectation:	1135	1135	1135
Bias:	0	0	0
Variance:	954,975.4	274,498.6	262,217.47
Standard Error:	977.23	523.93	512.07

Here $n_1 = 3$, for a population of six units with y-values 4, 20, 1, 10, 1000, 100, in which the neighborhood of each unit consists of itself plus adjacent units. The initial sample of three units is selected by simple random sampling without replacement. The bottom lines of the table give expected values, variances, and standard errors for each estimator.

units in the neighborhood of that added unit are also added, and so on. The 20 possible samples, with estimates and calculations for each, are listed in Table 6.3.

Note that for some samples, the units with values (4, 20) form a network, whereas in other samples, each of those is in a network by itself. The units with values (10, 1000, 100) form a network in some samples, whereas in other samples, the units with values (1000, 100) form a network whereas the one with value 10 forms a network by itself. Intersection of a network more than once by the initial sample is reflected in the coefficients in the calculations, as in the first row of the table in which the factor 2 before the parentheses in the calculation indicates that the network with values (4, 20) was intersected twice. Note also that for some samples, every unit in the population is included yet the estimator does not use the true population total. This is a price of unbiasedness, compensating for other samples giving too low an estimate. In practice, with a census sample anyone would use the true value, but the procedure would then not be precisely unbiased over every possible sample. In fact, the problem is due to the artificial smallness for illustrative purposes of the population relative to the sample size. In realistic situations, the sample would be very much smaller than the whole population.

CHAPTER 7

Adaptive Allocation in Stratified Sampling

7.1 INTRODUCTION

In conventional stratified random sampling, the population is partitioned into regions or strata and a simple random sample is selected in each stratum, with the selections in one stratum being independent of selections in every other. To obtain the best estimate of the population total with a given total sample size or total survey cost, or to achieve a desired precision with minimum cost, optimal allocation of sample size among the strata involves using larger sample sizes in strata that are larger, more variable, and less costly to sample.

In practice, one may not have good prior knowledge of the stratum variances. In such cases, it is natural to consider computing sample variances from an initial part of the stratified sample and using those estimates to adaptively allocate the remaining sample size among the strata. Alternatively, the remaining sample sizes may be based on initial sample means rather than sample variances, since with many natural populations high means are associated with high variances. Designs such as these are referred to as *adaptive allocation* designs. The standard stratified sampling estimator gives an unbiased estimate of the population total with conventional stratified random sampling but is not in general unbiased with adaptive allocation designs.

In this chapter, we first describe designs that approximate the optimal allocation based on prior knowledge by using sample variances or means from initial phases of the survey. We then examine unbiased estimation methods based on the Rao–Blackwell method and on fixed-weight strategies. After this we briefly consider model-based methods in adaptive allocation. This is followed by consideration of adaptive allocation strategies with sample sizes based on previous strata, with which the classical estimator is design or model unbiased.

The literature on adaptive stratification and adaptive allocation is quite extensive. A wide variety of strategies in which the population is stratified adaptively or the allocation to strata is adaptive is reviewed in Solomon and Zacks [1970]. Related issues are dealt with in the sequential statistics literature (cf. Wald [1947] and Siegmund [1985]) and in the literature on sequential allocation in experiments (cf. Robbins

[1952] and Barry and Fristedt [1985]), though the presence of unit labels as part of the sampling data and the combination of design-based and model-based methods offer a wider set of possibilities than has usually been associated with those topics. More recent studies are cited throughout this chapter.

7.2 SAMPLE SIZES BASED ON INITIAL OBSERVATIONS IN EACH STRATUM

7.2.1 Approximately Optimal Allocation

In conventional stratified random sampling, simple random samples of fixed sizes $n_1, n_2, \ldots, n_H$ are selected independently in each of the H strata. An estimate unbiased for the population total $\tau = \sum_{h=1}^{H} \sum_{i=1}^{N_h} y_{hi}$ under the conventional design is provided by

$$\hat{\tau}_{st} = \sum_{h=1}^{H} N_h \bar{y}_h,$$

where N_h is the number of units in stratum h and $\bar{y}_h = \sum_{i=1}^{n_h} y_{hi}/n_h$ is the sample mean in stratum h. The variance of $\hat{\tau}_{st}$ is

$$\text{var}[\hat{\tau}_{st}] = \sum_{h=1}^{H} N_h^2 \left(\frac{N_h - n_h}{N_h n_h} \right) \sigma_h^2,$$

where $\sigma_h^2 = \sum_{i=1}^{N_h} (y_{hi} - \mu_h)^2/(N_h - 1)$ is the population variance in stratum h, and $\mu_h = \sum_{i=1}^{N_h} y_{hi}/n_h$ is the population mean in stratum h. An unbiased estimate of the variance of $\hat{\tau}_{st}$ is

$$\widehat{\text{var}}[\hat{\tau}_{st}] = \sum_{h=1}^{H} N_h^2 \left(\frac{N_h - n_h}{N_h n_h} \right) s_h^2,$$

where $s_h^2 = \sum_{i=1}^{n_h} (y_{hi} - \bar{y}_h)^2/(n_h - 1)$ is the sample variance in stratum h.

For a fixed total sample size $n = \sum_{h=1}^{H} n_h$, the smallest variance of $\hat{\tau}_{st}$ is obtained by allocating to stratum h sample size

$$n_h = n \frac{N_h \sigma_h}{\sum_{h'=1}^{H} N_{h'} \sigma_{h'}}, \tag{7.1}$$

for $h = 1, \ldots, H$. This is the well-known optimal or Neyman allocation formula (cf. Cochran [1977, p. 97] and Thompson [1992]).

Since the optimal allocation formula depends on population variances that are generally unknown, practical recommendations include substituting sample variances from past survey data or a pilot survey. Another approach that is natural to consider is to do the survey in phases, using variance estimates from initial stages to adaptively allocate sample size in subsequent stages, thus seeking to approximate the optimal allocation based on prior knowledge of the actual variances. In this section we describe some recent investigations and uses of such procedures.

In Example 1.5 (Section 1.5), we described a two-phase procedure used by Francis [1984] in which the phase-2 sample was carried out in a sequential fashion. The details of the method are now described.

Suppose that n_{h1} phase-1 tows of length l_h are carried out in stratum h yielding fish weights of y_{hj} $(j = 1, 2, \ldots, n_{h1})$ with sample mean $\bar{y}_{h1}$ and sample variance s_{h1}^2. If b is the *effective* fishing width of the net, then the area swept by the net is $a_h = l_h b$. The estimate of the total weight of fish (biomass) in stratum h is $\bar{y}_{h1} A_h / a_h$ with variance estimate [cf.(1.28)] $s_{h1}^2 A_h^2 \left[\frac{1}{n_{h1}} - \frac{1}{N_h} \right] / a_h^2$. Often $1/N_h$ is negligible in the latter expression. If some prior information is available, then the phase-1 sample can be allocated among the strata using an approximate Neyman optimal allocation.

Looking now at the phase-2 tows, we see that if an additional tow was carried out in stratum h, then, using the same estimate s_{h1}^2, the reduction in the estimated variance of the biomass estimate is

$$G_h = \frac{s_{h1}^2 A_h^2}{a_h^2} \left(\frac{1}{n_{h1}} - \frac{1}{n_{h1} + 1} \right) = \frac{s_{h1}^2 A_h^2}{a_h^2 n_{h1}(n_{h1} + 1)}. \tag{7.2}$$

This formula is now used to determine phase-2 allocations sequentially. The first tow of phase 2 is chosen at random in the stratum for which G_h is the greatest. Suppose this is stratum j. Then G_j is recalculated as $s_{j1}^2 A_j^2 / a_j^2 (n_{j1} + 1)(n_{j1} + 2)$. The next tow is selected in the stratum for which G_h is a maximum, and so on. Since s_{h1}^2 from the phase-1 tows is used throughout, all the second phase tows can be determined before further fishing is carried out. These additional tows can then be planned geographically.

In practice, the tows in a stratum may differ in length because of logistical problems of laying and picking up the net. Let l_{hj} be the length of the jth tow in stratum h. Francis worked with the catch rates (catch weight per unit distance traveled), $c_{hj} = y_{hj}/l_{hj}$, and computed the mean $\bar{c}_{h1}$ and sample variance s_{ch1}^2 for these rates. The biomass estimate for stratum h is the density estimate $\bar{c}_{h1}/b$ times the area, or $\bar{c}_{h1} A_h / b$, with variance estimate approximately $s_{ch1}^2 A_h^2 / b^2 n_{h1}$ for large N_h. Also

$$G_h = \frac{s_{ch1}^2 A_h^2}{b^2 n_{h1}(n_{h1} + 1)}.$$

Francis then combined the phase-1 and phase-2 samples and used the usual stratified estimate for total biomass, namely

$$\hat{B} = \sum_{h=1}^{H} \frac{\bar{c}_h A_h}{b}, \tag{7.3}$$

with variance estimate

$$\widehat{\text{var}}[\hat{B}] = \sum_{h=1}^{H} \frac{s_{ch}^2 A_h^2}{b^2 n_h}. \tag{7.4}$$

Here $\bar{c}_h$ and s_{ch}^2 are now the corresponding estimates for the total stratum sample size $n_h = n_{h1} + n_{h2}$.

As the procedure is adaptive, (7.3) and (7.4) are biased. Francis carried out some simulation experiments, using parameter values based on data from mackerel and orange roughy surveys, to evaluate the extent of the bias and to compare the relative efficiencies of a conventional and a two-phase design. The estimate $\hat{B}$ was found to be negatively biased. Francis explained the downward bias as follows. In a conventional survey, low catch rates in some strata will tend to be balanced by high catches in others so that $\hat{B}$ is unbiased. However, in a two-phase survey, extra tows are more likely to be added to strata with high catch rates thus lowering the average catch rates in these strata and thereby biasing $\hat{B}$.

Francis also found that the size of the bias in $\hat{B}$ decreases as the ratio $r = n_1/n$ of the total phase-1 sample size to the total sample size increases. This is not surprising as the bias is zero when $r = 1$. However, in contrast to this trend, the efficiency of the two-phase design relative to the conventional design, as measured by the mean absolute bias, goes down as r increases. These opposing trends have to be balanced, and Francis suggests, as a working rule, that $r \approx 0.75$.

Using data from the mackerel and orange roughy surveys, Francis fitted a log-normal model and simulated the adaptive and the conventional designs for a variety of sample size ratios r. In the case of mackerel, allocation for the phase-1 sample was proportional to stratum area, while for roughy it was based on commercial catch rate data. However, better results were obtained by using $\bar{c}_{h1}^2$, the sample mean in stratum h, in place of the sample variance s_{ch1}^2 in the allocation formula (7.2). This was because, with the fish data, standard deviations tended to be proportional to means and sample means were more stable than sample variances. In all cases, the adaptive strategy gave a lower mean absolute error than the the conventional stratified design with the same total number of units. Mean absolute error was used rather than mean-square error in order to have a measure less sensitive to the skewness of the fish abundance distributions. Relative efficiency of the adaptive allocation strategy, measured as the ratio of mean-absolute errors, ranged up to 127% for mackerel and 180% for orange roughy.

Francis notes that the two-phase design gives much more flexibility in determining the total sample size n. In a conventional design, all n tows are initially allocated to strata and the strata will be visited in an order determined by location. However, if there is bad weather or problems with the gear, some strata will end up being under-sampled or not sampled at all. In contrast, a two-phase design can be open ended. Once the first phase is complete there are no serious problems if not all the allocated phase-2 tows are made. Furthermore, the procedure for allocating the phase-2 tows also gives an estimate of the gains to be made in each stratum that can be taken into account if the vessel is running out of time. On the other hand, if extra time is available, the two-phase design provides a rational method for allocating further tows. For further details and applications to orange roughy and mackerel populations, the reader is referred to Francis [1984].

It should be noted that the catch rate c_{hj}/b per unit area swept is the ratio of two random variables—biomass y_{hj} and area swept $a_{hj} = bl_{hj}$—so that a ratio estimator can also be used. In this case, it is more convenient to work with density so that we define $D_h = B_h/A_h$ to be the biomass density in stratum h and $D = B/A$ to be the overall population density. Then a (biased) estimate of D_h is the ratio estimate (cf. Cochran [1977, p. 153])

$$\hat{D}_h = \frac{\sum_{j=1}^{n_h} y_{hj}}{\sum_{j=1}^{n_h} bl_{hj}} = \frac{\bar{y}_h}{\bar{a}_h}, \tag{7.5}$$

where $\bar{y}_h$ and $\bar{a}_h$ are the sample means of catch and area swept, respectively, in stratum h. The variance is

$$\text{var}[\hat{D}_h] = \frac{(1 - f_h)}{n_h \mu_{ah}^2} \frac{\sum_{j=1}^{N_h} (y_{hj} - D_h a_{hj})^2}{N_h - 1}$$

$$= \frac{1 - f_h}{n_h} \sigma_{Dh}^2 \quad \text{say,} \tag{7.6}$$

where μ_{ah} is the population mean of the a_{hj}, N_h is the number of possible tows in stratum h, and $f_h = n_h/N_h$ is the sampling fraction in stratum h. In practical situations, f_h may be difficult to determine; fortunately it is usually negligible and can be ignored. The above variance can be estimated by

$$\widehat{\text{var}}[\hat{D}_h] = \frac{(1 - f_h)}{n_h \bar{a}_h^2} \frac{\sum_{j=1}^{n_h} (y_{hj} - \hat{D}_h a_{hj})^2}{n_h - 1}$$

$$= \frac{1 - f_h}{n_h(n_h - 1)} \sum_{j=1}^{n_h} \frac{a_{hj}^2}{\bar{a}_h^2} \left(\frac{y_{hj}}{a_{hj}} - \hat{D}_h \right)^2$$

$$= \frac{1 - f_h}{n_h(n_h - 1)} \sum_{j=1}^{n_h} w_{hj}^2 (d_{hj} - \hat{D}_h)^2$$

$$= \frac{1 - f_h}{n_h} s_{Dh}^2, \quad \text{say,} \tag{7.7}$$

where $d_{hj} = y_{hj}/a_{hj}$, the observed density on the jth unit, and

$$w_{hj} = \frac{a_{hj}}{\bar{a}_h} = \frac{l_{hj}}{\bar{l}_h}.$$

Here $\hat{D}_h$ also takes the form

$$\hat{D}_h = \frac{1}{n_h} \sum_{j=1}^{n_h} w_{hj} d_{hj}. \tag{7.8}$$

These densities can then be combined to give

$$\hat{D} = \frac{1}{A} \sum_{h=1}^{H} A_h \hat{D}_h \tag{7.9}$$

with variance estimate

$$\widehat{\text{var}}[\hat{D}] = \frac{1}{A^2} \sum_{h=1}^{H} A_h^2 \widehat{\text{var}}[\hat{D}_h]. \tag{7.10}$$

The above theory still applies if b is replaced by b_h (i.e., $a_{hj} = b_h l_{hj}$) as, like b, it cancels out of w_{hj}.

Jolly and Hampton [1990] proposed a related two-phase procedure. Their theory is similar to that given for the estimator (7.5) but is based on transect samples of length L_{hj} and width b_h. Thus, $a_{hj} = b_h L_{hj}$ and, as before [cf. (7.6)],

$$\text{var}[\hat{D}_h] = \frac{1 - f_h}{n_h} \sigma_{Dh}^2.$$

To find the optimal allocation, we wish to minimize [cf. (7.9)]

$$\text{var}[\hat{D}] = \frac{1}{A^2} \sum_{h=1}^{H} A_h^2 \frac{(1 - f_h)}{n_h} \sigma_{Dh}^2 \tag{7.11}$$

subject to a fixed expected distance traveled E[d], where

$$E[d] = E\left[\sum_h \sum_j L_{hj}\right] = \sum_h n_h E[\bar{L}_h].$$

The method of Lagrange multipliers gives the solution

$$n_h \propto \frac{A_h \sigma_{Dh}}{\sqrt{E[\bar{L}_h]}}. \tag{7.12}$$

The solution may alternatively be obtained directly from the standard optimal allocation formula incorporating cost (cf. Cochran [1977, Theorem 5.6, p. 97]) by interpreting E[$\bar{L}_h$] as the cost of including a unit in stratum h.

This procedure was used in a two-phase acoustic survey of South African anchovy. In the first phase, the research vessel was outward going, traversing the study region in an eastward direction. The second phase comprised the return, or westward journey. In the first phase, sample transects were allocated to strata based on expected density from past survey data. In the second phase, additional transects were allocated to the strata based on the actual densities found in the first phase.

A total of n transects are selected, where n depends on the time available for sampling. The number n_1 selected for surveying in phase 1 is chosen to be approximately $0.75n_1$. The size n_{h1} of the phase-1 sample in each stratum h is determined by (7.12), with n_h replaced by n_{h1}, where the population standard deviation σ_{Dh} is estimated from a previous survey. After the first-phase survey is carried out, we then have estimates s_{Dh1} [cf. (7.7)] and $\bar{L}_{h1}$ of σ_{Dh} and $E[\bar{L}_h]$, respectively, which we use back in (7.12) to determine an approximately optimal allocation for the whole sample (with the two phases combined), namely

$$
n_h = n \frac{A_h s_{Dh1}/\sqrt{\bar{L}_{h1}}}{\sum_{h'} A_{h'} s_{Dh'1}/\sqrt{\bar{L}_{h'1}}}.
\tag{7.13}
$$

The stratum allocations in the phase-2 sample are then given by $n_{h2} = n_h - n_{h1}$ ($h = 1, 2, \ldots, H$). If $n_h \leq n_{h1}$, then no further sampling is carried out in stratum h and the other n_{h2} are allocated in the proportions as originally intended.

Analysis of the anchovy data indicated that the variance was more nearly proportional to the square of density than to density itself, so that substitution of the first-phase sample mean for the sample standard deviation s_{Dh1} in (7.13) was suggested (Jolly and Hampton [1990]).

Because of the adaptive allocation, the estimates (7.9) and (7.10) will also be biased (Francis [1991]). However, because the major proportion of the sampling is carried out in the phase-1 stage, we can expect the biases to be small. This was noted in the reply by Jolly and Hampton [1991]. However, Jolly [1993] has recently proposed a method for estimating the bias.

For two strata, Francis [1991] showed that the two methods lead to similar results. The relationship is more than coincidental since the Lagrange method with a linear constraint leads to the condition that all partial derivatives of the variance of the estimator with respect to the sample sizes n_h must be equal at the constrained relative minimum. An algorithm for finding such a solution would be to add an increment to n_h of the stratum having the derivative of largest magnitude, thus reducing its magnitude, and continuing until all the derivatives are equal and the total sample size is as desired. Francis' procedure adds a unit increment to the stratum that would result in the most change to the variance. The optimality of each procedure is only approximate not just because variances are being estimated but also because of the discreteness of the sample size variables and the fact that the theoretically optimal sample size for a stratum could be either less than the phase-1 sample size already committed or greater than the size of the stratum.

The above description of the method of Jolly and Hampton [1990] is a simplification of their procedure and does not deal with some of the practicalities involved in carrying out the procedure, such as the transect layout and the transect subsampling that they use.

A multiphase adaptive allocation design is used for surveying moose in interior Alaska (Gasaway et al. [1986]). Moose are surveyed from aircraft flying over selected spatial units. The study region is divided into three strata, representing areas of expected high, medium, and low moose densities. Because estimates of stratum variances obtained prior to the survey are unreliable due to mobility of the animals and other factors, sample variances are computed at the end of each day during the survey, and an optimal allocation for the next day is computed with those estimates.

Sample sizes are initially computed to be proportional to area of the stratum times sample standard deviation from the initial data, divided by the sample mean of the areas of the sample units. Adjustments are then made to provide for required minimum sample sizes, integer-valued sample sizes, and maximum possible sizes. Total sample size is limited by time and other constraints. The adaptive procedure also accommodates unforeseen difficulties with weather and aircraft. An additional aspect to the moose survey is the estimation of and adjustment for visibility bias (imperfect detectability) using a double sampling procedure in which a portion of a subsample of units is searched more intensively. An additional visibility bias correction is based on observations from the air of a known number of marked animals. An adaptive procedure is also used to allocate effort between standard and intensive searches.

7.2.2   Unbiased Estimation with the Rao–Blackwell Method

With adaptive allocation procedures, conventional estimators tend to be biased, as in the previous section. One approach to finding estimators that are design unbiased with an adaptive allocation procedure is to use the Rao–Blackwell method. In this section, we consider a simple two-phase design. In the first phase, a simple random selection of units is made in each stratum. In the second phase, one returns to those strata with, say, the largest y-values and samples more units.

Suppose there are N_h units in stratum h ($h = 1, 2, \ldots, H$), and $N = \sum_h N_h$ units altogether. In phase r ($r = 1, 2$), a simple random sample of size n_{hr} is taken without replacement from stratum h. Let $n_r = \sum_h n_{hr}$. Then the n_{h1} and their sum n_1 are fixed but the n_{h2} are random as they depend on the allocation or stopping rule C. One criterion might be: if $\bar{y}_{h1} > c$, where $\bar{y}_{h1}$ is the sample mean for the phase-1 sample in stratum h, then stratum h undergoes second-phase sampling. Our aim is to estimate each μ_h, the population mean of the y-values for stratum h, and then combine the estimates to give an estimate of μ, the overall mean of the N units.

We note at this stage that if we used just phase-1 samples, then the usual unbiased stratified estimate of μ is

$$\bar{y}_{st1} = \sum_{h=1}^{H} \frac{N_h}{N} \bar{y}_{h1}, \tag{7.14}$$

and an unbiased estimate of $\text{var}[\bar{y}_{st1}]$ is

$$\widehat{\text{var}}[\bar{y}_{st1}] = \sum_{h=1}^{H} \frac{N_h^2}{N^2} s_{h1}^2 \left(\frac{1}{n_{h1}} - \frac{1}{N_h} \right), \tag{7.15}$$

where

$$s_{h1}^2 = \frac{1}{n_{h1} - 1} \sum_{j=1}^{n_{h1}} (y_{hj} - \bar{y}_{h1})^2 \tag{7.16}$$

and y_{hj} is the y-value for the jth unit in stratum h.

Since unit labels within the strata are not used for estimation, we have from Theorem 2.8 in Section 2.7 that $\mathbf{y}_{hR}$, the set of y-values corresponding to the unordered distinct units for the total sample (phase 1 and 2) in stratum h, is equivalent to $\mathbf{y}_{h,rank}$ and is therefore a complete sufficient statistic for μ_h. We note that some of the y-values may be the same. Then, using the unbiased stratum estimates $\bar{y}_{h1}$ and s_{h1}^2, we can apply the Rao–Blackwell method to construct the minimum variance unbiased estimators $\hat{\mu}_{hRB} = \mathrm{E}[\bar{y}_{h1} \mid \mathbf{y}_{hR}]$ and $s_{hRB}^2 = \mathrm{E}[s_{h1}^2 \mid \mathbf{y}_{hR}]$ to estimate μ_h and the stratum variance $\sigma_h^2 = \sum_{i=1}^{N_h}(y_{hi} - \mu_h)^2/(N_h - 1)$, respectively, when $n_{h2} > 0$.

In applying the method, it is helpful to simplify the notation by dropping the subscript h and working with just one stratum for the time being. Let $n = n_1 + n_2$ be the total sample size for the stratum and consider the case $n_2 > 0$. Then, for every one of the $G = n!$ permutations of $\mathbf{y}_R$, we can compute the mean $\bar{y}_{1g}$ of the first n_1 units using permutation g ($g = 1, 2, \ldots, G$). Let I_g be an indicator variable taking the value of 1 when C is satisfied, and 0 otherwise. Since all G permutations are equally likely, it follows that the $\xi = \sum_{g=1}^{G} I_g$ permutations satisfying C are also equally likely and

$$\hat{\mu}_{RB} = \mathrm{E}[\bar{y}_1 \mid \mathbf{y}_R]$$

$$= \begin{cases} \dfrac{1}{\xi} \displaystyle\sum_{g=1}^{G} \bar{y}_{1g} I_g, & \xi > 1, \\[2mm] \bar{y}_1, & \xi = 1. \end{cases} \tag{7.17}$$

To find an unbiased estimator of $\mathrm{var}[\hat{\mu}_{RB}]$, we note that (7.17) takes the same form as (4.33) so that the theory of Section 4.5 can be used here. From (4.31) we have

$$\mathrm{var}[\hat{\mu}_{RB}] = \mathrm{var}[\bar{y}_1] - \mathrm{E}\{\mathrm{var}[\bar{y}_1 \mid \mathbf{y}_R]\}, \tag{7.18}$$

and this has an unbiased estimator

$$\widehat{\mathrm{var}}[\hat{\mu}_{RB}] = \mathrm{E}\{\widehat{\mathrm{var}}[\bar{y}_1] \mid \mathbf{y}_R\} - \mathrm{var}[\bar{y}_1 \mid \mathbf{y}_R]$$

$$= \mathrm{E}[s_1^2 \mid \mathbf{y}_R]\left(\frac{1}{n_1} - \frac{1}{N}\right) - \mathrm{var}[\bar{y}_1 \mid \mathbf{y}_R]$$

$$= \begin{cases} \dfrac{1}{\xi} \displaystyle\sum_{g=1}^{\xi}\left\{s_{1g}^2\left(\dfrac{1}{n_1} - \dfrac{1}{N}\right) - \left(\bar{y}_{1g} - \hat{\mu}_{RB}\right)^2\right\}, & \xi > 1, \\[2mm] s_1^2\left(\dfrac{1}{n_1} - \dfrac{1}{N}\right), & \xi = 1, \end{cases} \tag{7.19}$$

where s_{1g}^2 is the value of s_1^2 calculated from the first n_1 observations using the permutation g.

The estimators (7.17) and (7.19) were obtained by Kremers [1987]. The idea of using the Rao–Blackwell method for unbiased estimation following a sequential stopping rule goes back to Blackwell [1947]. Kremers noted that in considering the

$n!$ permutations of the n y-values, the permutations within the first n_1 elements and within the final n_2 elements have no effect on the sampling rule and therefore need not be considered in the calculation of $\hat{\mu}_{RB}$. The $G_0 = \binom{n}{n_1}$ partitions that divide the sample into two sets of sizes n_1 and n_2, not all of which need be distinct, are all that is necessary for the description of the distribution of $\bar{y}_1$ given $\mathbf{y}_R$ and $n_2 > 0$. Since all G_0 partitions are equally likely, the ξ_0 of these partitions which give rise to $\bar{y}_1 > c$ (given $n_2 > 0$) are also equally likely. This means that the ξ combinations of (7.17) contain $n_1!n_2!$ replications of the ξ_0 partitions. We can therefore replace ξ by ξ_0 and G by G_0 in (7.17) and (7.19) as a factor of $n_1!n_2!$ cancels out.

Since second-phase sampling concentrates on strata with higher y-values, we might expect, at first sight, that the mean of the whole sample, $\bar{y}$, will be positively biased. However, the reverse is actually true. To begin with, $\bar{y}$ is the average over *all* the G_0 partitions including those for which $\bar{y}_1 \leq c$, that is,

$$\mathrm{E}[\bar{y}_1 \mid \bar{y}] = \bar{y} \qquad (7.20)$$

since, conditionally, $\bar{y}_1$ is the mean of a simple random sample from a "population" with mean $\bar{y}$. If $n_2 = 0$, then $\bar{y} = \bar{y}_1$. However, if $n_2 > 0$, then, since $\hat{\mu}_{RB}$ is the average over all ξ_0 partitions such that $\bar{y}_{1g} > c$, we must have $\hat{\mu}_{RB} \geq \bar{y}$ with strict inequality if $\xi_0 < G_0$. Thus, $\mu = \mathrm{E}[\hat{\mu}_{RB}] \geq \mathrm{E}[\bar{y}]$, with strict inequality in the usual case of $P(\xi_0 < G_0) \neq 0$, and $\bar{y}$ is negatively biased.

Once estimators have been calculated for each stratum, they can be combined as in (7.14) and (7.15) to give overall estimators

$$\hat{\mu}_{RB} = \sum_{h=1}^{H} \frac{N_h}{N} \hat{\mu}_{hRB} \qquad (7.21)$$

and

$$\widehat{\mathrm{var}}[\hat{\mu}_{RB}] = \sum_{h=1}^{N} \frac{N_h^2}{N^2} \widehat{\mathrm{var}}[\hat{\mu}_{hRB}], \qquad (7.22)$$

using (7.18) and (7.19) for each stratum.

One of the reasons why we chose the rule "sample further if $\bar{y}_{h1} > c$" is that we have in mind the common fisheries situation where stratum variances tend to increase with stratum means. The sampling rule then implies the intuitive notion that we should take further samples in the strata with greater variability.

Example 7.1   Two-Phase Sampling.   The following example is taken from Kremers [1987]. Suppose we have a single stratum with $N = 50$ units and an initial simple random sample of size $n_1 = 2$ is taken. If $\bar{y}_1 > 1.5$, or equivalently *sum* > 3, then a further $n_2 = 6$ units are sampled. Suppose that a final sample yielded y-values in rank order of $\mathbf{y}_{rank} = (1, 1, 2, 4, 6, 8, 9, 9)$. Then the only initial samples of size 2 which are incompatible with this final sample are $\{1, 1\}$ and $\{1, 2\}$, with the latter occurring twice as there are two units with y-value equal to 1. Thus, $G_0 = \binom{8}{2} = 28$ and $\xi_0 = 28 - 3 = 25$.

To find $\hat{\mu}_{RB}$, we have to calculate $\bar{y}_1$ for all the 25 compatible initial samples and then take the average. This can be done by summing the observations for each pair, then summing the 25 compatible values and dividing by 50. Following Kremers we can set the calculations out in an upper triangular table (Table 7.1) showing the sums. From the table we have

$$\hat{\mu}_{RB} = \frac{272}{50} = 5.44,$$

which is greater than the mean of the whole sample $\bar{y} = 5$, as expected. The computation of $\widehat{\text{var}}[\hat{\mu}_{RB}]$ is more complicated and, for this simple example, can be achieved by forming further triangular arrays involving the squares of the sums and the differences for each initial sample [since $\sum(y_i - \bar{y})^2 = \frac{1}{2}(y_1 - y_2)^2$ for samples of size 2]. We have

$$\frac{1}{\xi_0}\left(\frac{1}{n_1} - \frac{1}{N}\right)\sum_{g=1}^{\xi_0} s_{1g}^2 = \frac{1}{25}\frac{48}{100}\frac{670}{2} = 6.4320,$$

since the sum of the squared differences is 670. Also, using the last column of Table 7.1 (where the first element, for example, comes from $355 = 5^2 + 7^2 + 9^2 + 10^2 + 10^2$),

$$\frac{1}{\xi_0}\sum_{g=1}^{\xi_0}\left(\bar{y}_{1g} - \hat{\mu}_{RB}\right)^2 = \frac{1}{\xi_0}\sum_{g=1}^{\xi_0}\bar{y}_{1g}^2 - \hat{\mu}_{RB}^2 = \frac{1}{25}\frac{3282}{4} - 5.44^2 = 3.2264.$$

We finally obtain

$$\widehat{\text{var}}[\hat{\mu}_{RB}] = 6.4320 - 3.2264 = 3.21.$$

Table 7.1. The Sum of the Order Statistics for All Initial Samples of Size 2

			Possible Samples						
1	1	2	4	6	8	9	9	Sum	Squares
1	—	2	3	5	7	9	10	10	41	355
1	—	—	3	5	7	9	10	10	41	355
2	—	—	—	6	8	10	11	11	46	442
4	—	—	—	—	10	12	13	13	48	582
6	—	—	—	—	—	14	15	15	44	646
8	—	—	—	—	—	—	17	17	34	578
9	—	—	—	—	—	—	—	18	18	324
9	—	—	—	—	—	—	—	—	—	—
								Total	272	3282

The same answers were given by Kremers using a much shorter method which involves working with the incompatible pairs, 3 in this case, rather than the 25 compatible pairs. His shortcut method works well here, as most of the pairs are compatible. If the rule was *sum* > 15, then there would only be three compatible pairs, {8, 9} twice and {9.9}, and calculations would be much shorter. In sparse, highly clustered populations, this second type of situation is usually the one that arises. A computer program can be readily written to perform the calculations for any n_1.

Before leaving this example, we note that if we calculate the usual estimate of var[$\bar{y}$] we get

$$s^2 \left(\frac{1}{n} - \frac{1}{N} \right) = 12 \left(\frac{1}{8} - \frac{1}{50} \right) = 1.26,$$

which is a lot less than $\widehat{\text{var}}[\hat{\mu}_{RB}] = 3.21$. There is one occasion, however, for $n_2 > 0$ when $\hat{\mu}_{RB} = \bar{y}$. This occurs when the final sample is such that *every* pair is compatible with the final sample, namely, $\xi_0 = G_0$. □

In applications of the above theory, the main emphasis is usually in estimating density rather than means or totals. If all the units in all the strata have the same area a, and A is the total population area, then the density $D = \tau/A \, (= \tau/Na = \mu/a)$ is estimated by $\hat{D} = \hat{\mu}_{RB}/a$ with unbiased variance estimate $\widehat{\text{var}}[\hat{D}] = (1/a^2)\widehat{\text{var}}[\hat{\mu}_{RB}]$ [cf. (7.21) and (7.22)]. However, if the units in stratum h each have area a_h, then the density estimator for stratum h is $\hat{D}_h = \hat{\mu}_{hRB}/a_h$. Combining these leads to

$$\hat{D} = \sum_{h=1}^{H} \frac{A_h}{A} \hat{D}_h, \tag{7.23}$$

where $A_h \, (= N_h a_h)$ is the area of stratum h. An unbiased estimate of the variance of $\hat{D}$ is

$$\widehat{\text{var}}[\hat{D}] = \frac{1}{A^2} \sum_{h=1}^{H} \frac{A_h^2}{a_h^2} \, \widehat{\text{var}}[\hat{\mu}_{hRB}]. \tag{7.24}$$

An important question is, how does one choose the phase-1 sample? Ignoring costs, the minimum variance (Neyman) allocation is to choose the number of units in stratum h proportional to the product of the total number of units in the stratum and the standard deviation, as in (7.1), namely

$$n_{h1} \propto N_h \sigma_h \tag{7.25}$$

or

$$n_{h1} = n_1 \frac{N_h \sigma_h}{\sum_{h'} N_{h'} \sigma_{h'}}. \tag{7.26}$$

Here n_1 would generally be determined by cost or effort considerations, and σ_h would have to be estimated from prior information such as a pilot or previous sample using (7.16).

7.2.3   Fixed-Weight Strategies

In classical stratified survey methodology (cf. Cochran [1977]), one is advised to base allocation in stratified sampling on a past survey. The new stratified survey is then carried out and the estimator of the population total based on it is unbiased, even though sample size allocation and perhaps stratum boundaries were based on the previous survey. The estimator of the previous survey was, we assume, also an unbiased one. Since both estimators are unbiased, their average is also unbiased (assuming the population has not changed between the two surveys).

The same idea can be applied to a single stratified random survey in two phases (as described above) or, more generally, to a survey in K phases. At the kth phase, a complete stratified random sample is selected, with sample sizes possibly depending on data from previous phases. Then the conventional stratified estimator, call it $\hat{\tau}_k$, based on the data from the kth phase is unbiased for the population total τ. Therefore the weighted average

$$\hat{\tau}_w = \sum_{k=1}^{K} w_k \hat{\tau}_k,$$

with any fixed set of weights for which $\sum_{k=1}^{K} w_k = 1$, is an unbiased estimator of the population total τ.

The key to design unbiasedness of such an estimator is that each of the estimators $\hat{\tau}_k$ is design unbiased and that the weights w_k are fixed in advance and do not depend on observations made during the survey. For an unbiased estimator $\hat{\tau}_k$ at the kth phase, each of the strata need to be sampled at that phase. However, stratum boundaries as well as sample sizes can change from phase to phase, being based on observations in the previous phases of the survey.

Choices for the weights w_k include having them all equal (with $w_k = 1/K$), having w_k proportional to the total sample size planned for phase k, or determining optimal weights to minimize the variance of $\hat{\tau}_w$ under an assumed model for the population.

As an example of a fixed-weight strategy, consider one in which stratified random sampling is used at each phase, with stratum sample sizes at phase k depending on data from phases previous to k. For the kth phase,

$$\hat{\tau}_k = \sum_{h=1}^{H} \frac{N_h}{n_{kh}} \sum_{j=1}^{n_{kh}} y_{khj},$$

is an estimator of τ, where n_{kh} is the sample size used in stratum h at phase k and y_{khj} is the value of the variable of interest for the jth unit in the sample in stratum h and phase k. The fixed-weight estimator for the population total is then

$$
\begin{aligned}
\hat{\tau} &= \sum_{k=1}^{K} w_k \hat{\tau}_k \\
&= \sum_{k=1}^{K} w_k \sum_{h=1}^{H} \frac{N_h}{n_{kh}} \sum_{j=1}^{n_{kh}} y_{khj},
\end{aligned}
$$

which we now show to be unbiased.

Let d_{k-1} denote the data from the first $(k-1)$ phases. Each sample size n_{kh} at the kth phase is a function of d_{k-1}, and the selection of units at the kth phase depends on d_{k-1} only through the sample sizes n_{kh}. Thus, the conditional expectation of $\hat{\tau}_k$ given d_{k-1} is $\mathrm{E}[\hat{\tau}_k \mid d_{k-1}] = \mathrm{E}[\hat{\tau}_k \mid n_{k1}, \ldots, n_{kH}]$. Given the sample sizes n_{kh}, the kth phase estimator $\hat{\tau}_k$ is, under stratified random sampling, unbiased for the population total τ, that is, $\mathrm{E}[\hat{\tau}_k \mid n_{k1}, \ldots, n_{kH}] = \tau$, for $k = 1, \ldots, K$. Therefore the unconditional expectation of $\hat{\tau}_k$ is also τ. The expectation of the weighted average $\hat{\tau}$ is $\mathrm{E}[\hat{\tau}] = \sum_{k=1}^{K} w_k \mathrm{E}[\hat{\tau}_k]$ so long as the weights w_k are fixed constants, not random variables. Since the $\hat{\tau}_k$ are unbiased at each phase and the fixed weights sum to 1, $\mathrm{E}[\hat{\tau}] = \tau \sum_{k=1}^{K} w_k = \tau$, so that the fixed-weight multiphase strategy is unbiased.

The variance of $\hat{\tau}_k$ depends on the sample sizes. Under simple random sampling, the conditional variance for the kth phase is

$$
\mathrm{var}[\hat{\tau}_k \mid n_{k1}, \ldots, n_{kH}] = \sum_{h=1}^{H} N_h^2 \frac{N_h - n_{kh}}{N_h n_{kh}} \sigma_h^2.
$$

The unconditional variance is

$$
\mathrm{var}[\hat{\tau}_k] = \mathrm{E}\left[\mathrm{var}(\hat{\tau}_k \mid n_{k1}, \ldots, n_{kH})\right] + \mathrm{var}\left[\mathrm{E}(\hat{\tau}_k \mid n_{k1}, \ldots, n_{kH})\right].
$$

But $\mathrm{E}(\hat{\tau}_k \mid n_{k1}, \ldots, n_{kH})$ is the constant τ, having variance zero, so that

$$
\mathrm{var}[\hat{\tau}_k] = \mathrm{E}\left[\mathrm{var}(\hat{\tau}_k \mid n_{k1}, \ldots, n_{kH})\right], \tag{7.27}
$$

which, with stratified random sampling, is

$$
\mathrm{E}\left[\mathrm{var}(\hat{\tau}_k \mid n_{k1}, \ldots, n_{kH})\right] = \mathrm{E}\left[\sum_{h=1}^{H} N_h^2 \frac{N_h - n_{kh}}{N_h n_{kh}} \sigma_h^2\right],
$$

where σ_h^2 is the population variance for stratum h.

The covariance of $\hat{\tau}_k$ and $\hat{\tau}_{k'}$, for $k' < k$, is

$$
\mathrm{cov}[\hat{\tau}_k, \hat{\tau}_{k'}] = \mathrm{E}\left[\mathrm{cov}(\hat{\tau}_k, \hat{\tau}_{k'} \mid d_{k'})\right] + \mathrm{cov}\left[\mathrm{E}(\hat{\tau}_k \mid d_{k'}), \mathrm{E}(\hat{\tau}_{k'} \mid d_{k'})\right]. \tag{7.28}
$$

Since $\hat{\tau}_{k'}$ is a constant given all the data $d_{k'}$ up to phase k', the conditional covariance $\text{cov}(\hat{\tau}_k, \hat{\tau}_{k'} \mid d_{k'})$ is zero, so that the first term on the right in (7.28) is zero. Also,

$$\begin{aligned} \text{E}(\hat{\tau}_k \mid d_{k'}) &= \text{E}\big[\text{E}(\hat{\tau}_k \mid d_{k'}, d_{k-1}) \mid d_{k'}\big] \\ &= \text{E}\big[\text{E}(\hat{\tau}_k \mid d_{k-1}) \mid d_{k'}\big] \\ &= \text{E}[\tau \mid d_{k'}] = \tau, \end{aligned}$$

which is constant, so the second term on the right is also zero. Thus, for $k \neq k'$, $\text{cov}[\hat{\tau}_k, \hat{\tau}_{k'}] = 0$. Therefore the variance of the fixed weight estimator $\hat{\tau}$ is

$$\text{var}[\hat{\tau}] = \sum_{k=1}^{K} w_k^2 \text{var}[\hat{\tau}_k]. \tag{7.29}$$

To estimate $\text{var}[\hat{\tau}]$, suppose that for each phase k there is an unbiased estimator of $\text{E}[\text{var}(\hat{\tau}_k \mid n_{k1}, \ldots, n_{kH})]$. With stratified random sampling, such an estimator is provided by

$$\widehat{\text{var}}[\hat{\tau}_k \mid n_{k1}, \ldots, n_{kH}] = \sum_{h=1}^{H} N_h^2 \frac{N_h - n_{kh}}{N_h n_{kh}} s_{kh}^2,$$

where s_{kh}^2 is the sample variance from the kth phase in stratum h. Then by (7.27) and (7.29) an unbiased estimator of $\text{var}[\hat{\tau}]$ is

$$\widehat{\text{var}}[\hat{\tau}] = \sum_{k=1}^{K} w_k^2 \widehat{\text{var}}[\hat{\tau}_k \mid n_{k1}, \ldots, n_{kH}],$$

which, for stratified random sampling, is

$$\widehat{\text{var}}[\hat{\tau}] = \sum_{k=1}^{K} w_k^2 \sum_{h=1}^{H} N_h^2 \frac{N_h - n_{kh}}{N_h n_{kh}} s_{kh}^2.$$

In short, conventional estimators are used for each phase and combined according to the fixed weights.

7.2.4 Model-Based Stratified Sampling

Consider the model for a stratified population in which the Y-values are independent normal random variables, with means and variances depending on stratum membership. That is, Y_{hi} has a normal (μ_h, σ_h^2) distribution, $h = 1, \ldots, H$, $i = 1, \ldots, N_h$. A sample is selected by any design of the type $p(s \mid \mathbf{y}) = p(s \mid \mathbf{y}_s)$, giving ν_h $(= \nu_h(s))$ distinct units in each of the H strata. In Section 3.13, Example 3.8, it was shown that the maximum likelihood predictor of the population total is the classical stratified sampling estimator

$$\hat{Z} = \sum_{h=1}^{L} N_h \bar{y}_h.$$

Under the assumed model, the standard predictor $\hat{Z}$ is a maximum likelihood predictor for the population total whether the design is conventional or adaptive, so long as the design does not depend on Y-values outside the sample or on unknown parameter values. The logical implication of this model-based approach is that the field workers could have quite a bit of discretion in how to allocate the effort during the survey and maximum likelihood prediction would not be affected.

Bayesian models have also been applied to the problem of adaptive allocation. Draper and Guttman [1968] describe two-phase allocation designs assuming normal distributions for the y-values, with prior distributions on the parameters μ and σ^2. Zacks [1970] discusses a two-phase allocation plan for estimating the proportion of defectives in quality control sampling. With prior independent binomial distributions assumed for the numbers of defectives in each stratum, the posterior risk was independent of the observations so that no gain was achieved with adaptive allocation. With prior independent discrete uniform distributions assumed, the posterior risk was lower with an adaptive allocation plan than with conventional allocation, but Zacks did not consider the gain to be substantial. Geiger [1994] describes a Bayes adaptive allocation design for estimating hatchery contribution to commercial salmon fisheries. A prior beta distribution is assumed for the binomial proportion of fish that are of hatchery, as opposed to wild, origin. The posterior distribution is shown to be insensitive to the exact choice of prior distribution. Solomon and Zacks [1970] describe thesis studies by Grosh [1969] that indicate that with a beta prior, the efficiency of adaptive allocation plans was only slightly higher than the optimal conventional allocation.

7.3 SAMPLE SIZES BASED ON OBSERVATIONS FROM PREVIOUS STRATA

7.3.1 Design-Unbiased Estimation

One of the problems with the above methods is that they require two "passes" over the population area. The first pass provides the phase-1 sample, and this determines the phase-2 sampling effort in the second pass. Another approach, that requires only a single pass, proceeds sequentially. Here the strata are sampled one at a time in a predetermined sequential fashion, and the level of sampling in a particular stratum depends on what happens in the previous stratum. By way of illustration, we consider a modification of sampling the shrimp in Example 1.4 of Section 1.5. If less than 50 lbs of shrimp are caught in a given stratum, then a simple random sample of m_1

1-mile tows are made in the next stratum. However, if more than 50 lbs are caught, then m_2 ($m_2 > m_1$) 1-mile tows are made. The process begins with m_1 1-mile tows in the first stratum. We now set up a general scheme which includes this example as a special case. The treatment, based on Thompson et al. [1992], focuses on estimation of density per unit area. The population total in the study region can be estimated by multiplying the estimate of density by the area of the study region.

Adaptive allocation based on previous strata is particularly appropriate for surveys in which strata are defined by time periods. For example, in one method for estimating the number of salmon migrating up a river, fish are counted using observers on the bank. During each time stratum, such as a day, observers count fish for a selection of sample time units, such as 15-minute periods. Increasing observed abundance may indicate the onset of the main part of the salmon run, so that sampling intensity may be increased for the next day.

The notation we use is similar to that given in Section 7.1. The study region is partitioned into H strata of area A_h ($h = 1, 2, \ldots, H$), and there are N_h units each of area a_h in stratum h so that $A_h = N_h a_h$. The total number of units is $N = \sum_h N_h$; and we wish to estimate the population density $D = \tau/A$, where τ is the sum of all the y-values and A is the total population area. Let τ_h be the sum of the y-values in stratum h and let $\mu_h = \tau_h/N_h$. A simple random sample of n_h units is taken from stratum h and y_{hj} ($j = 1, 2, \ldots, n_h$) is the associated y-value (e.g., number of animals or biomass) in unit j of the sample. Let $y_{h.} = \sum_j y_{hj}$ and let $\bar{y}_h = y_{h.}/n_h$, the sample mean for stratum h. The stratum density $D_h = \tau_h/A_h$ ($= \mu_h/a_h$) can be estimated by the sample density $\hat{D}_h = \bar{y}_h/a_h$.

The sampling design is as follows. Take a simple random sample of n_1 units from the first stratum. Next take a simple random sample of n_2 units from the second stratum, but with n_2 depending on $\hat{D}_1$. We continue in this fashion selecting n_h units from stratum h with n_h depending on the observed density $\hat{D}_{h-1}$ in the preceding stratum. The appropriate stratified estimate of D is then

$$\hat{D} = \sum_{h=1}^{H} W_h \hat{D}_h, \qquad (7.30)$$

where $W_h = A_h/A$. If c_h denotes "the set of all $\hat{D}_k$ for $k < h$", then, since c_h determines n_h, it follows from the theory of simple random sampling with fixed n_h that

$$\mathrm{E}[\hat{D}_h \mid c_h] = \frac{1}{a_h}\mathrm{E}[\bar{y}_h \mid n_h] = D_h, \qquad (7.31)$$

which implies

$$\mathrm{E}[\hat{D}_h] = D_h \qquad (7.32)$$

and

$$E[\hat{D}] = \sum_h W_h E[\hat{D}_h]$$

$$= \sum_h W_h D_h \qquad (7.33)$$

$$= \sum_h \frac{A_h D_h}{A}$$

$$= D. \qquad (7.34)$$

Also

$$\text{var}[\hat{D}_h] = E_{c_h}\left[\text{var}(\hat{D}_h \mid c_h)\right] + \text{var}_{c_h}\left[E(\hat{D}_h \mid c_h)\right] \qquad (7.35)$$

with the second term being zero by (7.31). Now for $h < i$, c_i determines $\hat{D}_h$ so that, by (7.31),

$$E\left[(\hat{D}_h - D_h)(\hat{D}_i - D_i) \mid c_i\right] = (\hat{D}_h - D_h)E[\hat{D}_i - D_i \mid c_i]$$

$$= 0.$$

Taking expectations with respect to c_i leads to

$$E\left[(\hat{D}_h - D_h)(\hat{D}_i - D_i)\right] = 0. \qquad (7.36)$$

Hence using (7.33) and (7.34), and applying (7.36) and (7.35), we have

$$\text{var}[\hat{D}] = E\left\{\left[\sum_{h=1}^{H} W_h \hat{D}_h - D)\right]^2\right\}$$

$$= E\left\{\left[\sum_{h=1}^{H} W_h(\hat{D}_h - D)\right]^2\right\}$$

$$= E\left[\sum_{h=1}^{H} W_h^2(\hat{D}_h - D_h)^2\right] + \sum_{h=1}^{H}\sum_{i \neq h} W_h W_i E\left[(\hat{D}_h - D_h)(\hat{D}_i - D_i)\right]$$

$$= \sum_h W_h^2 \text{var}[\hat{D}_h] \qquad (7.37)$$

$$= \sum_h W_h^2 E_{c_h}\left[\text{var}(\hat{D}_h \mid c_h)\right] \qquad (7.38)$$

$$= \sum_h W_h^2 E_{n_h}\left\{\frac{1}{a_h^2}\left[\text{var}(\bar{y}_h \mid n_h)\right]\right\}. \qquad (7.39)$$

Defining

$$s_h^2 = \frac{1}{n_h - 1} \sum_{j=1}^{n_h} (y_{hj} - \bar{y}_h)^2, \tag{7.40}$$

we obtain the usual unbiased estimate of $\text{var}[\bar{y}_h \mid n_h]$ from simple random sampling. Hence

$$\widehat{\text{var}}[\hat{D}] = \sum_h W_h^2 \frac{s_h^2}{a_h^2 n_h} \left(1 - \frac{n_h}{N_h}\right)$$

$$= \sum_h W_h^2 v_h, \quad \text{say}, \tag{7.41}$$

is an unbiased estimate of $\text{var}[\hat{D}]$. This leads to the somewhat surprising result that the usual formulas (7.30) and (7.41) for nonadaptive stratified sampling are still unbiased for the adaptive allocation method.

The preceding theory can be extended. If we now assume that the size a_h of the units in stratum h also depends on c_h, then, from (7.31),

$$E[\hat{D}_h \mid c_h] = E[\hat{D}_h \mid n_h, a_h] = D_h,$$

and $\hat{D}$ is still unbiased. Also $\text{var}[\hat{D}]$ is still given by (7.37), and (7.39) holds if the expectation is with respect to both a_h and n_h. With this change, it then follows from (7.39) and

$$E[v_h] = EE[v_h \mid n_h, a_h]$$

that $\widehat{\text{var}}[\hat{D}]$ of (7.41) is still unbiased.

In applying the above theory to the shrimp example, a possible scheme is the following. Each stratum is divided into a grid of primary plots and a plot is chosen at random. Tows are then made at random points in the plot. The area a_h swept out by the tow is then the length of the tow multiplied by the effective fishing width of the net.

7.3.2 Model-Unbiased Estimation

In some survey situations it may not be practical to carry out the random sampling within each stratum that makes the adaptive allocation procedure of the previous section design unbiased. For example, a centered systematic sample may be desired for reasons of travel efficiency or presumed representativeness. In such cases, model unbiasedness of estimates may be obtained with a minimum of assumptions about the population itself.

An example is given by Example 1.4 where just a single-mile or 1/2-mile tow is used in each stratum. If 50 lbs of shrimp per mile are caught in a stratum, then a single-mile tow is made in the next stratum. If less than 50 lbs are caught, then a single 1/2-mile tow is made. The process begins with a 1-mile tow. One approach to that example is to let $n_h = 1$ for each stratum so that a_h is random, being either one or two units of area. However, an alternative approach is to assume a_h represents a 1/2-mile tow for every stratum so that a_h is now constant but n_h is random, being either 1 or 2. Unfortunately the two 1/2-mile tows making up the 1-mile tow corresponding to $n_h = 2$ are not a simple random sample of two units, as they are contiguous. In developing a theory we adopt the second approach assuming fixed a_h but random n_h.

To get around the problem of nonrandom sampling, we use a model-based approach and develop a model for the spatial distribution of the animals. Here the y_{hj} now refer to numbers of animals. We assume that the population is clustered but that the clustering is due to variations in density from stratum to stratum and not to clustering within a stratum. Such an assumption could hold when the animal "patches" are much larger than a stratum. We therefore treat this model as an "apparent" rather than a "true" contagious model (Seber [1982, p. 488]). With this in mind, we assume that in stratum h we have a spatial Poisson process with intensity (mean per unit area) λ_h. Then, conditional on λ_h and n_h, the y_{hj} $(j = 1, 2, \ldots, N_h)$ are independent Poisson($\lambda_h a_h$), the stratum total τ_h is Poisson($N_h \lambda_h a_h$), and the sample total $y_{h\cdot}$ is Poisson($n_h \lambda_h a_h$), as the sum of independent Poisson variables is also Poisson. Furthermore, the λ_h $(h = 1, 2, \ldots, H)$ are assumed to represent observations from some spatial process with mean λ. For two different strata h and i, λ_h and λ_i are not necessarily independent. However, conditional on λ_h and λ_i, the y_{hj} (and therefore τ_h) are independent of the y_{ik} and τ_i. Also, under this model, the grand total τ and the density $D = \tau/A$ are now random variables.

The sample size n_h for stratum h is again assumed to depend on the observed density in the preceding stratum. However, we now need not assume that the units are randomly selected. They may be systematically or purposively located, as randomness is now imposed by the model.

We shall use the same density estimate $\hat{D}$ as (7.30). From the Poisson assumptions we have

$$E[\hat{D}_h \mid \lambda_h, n_h] = E\left[\frac{y_{h\cdot}}{n_h a_h} \mid \lambda_h, n_h\right] = \lambda_h, \tag{7.42}$$

$$E[\hat{D}_h] = E[\lambda_h] = \lambda$$

and

$$E[\hat{D}] = \sum_h W_h E[\hat{D}_h] = \sum_h W_h \lambda = \lambda.$$

Also

$$E[D_h \mid \lambda_h] = E\left[\frac{\tau_h}{N_h a_h} \mid \lambda_h\right] = \lambda_h \tag{7.43}$$

and

$$E[D] = \lambda.$$

Depending on whether we are estimating the parameter λ or the random variable D we need to determine $\text{var}[\hat{D}]$ or the mean-square error $E[(\hat{D} - D)^2]$, respectively (Thompson and Ramsey [1987]). The latter would seem more appropriate here, and we now proceed to find this mean-square error.

To begin with,

$$\hat{D}_h - D_h = \frac{y_{h\cdot}}{n_h a_h} - \frac{\tau_h}{N_h a_h}$$

$$= \frac{1}{a_h}\left[y_{h\cdot}\left(\frac{1}{n_h} - \frac{1}{N_h}\right) - \frac{\tau_h - y_{h\cdot}}{N_h}\right] \tag{7.44}$$

$$= \sum_{j=1}^{N_h} d_j y_{hj}, \quad \text{say;} \tag{7.45}$$

and from (7.42) and (7.43),

$$E[\hat{D}_h - D_h | \lambda_h, n_h] = 0. \tag{7.46}$$

Also, for fixed λ_h and n_h, $y_{h\cdot}$ and $\tau_h - y_{h\cdot}$ in (7.44) are independent Poisson variables, so that from (7.46)

$$E\left[(\hat{D}_h - D_h)^2 \mid \lambda_h, n_h\right] = \text{var}[\hat{D}_h - D_h \mid \lambda_h, n_h]$$

$$= \frac{1}{a_h^2}\left[\left(\frac{1}{n_h} - \frac{1}{N_h}\right)^2 n_h \lambda_h a_h + \frac{(N_h - n_h)}{N_h^2}\lambda_h a_h\right]$$

$$= \frac{\lambda_h}{a_h}\left[\frac{(N_h - n_h)^2}{n_h N_h^2} + \frac{N_h - n_h}{N_h^2}\right]$$

$$= \frac{\lambda_h}{a_h}\left(\frac{1}{n_h} - \frac{1}{N_h}\right).$$

Hence

$$E[(\hat{D}_h - D_h)^2] = E\left[\frac{\lambda_h}{a_h}\left(\frac{1}{n_h} - \frac{1}{N_h}\right)\right]. \tag{7.47}$$

Now, by (7.45), we have the representation

$$\hat{D}_i - D_i = \sum_{k=1}^{N_i} e_k y_{ik}.$$

Conditional on λ_h, λ_i, n_h, and n_i, the coefficients d_j and e_k are constants (as n_h and n_i are fixed) and $\hat{D}_h - D_h$ and $\hat{D}_i - D_i$ are statistically independent as the y_{hj} and y_{ik}

are statistically independent for fixed λ_h and λ_i. Hence, for $h < i$,

$$
\mathrm{E}\big[(\hat{D}_h - D_h)(\hat{D}_i - D_i) \mid \lambda_h, \lambda_i, n_h, n_i\big]
$$
$$
= \mathrm{E}[\hat{D}_h - D_h \mid \lambda_h, \lambda_i, n_h, n_i]\mathrm{E}[\hat{D}_i - D_i \mid \lambda_h, \lambda_i, n_h, n_i].
$$

Furthermore, as the distribution of $\hat{D}_i - D_i$ is solely determined by λ_i and n_i, we have, by (7.46),

$$
\mathrm{E}[\hat{D}_i - D_i \mid \lambda_h, \lambda_i, n_h, n_i] = \mathrm{E}[\hat{D}_i - D_i \mid \lambda_i, n_i] = 0,
$$

Taking further expectations leads to

$$
\mathrm{E}\big[(\hat{D}_h - D_h)(\hat{D}_i - D_i)\big] = 0, \tag{7.48}
$$

as in (7.36). Using the above equation and arguing as in (7.37) leads to

$$
\mathrm{E}[(\hat{D} - D)^2] = \sum_h W_h^2 \, \mathrm{E}[(\hat{D}_h - D_h)^2]
$$
$$
= \sum_h W_h^2 \, \mathrm{E}\left[\frac{\lambda_h}{a_h}\left(\frac{1}{n_h} - \frac{1}{N_h}\right)\right],
$$

by (7.47). Since $\mathrm{E}[\bar{y}_h \mid \lambda_h, n_h] = a_h\lambda_h$, we have

$$
\mathrm{E}\mathrm{E}\left[\frac{\bar{y}_h}{a_h^2}\left(\frac{1}{n_h} - \frac{1}{N_h}\right) \mid \lambda_h, n_h\right] = \mathrm{E}\left[\frac{\lambda_h}{a_h}\left(\frac{1}{n_h} - \frac{1}{N_h}\right)\right],
$$

and an unbiased estimate of $\mathrm{E}[(\hat{D} - D)^2]$ is

$$
v_1 = \sum_{h=1}^{H} W_h^2 \, \frac{\bar{y}_h}{a_h^2}\left(\frac{1}{n_h} - \frac{1}{N_h}\right). \tag{7.49}
$$

It is of interest to note that, for the Poisson model,

$$
\mathrm{E}[s_h^2 \mid \lambda_h, n_h] = \mathrm{E}[\bar{y}_h \mid \lambda_h, n_h]
$$

so that $\widehat{\mathrm{var}}[\hat{D}]$ of (7.41) is also an unbiased estimate of $\mathrm{E}[(\hat{D} - D)^2]$.

A slightly different approach to the above problem is to regard the stratum totals τ_h (and hence the D_h) as fixed. Then, under the above Poisson assumptions, the joint distribution of the y_{hj} $(j = 1, 2, \ldots, N_h)$ is multinomial with τ_h "trials" and individual probabilities $1/N_h$. The conditional distribution of y_h. given n_h is $\mathrm{Bin}(\tau_h, n_h/N_h)$. With these distributions now replacing the Poisson assumptions, and conditioning on c_h the set of all $\hat{D}_k$ for $k < h$, we have

$$
\mathrm{E}[\hat{D}_h \mid c_h] = \frac{1}{a_h}\mathrm{E}[\bar{y}_h \mid n_h] = \frac{1}{a_h} \cdot \frac{\tau_h}{N_h} = D_h,
$$

as c_h determines n_h, and finally $E[\hat{D}_h] = D_h$. Thus, (7.31) and (7.34) still hold and the following theory there leading up to (7.39) still applies, so that

$$\text{var}[\hat{D}] = \sum_h W_h^2 \, E_{n_h} \left\{ \frac{1}{a_h^2} \left[\text{var}(\bar{y}_h \mid n_h) \right] \right\}.$$

Now, since $y_{h\cdot}$ is conditionally binomial,

$$\text{var}[\bar{y}_h \mid n_h] = \frac{1}{n_h^2} \tau_h \frac{n_h}{N_h} \left(1 - \frac{n_h}{N_h} \right)$$

$$= D_h a_h \left(\frac{1}{n_h} - \frac{1}{N_h} \right), \tag{7.50}$$

so that replacing D_h by its unbiased estimate $\hat{D}_h = \bar{y}_h / a_h$ shows that v_1 of (7.49) is an unbiased estimate of $\text{var}[\hat{D}]$.

Suppose that $n_h > 1$. We now consider (7.40). If $\mathbf{Y} = (y_{h1}, y_{h2}, \dots, y_{hn_h})'$ with mean vector $\mathbf{m}$ and variance covariance matrix $\mathbf{V}$, and $\mathbf{Y}^T \mathbf{A Y}$ is a quadratic form, then (Seber [1977, Theorem 1.7])

$$E[\mathbf{Y}'\mathbf{A Y}] = \text{trace}(\mathbf{A V}) + \mathbf{m}'\mathbf{A m}. \tag{7.51}$$

Here all the elements of $\mathbf{m}$ are equal to τ_h / N_h and, from the properties of the multinomial distribution, the (i, j)th element of $\mathbf{V}$ is $\tau_h (\delta_{ij} N_h - 1)/N_h^2$, where $\delta_{ij} = 1$ when $i = j$ and 0 otherwise. Applying (7.51) to s_h^2 of (7.40) we find that the (i, j)th element of $\mathbf{A}$ is $(\delta_{ij} - n_h^{-1})/(n_h - 1)$ so that $\mathbf{A m} = \mathbf{0}$ and

$$E[s_h^2] = D_h a_h = E[\bar{y}_h].$$

Hence, if each $n_h > 1$, we find that (7.41) is still an unbiased estimate of $\text{var}[\hat{D}]$. Thus, if each $n_h > 1$, we find that under the above assumptions, (7.41) is both a design and model-unbiased estimate of $\text{var}[\hat{D}]$.

CHAPTER 8

Multivariate Aspects of
Adaptive Sampling

8.1 INTRODUCTION

In many of the survey situations in which adaptive cluster sampling could be beneficial, there is more than one variable of interest. For example, when surveying pollution, concentrations of more than one chemical substance are typically measured at each site. With surveys of animal species, the abundances of several species are often recorded for each site. In addition, it may be desirable to base adaptive sampling on one variable, such as a rapid assessment of abundance, in order to estimate the population total of another variable such as actual abundance.

In Chapter 4, where we discussed adaptive cluster sampling, the condition for additional sampling depended on some function of a single variable. However, in the multivariate situation, the condition may depend on any one of the variables or on all of them. For instance, in an environmental survey, extra sampling might be carried out when a key indicator variable exceeds a threshold value, or whenever any one of the variables exceeds its respective threshold value, or when the sum (or other function) of all the variables exceeds a specified value. Similarly in an ecological survey, the condition for additional sampling could be based on just the one species of primary interest or on all of the species. We shall find that design-unbiased estimates exist for the population mean (or total) of each variable irrespective of the nature of the condition. Furthermore, an unbiased estimate of the variance-covariance matrix of these estimates also exists. Not unexpectedly we find that, for a given variable, adaptive cluster sampling works best in terms of relative efficiency with respect to conventional sampling when, for the networks determined by the condition, the within-network variability is high for that variable.

Multivariate considerations in conventional sampling designs involve questions of sample size and survey design when there is more than one variable of interest (see Cochran [1977, pp. 81–82]). Also there is the question of allocation in stratified and two-stage sampling in the multivariate case (Cochran [1977, pp. 119–123], Bethel [1989], and Kokan and Khan [1967]). For adaptive cluster sampling, the results on allocation for conventional stratified sampling apply directly to the initial sample design of stratified adaptive cluster sampling (cf. Thompson [1991b]). The conventional advice that, when each variable has a sporadic distribution concentrated in a small part of the study region, a different sampling design is needed for each variable (Cochran [1977, p.82]) may not need to be adhered to when adaptive cluster sampling is used. With adaptive cluster sampling, where the course of the survey depends on the observed values, it is possible to accommodate in a single survey the sporadic distributions of several variables of interest.

We now introduce some notation. The population again consists of N units but the measurement associated with unit i is now a p-dimensional vector $\mathbf{y}_i = (y_{i1}, y_{i2}, \ldots, y_{ip})'$. Here y_{ij} is the value of the jth variable for unit i, and the population matrix for these "parameters" is

$$
\boldsymbol{\Theta} = \begin{pmatrix} \mathbf{y}_1' \\ \mathbf{y}_2' \\ \vdots \\ \mathbf{y}_N' \end{pmatrix} = \begin{pmatrix} y_{11} & y_{12} & \cdots & y_{1p} \\ y_{21} & y_{22} & \cdots & y_{2p} \\ \vdots & \vdots & \ddots & \vdots \\ y_{N1} & y_{N2} & \cdots & y_{Np} \end{pmatrix}.
$$

In general we wish to estimate some vector function of $\boldsymbol{\Theta}$, say $\boldsymbol{\phi} = (\phi_1, \phi_2, \ldots, \phi_q)'$, using an estimator $\widehat{\boldsymbol{\phi}}$ whose jth component $\widehat{\phi}_j$ is design unbiased for ϕ_j ($j = 1, 2, \ldots, q$). We would also like a design-unbiased estimate $\widehat{\text{cov}}[\widehat{\boldsymbol{\phi}}]$ for the variance-covariance matrix $\text{cov}[\widehat{\boldsymbol{\phi}}]$.

For a loss function in the multivariate sampling situation, one approach has been to use a linear combination of the variances of the q estimators (for reviews of this approach, see Bethel [1989] and Cochran [1977, pp. 121–122]). With this approach, a low value of $\sum_{j=1}^{q} b_j \text{var}[\widehat{\phi}_j]$ is desired, where b_j is a specified weight for the jth variable.

Population characteristics commonly of interest include the vector of population totals

$$
\boldsymbol{\tau} = (\tau_1, \tau_2, \ldots, \tau_p)',
$$

where $\tau_j = \sum_{i=1}^{N} y_{ij}$, or the vector of population means

$$
\boldsymbol{\mu} = \boldsymbol{\tau}/N = (\mu_1, \mu_2, \ldots, \mu_p)'.
$$

8.2   MULTIVARIATE CONDITIONS FOR ADDING NEIGHBORHOODS

The sampling scheme we follow is that described in Section 4.2. An initial sample of n_1 units is selected at random *without* replacement. Whenever a unit in the sample satisfies a specified condition, the units in the neighborhood of that unit are added to the sample. In the multivariate case, the condition is specified by a region C in p-dimensional space. Unit i satisfies the condition if $\mathbf{y}_i \in C$. Examples of such conditions in the multivariate setting include: (1) $y_{i1} > c$, in which case extra sampling is carried out whenever the value of the first variable exceeds some constant c; (2) $y_{ij} > c_j$ for all j, in which case extra sampling is carried out only if each of the variables exceeds a specified value; (3) $y_{ij} > c_j$ for some j, in which case extra sampling is carried out if any one or more of the variables exceed the specified value; and (4) $\sum_{j=1}^{p} y_{ij} > c$, requiring that the sum of the p-variables exceed a specified value. Many other conditions can be specified as functions of the p observed values.

With the adaptive procedure based on one of the variables y_{ij} or on a function of all of the variables in y_i, what is the effect of the adaptive procedure on the estimator $\hat{\tau}_j$ of the population total τ_j for a possibly different variable j? Two aspects of that question will be answered in the following sections.

8.3 DESIGN-UNBIASED ESTIMATION

For the univariate situation described in Chapter 4, we saw that the estimate of the population mean or total does not depend directly on the condition C. The condition simply determines the network structure before any sampling takes place. After the sampling we have two methods of estimation. The first uses the intersection probabilities α_k that the initial sample intersects the kth network. The second uses the numbers of initial intersections which in turn depends on the network means.

Turning to the multivariate case, it is readily shown that the univariate results of Chapter 4 apply to each of the p-variables in turn, even though the condition for extra sampling may depend on other variables (Thompson [1993]). Thus, unbiased estimator and variance estimators for the population mean or total of each variable are obtained as in Chapter 4, though networks are devined by the multivariate condition specified. The new feature that needs to be considered, however, is finding and estimating the covariance of the estimates for any pair of variables so that the full variance–covariance matrix can be estimated.

8.3.1 Estimator Based on Intersection Probabilities

Using the same notation as in Chapter 4, let the K distinct networks in the population be denoted by $B_1, B_2, \ldots, B_K$. The structure of these networks will depend on the condition C. Defining x_k to be the number of units in B_k, we recall that α_k, the probability that the initial sample intersects B_k, is

$$\alpha_k = 1 - \left[\binom{N - x_k}{n_1} \Big/ \binom{N}{n_1} \right].$$

Let y_{kj}^* denote the total of the values of the jth variable of interest in the kth network; that is

$$y_{kj}^* = \sum_{i \in B_k} y_{ij}.$$

Define the indicator variable I_k by $I_k = 1$ if the initial sample intersects network k, and $I_k = 0$ otherwise. From (4.6), an unbiased estimator of μ_j is

$$\hat{\mu}_j = \frac{1}{N} \sum_{k=1}^{K} \frac{y_{kj}^* I_k}{\alpha_k}.$$

The variance of $\hat{\mu}_j$ is given by [cf. (4.10) and (4.8)]

$$\text{var}[\hat{\mu}_j] = \frac{1}{N^2} \sum_{k=1}^{K} \sum_{k'=1}^{K} y_{kj}^* y_{k'j}^* \left(\frac{\alpha_{kk'} - \alpha_k \alpha_{k'}}{\alpha_k \alpha_{k'}} \right),$$

where

$$\alpha_{kk'} = \alpha_k + \alpha_{k'} - 1 + \binom{N - x_k - x_{k'}}{n_1} \bigg/ \binom{N}{n_1}$$

is the probability that the initial sample intersects both networks k and k' and $\alpha_{kk} = \alpha_k$. The covariance of $\hat{\mu}_j$ and $\hat{\mu}_{j'}$, for $j \neq j'$, is

$$\text{cov}[\hat{\mu}_j, \hat{\mu}_{j'}] = \frac{1}{N^2} \sum_{k=1}^{K} \sum_{k'=1}^{K} y_{kj}^* y_{k'j'}^* \left(\frac{\alpha_{kk'} - \alpha_k \alpha_{k'}}{\alpha_k \alpha_{k'}} \right).$$

This follows from the fact that

$$\text{cov}\left[\sum_{k=1}^{K} \frac{y_{kj}^* I_k}{\alpha_k}, \sum_{k'=1}^{K} \frac{y_{k'j'}^* I_{k'}}{\alpha_{k'}} \right] = \sum_{k=1}^{K} \sum_{k'=1}^{K} \frac{y_{kj}^* y_{k'j'}^*}{\alpha_k \alpha_{k'}} \text{cov}[I_k, I_{k'}]$$

and

$$\text{cov}[I_k, I_{k'}] = E[I_k I_{k'}] - E[I_k] E[I_{k'}]$$

$$= \alpha_{kk'} - \alpha_k \alpha_{k'}.$$

An unbiased estimator of $\text{var}[\hat{\mu}_j]$ is

$$\widehat{\text{var}}[\hat{\mu}_j] = \frac{1}{N^2} \sum_{k=1}^{K} \sum_{k'=1}^{K} y_{kj}^* y_{k'j}^* \left(\frac{\alpha_{kk'} - \alpha_k \alpha_{k'}}{\alpha_k \alpha_{k'} \alpha_{kk'}} \right) I_k I_{k'}, \tag{8.1}$$

and an unbiased estimator of $\text{cov}[\hat{\mu}_j, \hat{\mu}_{j'}]$, for $j \neq j'$, is

$$\widehat{\text{cov}}[\hat{\mu}_j, \hat{\mu}_{j'}] = \frac{1}{N^2} \sum_{k=1}^{K} \sum_{k'=1}^{K} y_{kj}^* y_{k'j'}^* \left(\frac{\alpha_{kk'} - \alpha_k \alpha_{k'}}{\alpha_k \alpha_{k'} \alpha_{kk'}} \right) I_k I_{k'}. \tag{8.2}$$

8.3.2 Estimator Based on Network Sizes

Let A_i denote the network that includes unit i, and let m_i be the number of units in that network. Define the new variable w_{ij} to be the average of the values of the jth variable of interest in A_i, that is,

$$w_{ij} = \frac{1}{m_i} \sum_{i' \in A_i} y_{i'j}.$$

From (4.15), a design-unbiased estimator of μ_j is

$$\tilde{\mu}_j = \frac{1}{n_1} \sum_{i=1}^{n_1} w_{ij}. \tag{8.3}$$

The variance of $\tilde{\mu}_j$ is [cf. (4.16)]

$$\text{var}[\tilde{\mu}_j] = \frac{N - n_1}{Nn_1(N - 1)} \sum_{i=1}^{N}(w_{ij} - \mu_j)^2, \tag{8.4}$$

and, from Cochran [1977, p.25], the covariance of $\tilde{\mu}_j$ and $\tilde{\mu}_{j'}$, for $j \neq j'$, is

$$\text{cov}[\tilde{\mu}_j, \tilde{\mu}_{j'}] = \frac{N - n_1}{Nn_1(N - 1)} \sum_{i=1}^{N}(w_{ij} - \mu_j)(w_{ij'} - \mu_{j'}).$$

From (4.17) an unbiased estimator of $\text{var}[\tilde{\mu}_j]$ is

$$\widehat{\text{var}}[\tilde{\mu}_j] = \frac{N - n_1}{Nn_1(n_1 - 1)} \sum_{i=1}^{n_1}(w_{ij} - \tilde{\mu}_j)^2, \tag{8.5}$$

and an unbiased estimator of $\text{cov}[\tilde{\mu}_j, \tilde{\mu}_{j'}]$, for $j \neq j'$, is

$$\widehat{\text{cov}}[\tilde{\mu}_j, \tilde{\mu}_{j'}] = \frac{N - n_1}{Nn_1(n_1 - 1)} \sum_{i=1}^{n_1}(w_{ij} - \tilde{\mu}_j)(w_{ij'} - \tilde{\mu}_{j'}). \tag{8.6}$$

To prove the unbiasedness of (8.6), we define $u_i = w_{ij} + w_{ij'}$ and let $\bar{u} = (1/n_1)\sum_{i=1}^{n_1} u_i$ and $s_u^2 = \sum_{i=1}^{n_1}(u_i - \bar{u})^2/(n_1 - 1)$. Then, from (8.3), we have

$$E[\bar{u}] = E[\tilde{\mu}_j + \tilde{\mu}_{j'}] = \mu_j + \mu_{j'},$$

and, applying the theory of simple random sampling to the u_i,

$$E[s_u^2] = \frac{1}{N - 1} \sum_{i=1}^{N}(u_i - E[\bar{u}])^2. \tag{8.7}$$

This equation may be written in terms of the w_{ij} as

$$E\left\{\frac{1}{n_1 - 1}\sum_{i=1}^{n_1}\left[(w_{ij} - \tilde{\mu}_j) + (w_{ij'} - \tilde{\mu}_{j'})\right]^2\right\}$$

$$= \frac{1}{N - 1}\sum_{i=1}^{N}\left[(w_{ij} - \mu_j) + (w_{ij'} - \mu_{j'})\right]^2.$$

Expanding the quadratic term on each side and using the fact that (8.5) is an unbiased estimator of (8.4) to cancel out the two squared terms, the cross-product terms give

$$E\left[\frac{1}{n_1 - 1}\sum_{i=1}^{n_1}(w_{ij} - \tilde{\mu}_j)(w_{ij'} - \tilde{\mu}_{j'})\right] = \frac{1}{N - 1}\sum_{i=1}^{N}(w_{ij} - \tilde{\mu}_j)(w_{ij'} - \tilde{\mu}_{j'}),$$

thus establishing the unbiasedness of (8.6).

8.3.3 Estimators Based on the Rao–Blackwell Method

Each of the unbiased estimators $\hat{\boldsymbol{\mu}}$ and $\tilde{\boldsymbol{\mu}}$ is in general not a function of the minimal sufficient statistic. As in Chapter 4, the minimal sufficient statistic D_R for Θ is the unordered set of distinct unit labels in the sample together with the corresponding values of the variables of interest. Both estimators $\hat{\boldsymbol{\mu}}$ and $\tilde{\boldsymbol{\mu}}$ depend on the order in which the sample is obtained, treating units in the initial sample differently from units subsequently added. Therefore, as in Section 4.5, each estimator can be improved by taking its conditional expectation given the value of D_R. The improved estimators have the form $\hat{\boldsymbol{\mu}}_{RB} = E[\hat{\boldsymbol{\mu}} \mid D_R = d_R]$ and $\tilde{\boldsymbol{\mu}}_{RB} = E[\tilde{\boldsymbol{\mu}} \mid D_R = d_R]$.

The estimator $\hat{\boldsymbol{\mu}}_{RB}$ is computed by applying the method of Section 4.5 simultaneously to each element $\hat{\mu}_j$ of $\hat{\boldsymbol{\mu}}$. We recall that if ν is the number of distinct units in the final sample, then $G = \binom{\nu}{n_1}$ is the number of possible initial samples of size n_1. These G combinations can be indexed in an arbitrary way by the label g ($g = 1, 2, \ldots, G$). Let $\hat{\mu}_{j,g}$ denote the value of $\hat{\mu}_j$ when the initial sample consists of combination g and let $\widehat{\text{var}}_g[\hat{\mu}_j]$ denote the value of the unbiased estimator $\widehat{\text{var}}[\hat{\mu}_j]$ of (8.5) when using the gth combination. If ξ is the number of combinations that could give rise to d_R (i.e., which are compatible with d_R), then, from (4.36),

$$\hat{\mu}_{j,RB} = E[\hat{\mu}_j \mid d_R]$$

$$= \frac{1}{\xi}\sum_{g=1}^{\xi}\hat{\mu}_{j,g},$$

which, in vector form, can be expressed as

$$\hat{\boldsymbol{\mu}}_{RB} = \frac{1}{\xi}\sum_{g=1}^{\xi}\hat{\boldsymbol{\mu}}_g.$$

An unbiased estimator for $\text{var}[\hat{\mu}_j \mid d_R]$ follows from (4.36), namely

$$\widehat{\text{var}}[\hat{\mu}_{j,RB}] = \frac{1}{\xi} \sum_{g=1}^{\xi} \left\{ \widehat{\text{var}}_g[\hat{\mu}_j] - (\hat{\mu}_{j,g} - \hat{\mu}_{j,RB})^2 \right\}.$$

To find an unbiased estimator of $\text{cov}[\hat{\mu}_{j,RB}, \hat{\mu}_{j',RB}]$, we use the result that, for random variables X and Y,

$$\text{cov}[X, Y] = \text{E}\left[\text{cov}(X, Y \mid d_R)\right] + \text{cov}\left[\text{E}(X \mid d_R), \text{E}(Y \mid d_R)\right]$$

or

$$\text{cov}\left[\text{E}(X \mid d_R), \text{E}(Y \mid d_R)\right] = \text{cov}[X, Y] - \text{E}\left[\text{cov}(X, Y \mid d_R)\right].$$

Thus,

$$\text{cov}[\hat{\mu}_{j,RB}, \hat{\mu}_{j',RB}] = \text{cov}[\hat{\mu}_j, \hat{\mu}_{j'}] - \text{E}\left[\text{cov}(\hat{\mu}_j, \hat{\mu}_{j'} \mid d_R)\right]. \tag{8.8}$$

Since $\text{cov}[\hat{\mu}_j, \hat{\mu}_{j'}]$ is a function of Θ and $\widehat{\text{cov}}[\hat{\mu}_j, \hat{\mu}_{j'}]$ of (8.2) is an unbiased estimator, we can apply the Rao–Blackwell theorem once again to obtain another unbiased estimator with smaller variance, namely

$$\widehat{\text{cov}}_{RB}[\hat{\mu}_j, \hat{\mu}_{j'}] = \text{E}\left[\widehat{\text{cov}}(\hat{\mu}_j, \hat{\mu}_{j'}) \mid d_R\right]$$

$$= \frac{1}{\xi} \sum_{g=1}^{\xi} \widehat{\text{cov}}_g[\hat{\mu}_j, \hat{\mu}_{j'}], \tag{8.9}$$

where $\widehat{\text{cov}}_g$ represents the unbiased estimator calculated using group g. Also

$$\text{cov}[\hat{\mu}_j, \hat{\mu}_{j'} \mid d_R] = \text{E}\{(\hat{\mu}_j - \text{E}[\hat{\mu}_j \mid d_R])(\hat{\mu}_{j'} - \text{E}[\hat{\mu}_{j'} \mid d_R]) \mid d_R\}$$

$$= \text{E}\{(\hat{\mu}_j - \hat{\mu}_{j,RB})(\hat{\mu}_{j'} - \hat{\mu}_{j',RB}) \mid d_R\}$$

$$= \frac{1}{\xi} \sum_{g=1}^{\xi} \left[(\hat{\mu}_{j,g} - \hat{\mu}_{j,RB})(\hat{\mu}_{j',g} - \hat{\mu}_{j',RB})\right], \tag{8.10}$$

is an unbiased estimator of its expected value with respect to D_R. Combining (8.9) and (8.10), we obtain from (8.8) the unbiased estimator

$$\widehat{\text{cov}}[\hat{\mu}_{j,RB}, \hat{\mu}_{j',RB}] = \frac{1}{\xi} \sum_{g=1}^{\xi} \left\{ \widehat{\text{cov}}_g[\hat{\mu}_j, \hat{\mu}_{j'}] - (\hat{\mu}_{j,g} - \hat{\mu}_{j,RB})(\hat{\mu}_{j',g} - \hat{\mu}_{j',RB}) \right\}.$$

8.4   ADAPTIVE CLUSTER SAMPLING COMPARED TO CONVENTIONAL SAMPLING

Given a partition $B_1, \ldots, B_K$ of the population into networks, the total sum of squares for the jth variable can be partitioned into between-network and within-network components as follows:

$$\sum_{i=1}^{N}(y_{ij} - \mu_j)^2 = \sum_{k=1}^{K}\sum_{i \in B_k}(y_{ij} - w_{ij})^2 + \sum_{k=1}^{K}\sum_{i \in B_k}(w_{ij} - \mu_j)^2.$$

This may be rewritten

$$\sum_{i=1}^{N}(y_{ij} - \mu_j)^2 = \sum_{k=1}^{K}\sum_{i \in B_k}(y_{ij} - w_{ij})^2 + \sum_{i=1}^{N}(w_{ij} - \mu_j)^2.$$

Thus, the variance of the estimator $\hat{\mu}_j$ with adaptive cluster sampling may be written [cf. (8.4)]

$$\mathrm{var}[\hat{\mu}_j] = \frac{N - n_1}{Nn_1(N - 1)}\left[\sum_{i=1}^{N}(y_{ij} - \mu_j)^2 - \sum_{k=1}^{K}\sum_{i \in B_k}(y_{ij} - w_{ij})^2\right].$$

With a conventional simple random sample of m units, say, the estimator

$$\bar{y}_j = \frac{1}{m}\sum_{i=1}^{m} y_{ij}$$

is unbiased for μ_j and has variance

$$\mathrm{var}[\bar{y}_j; srs] = \frac{N - m}{Nm(N - 1)}\sum_{i=1}^{N}(y_{ij} - \mu_j)^2.$$

The relative efficiency of a conventional simple random sampling design of sample size m, with the unbiased estimator $\bar{y}_j$, to the adaptive cluster sampling strategy of initial sample size n_1, with the estimator $\tilde{\mu}_j$, is

$$\frac{\mathrm{var}[\tilde{\mu}_j; acs]}{\mathrm{var}[\bar{y}_j; srs]} = \frac{m}{n_1}\left(\frac{N - n_1}{N - m}\right)\left[1 - \frac{\sum_{k=1}^{K}\sum_{i \in B_k}(y_{ij} - w_{ij})^2}{\sum_{i=1}^{N}(y_{ij} - \mu_j)^2}\right].$$

The adaptive cluster sampling strategy will be efficient relative to the conventional strategy if the within-network variation is large relative to the overall variation. The relative efficiency is dependent on the partitioning of the population into networks, which depends on the condition C specified and on the definition of neighborhoods.

For any of the p-variables of interest, efficiency is gained with a partitioning that leads to high within-network variation in the population.

Consider now the approach to multivariate sampling using as the loss function a linear combination of the variances of the p-estimators. Let V_j be the variance of the estimator of μ_j under whichever sampling strategy is used. With this approach, a low value of $\sum_{j=1}^{p} b_j V_j$ is desired, where b_j is the specified weight for the jth variable. Then the relative efficiency of simple random sampling with sample size m to adaptive cluster sampling with initial sample size n_1, in terms of $\sum_{j=1}^{p} b_j V_j$, is

$$\frac{\sum_{j=1}^{p} b_j \mathrm{var}[\hat{\mu}_j; acs]}{\sum_{j=1}^{p} b_j \mathrm{var}[\bar{y}_j; srs]} = \frac{m}{n_1} \left(\frac{N - n_1}{N - m} \right) \left[1 - \frac{\sum_{j=1}^{p} b_j \sum_{k=1}^{K} \sum_{i \in B_k} (y_{ij} - w_{ij})^2}{\sum_{j=1}^{p} b_j \sum_{i=1}^{N} (y_{ij} - \mu_j)^2} \right].$$

In terms of the linear combination of the variances, adaptive cluster sampling is relatively efficient when the weighted sum of the within-cluster sums of squares is high.

8.5 RAPID ASSESSMENT VARIABLES

The preceding results on multivariate variables of interest show that unbiased estimates for any of the variables are obtained even though the condition for adaptive sampling may be based on another of the variables. These results apply equally well to variables that are not directly of interest but are useful in deciding whether to add extra units. For example, in surveys of freshwater mussels, it can be very time consuming to sample a stream plot for mussels, but a rapid assessment of whether mussels are present can be readily made (D. R. Smith, personal communication). Let y_i be the number of mussels in unit (plot) i. Define the rapid assessment variable x_i to be 1 if mussels are assessed to be present in unit i, and 0 otherwise. For plot i in the sample, if $x_i = 1$, units in the neighborhood of unit i are added to the sample. If $x = 1$ for any of the added units, still more units are added, and so on. The design is adaptive cluster sampling, with the object of estimating the population total of the variable y, but with the condition for adaptive sampling based on the variable x. Note that the value of x is known only for units in the sample, not for all units as is usually assumed for auxiliary variables in sampling.

The usual estimators used in adaptive cluster sampling do not use the y-values for edge units, that is, for units not satisfying the condition and not in the initial sample but picked up only at the edge of networks of units satisfying the condition. The addition of such units to the sample adds to the cost of sampling while not contributing to the precision of the estimator. The key to cost savings with the rapid assessment variable x is that when $x_i = 0$ for any adaptively added unit, the more expensive variable of interest y_i does not have to be measured for that unit. That is, the variable of interest does not need to be measured for edge units. However, the variable of interest should be measured accurately for every unit in the initial sample, whatever the value of the rapid assessment variable. An example of a cluster of units

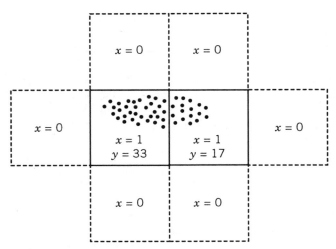

Figure 8.1. Adaptive cluster sampling using a rapid-assessment variable x. The variable of interest is y. The unit with $y = 33$ was included in the initial sample. Whenever $x = 1$, the neighboring units are added.

adaptively added when an initial unit strikes positive assessed abundance is shown in Figure 8.1.

Unbiased estimation of the total of y does not depend on any assumption of the relationship between x and y. For example, it may be that for some plots $y_i > 0$ even though $x_i = 0$, so that the rapid assessment missed the presence of the species. Efficiency of the strategy does rely on some correspondence between x and y, however. In particular, the strategy will be efficient if the variability of y is high within networks defined by x.

CHAPTER 9

Detectability in Adaptive Sampling

9.1 INTRODUCTION

Imperfect detectability is a prevalent source of nonsampling error in many surveys of natural and human populations. Even when a unit (such as a spatial plot) is included in the sample, not all individuals in the selected unit may be detected by the observer. For example, in an aerial survey of moose, a vessel survey of whales, a trawl survey of fish, a feasibility survey for mining diamonds, a survey of artifacts in an archaeological site, or a survey of homeless people, some individuals in the selected units may be missed. To estimate the population total in a survey with imperfect detectability, both the sampling design and the detection probabilities must be taken into account.

General results on dealing with detectability in conventional sampling designs are given in Steinhorst and Samuel [1989], Thompson and Ramsey [1987], and Thompson [1992, Chapter 16]. General detectability methods for adaptive as well as conventional designs were given in Thompson and Seber [1994]. Vast literatures exist in various scientific fields for dealing with problems of visibility, catchability, audibility, and other forms of detectability for specific species or types of objects. Methods for dealing with nonsampling errors such as nonresponse and measurement errors in surveys are discussed in Biemer et al. [1991], Lessler and Kalsbeek [1992], Rubin [1987], Särndal et al. [1992], and elsewhere. In the case of nonresponse, the number of people surveyed is known, but there may be no response from some people (unit nonresponse) or missing information from some of the questionnaire replies (item nonresponse).

In Chapter 2, detectability was briefly discussed as a source of nonsampling errors. In this chapter, available results on adjusting for detectability are simplified, generalized, and improved, and the theory is extended to include adaptive as well as conventional sampling designs. The case in which each object has the same detectability is treated separately (Sections 9.2 and 9.3) from the more general case in which objects have unequal detection probabilities (Sections 9.4 and 9.5).

Imperfect detectability, if not taken into account, will lead to underestimates of the population total. Adjustment for constant detectability consists of dividing an ordinary estimate by the constant probability of detection. In the case of unequal detection probabilities, the method developed below consists of dividing the variable of interest associated with a detected object by the detection probability for that object and then proceeding to use the estimation method that would ordinarily be used under the design if there were no detectability problems. Similarly, for variance and variance estimation of estimators, the ordinary methods that would be used in the absence of detectability problems serve as a starting point.

With conventional sampling designs, imperfect detectability changes what is observed with a given sample. With adaptive designs, the detectability alters not only what is seen with a given sample but it also changes the inclusion probabilities for different possible samples, even affecting which samples are possible under a given design. Because estimators may depend on such inclusion and selection probabilities, establishing results on detectability for adaptive designs offers a special challenge.

In the case of nonadaptive designs with detectability, the mean and variance of an adjusted estimator can be conveniently derived by conditioning on the sample s selected (as in Steinhorst and Samuel [1989] and Thompson [1992]). One takes expectations and variances, first with respect to visibility, given the sample of units, and second with respect to the sampling design. However, for obtaining an estimator of the variance and for the more general class of designs including adaptive sampling designs, the method of conditioning on the sample is not viable. This is because the distribution of the detected numbers given the sample may now depend on the design, and also because the distribution of the detected value in one unit may depend on other units in the sample, making the method very complicated. For the general case, a more useful approach is to reverse the order. We first take expectations and variances with respect to the initial sample conditional on the detectability, and then we take expectations with respect to detectability.

9.2 CONSTANT DETECTABILITY

In this section, the probability of detection g is assumed to be the same for each object in the population. Detections of different objects in any sample of units are assumed to be independent.

Let τ_i be the number of objects in unit i and let y_i be the number of objects detected by the observer in unit i. Using the y_i's from the sample, we wish to estimate $\tau = \sum_{i=1}^{N} \tau_i \, (= N\mu)$, the total number of objects in the study region. Although y_i is, in practice, observed only for units in the sample, it is mathematically convenient to define the vector $\mathbf{y} = (y_1, y_2, \ldots, y_N)'$, thus recognizing the hypothetical possibility of observing every unit in the population. Let $Y = \sum_{i=1}^{N} y_i$ be the total over all units in the population of the hypothetically detected objects.

Suppose that we have an estimator $\hat{\tau}_0$ which would be unbiased under the sampling design for the population total τ if detectability was perfect. Further, suppose that, under perfect detectability, we have an expression V_0 for the variance of $\hat{\tau}_0$ and an unbiased estimate v_0 of this variance. With perfect detectability, the observed value y_i is the actual number τ_i of objects in unit i, and the usual results of finite population sampling apply. Thus, V_0 is a function of the population y-values, while v_0 is a function of the sample y-values.

Now suppose that detectability is not perfect, that is, $g < 1$. Conditional on a realization of the detection vector $\mathbf{y}$, a fixed value y_i is recorded if the unit i is included in the sample. The estimator $\hat{\tau}_0$, applied to the detected y-values, is unbiased under the design for Y, that is, $E[\hat{\tau}_0 \mid \mathbf{y}] = Y$. Similarly, $\text{var}[\hat{\tau}_0 \mid \mathbf{y}] = V_0(\mathbf{y})$, and $E[v_0 \mid \mathbf{y}] = V_0(\mathbf{y})$.

The following results, given by Equations (9.1), (9.2), and (9.3), describe how to adjust for imperfect detectability by dividing the nominal estimate by the probability of detection and give the variance and variance estimate for this adjusted estimator. As we have mentioned above, the starting point for applying the method is to choose an estimator of the population total that would be design unbiased if there were no detectability problems. For the moment, the probability g of detection will be assumed to be known; the effect of estimating g will be considered in the next section. The result applies to both adaptive and nonadaptive designs.

Define the estimator

$$\hat{\tau} = \frac{\hat{\tau}_0}{g}. \tag{9.1}$$

Then, as we shall show below, $\hat{\tau}$ is an unbiased estimator of τ, with variance

$$\text{var}[\hat{\tau}] = \frac{E_{\mathbf{y}}[V_0]}{g^2} + \frac{\tau(1 - g)}{g}, \tag{9.2}$$

where $E_{\mathbf{y}}$ means expectation with respect to $\mathbf{y}$, the "visibility" vector. Also, an unbiased estimator of this variance is

$$\widehat{\text{var}}[\hat{\tau}] = \frac{v_0}{g^2} + \frac{\hat{\tau}(1 - g)}{g}. \tag{9.3}$$

Each of the above results is obtained by conditioning first on $\mathbf{y}$. For unit i, the observed number of objects y_i is a binomial random variable with parameters τ_i and g. Hence $E[y_i] = \tau_i g$, $\text{var}[y_i] = \tau_i g(1 - g)$, and the total Y has expected value $E[Y] = \sum_{i=1}^{N} E[y_i] = \tau g$. Also the y_i are mutually independent since the τ_i are considered fixed and detections of different objects are assumed independent. Hence

$$\text{var}[Y] = \sum_{i=1}^{N} \text{var}[y_i] = \tau g(1 - g).$$

The design-unbiasedness of $\hat{\tau}_0$ with perfect detectability gives $E[\hat{\tau}_0 \mid \mathbf{y}] = Y$. Unconditionally, then, $E[\hat{\tau}_0] = E[Y] = \tau g$. Therefore, $E[\hat{\tau}] = \tau g / g = \tau$, so that $\hat{\tau}$ is unbiased for τ.

The variance of $\hat{\tau}_0$ can be decomposed into

$$\begin{aligned}
\text{var}[\hat{\tau}_0] &= E[\text{var}(\hat{\tau}_0 \mid \mathbf{y})] + \text{var}[E(\hat{\tau}_0 \mid \mathbf{y})] \\
&= E[V_0(\mathbf{y})] + \text{var}[Y] \\
&= E[V_0(\mathbf{y})] + \tau g(1 - g).
\end{aligned} \tag{9.4}$$

Since $\hat{\tau} = \hat{\tau}_0/g$, its variance is $\text{var}[\hat{\tau}] = \text{var}[\hat{\tau}_0]/g^2$, or

$$\text{var}[\hat{\tau}] = \frac{\text{E}_y[V_0]}{g^2} + \frac{\tau(1-g)}{g},$$

which establishes (9.2).

From $\text{E}[v_0 \mid y] = V_0$, it follows that $\text{E}[v_0] = \text{E}_y[V_0]$. Also, since $\hat{\tau}$ is unbiased, $\text{E}[\hat{\tau}g(1-g)] = \tau g(1-g)$. Combining the two terms, (9.4) gives

$$\text{E}[v_0 + \hat{\tau}g(1-g)] = \text{var}[\hat{\tau}_0]$$

and, finally,

$$\text{E}[\widehat{\text{var}}(\hat{\tau})] = \text{E}_y\left[\frac{v_0}{g^2} + \frac{\hat{\tau}(1-g)}{g}\right] = \text{var}[\hat{\tau}],$$

showing that the variance estimator (9.3) is unbiased.

Example 9.1 Simple Random Sampling. Suppose that the population consists of N units or plots and a simple random sample of n of the units is selected. The number y_i of objects in unit i is recorded for each unit in the sample. With perfect detectability, an unbiased estimator of the total number of objects in the population would be

$$\hat{\tau}_0 = N\bar{y},$$

where $\bar{y} = \sum_{i=1}^n y_i/n$ is the sample mean. From familiar sampling results [cf. (1.27)],

$$V_0 = V_0(\mathbf{y}) = \frac{N(N-n)}{n} \frac{\sum_{i=1}^N (y_i - \mu_y)^2}{N-1},$$

where $\mu_y = Y/N$, and its unbiased estimator is

$$v_0 = \frac{N(N-n)}{n} \frac{\sum_{i=1}^n (y_i - \bar{y})^2}{n-1}.$$

Now we suppose that objects in sample units are detected with known probability g. Then, by (9.1),

$$\hat{\tau} = \frac{N\bar{y}}{g},$$

with variance given by (9.2), namely

$$\text{var}[\hat{\tau}] = \frac{N(N-n)}{g^2 n} \text{E}\left[\frac{\sum_{i=1}^N (y_i - \mu_y)^2}{N-1}\right] + \frac{\tau(1-g)}{g}. \tag{9.5}$$

By (9.3), an unbiased estimator of the above variance is

$$\widehat{\text{var}}[\hat{\tau}] = \frac{N(N-n)}{g^2 n} \frac{\sum_{i=1}^{n}(y_i - \bar{y})^2}{n-1} + \frac{\hat{\tau}(1-g)}{g}. \tag{9.6}$$

With a conventional design such as this, the variance of the estimator could alternatively be decomposed (cf. Thompson [1992]) by conditioning on the sample s selected. Thus, using $E[y_i] = \tau_i g$ and $\text{var}[y_i] = \tau_i g(1-g)$, we have

$$\text{var}[\hat{\tau}] = \text{var}\big[E(\hat{\tau}\,|\,s)\big] + E\big[\text{var}(\hat{\tau}\,|\,s)\big] \tag{9.7}$$

$$= \text{var}\left[E\left(\frac{N\bar{y}}{g}\,\Big|\,s\right)\right] + E\left[\text{var}\left(\frac{N\bar{y}}{g}\,\Big|\,s\right)\right]$$

$$= \text{var}_s\left[\frac{N}{n}\sum_{i=1}^{n}\tau_i\right] + E_s\left[\frac{N^2}{g^2 n^2}\sum_{i=1}^{n}\tau_i g(1-g)\right]$$

$$= \frac{N^2}{n}\frac{\sum_{i=1}^{N}(\tau_i - \mu)^2}{N-1}\left(1 - \frac{n}{N}\right) + \frac{N^2}{gn}(1-g)\frac{\sum_{i=1}^{N}\tau_i}{N}$$

$$= \frac{N(N-n)}{n}\frac{\sum_{i=1}^{N}(\tau_i - \mu)^2}{N-1} + \frac{N\tau(1-g)}{gn}, \tag{9.8}$$

where $\mu = \tau/N$. The first term can be interpreted as the sampling variance and the second term as the variance due to imperfect detectability. The second term of (9.8) could be unbiasedly estimated by

$$v_d = \frac{N\hat{\tau}(1-g)}{gn}.$$

An unbiased estimate v_s of the first term of (9.8) can be obtained by subtracting v_d from $\widehat{\text{var}}[\hat{\tau}]$ of (9.6) to give

$$v_s = \frac{N(N-n)}{g^2 n}\frac{\sum_{i=1}^{n}(y_i - \bar{y})^2}{n-1} - \left(\frac{N}{n} - 1\right)\frac{\hat{\tau}(1-g)}{g}.$$

Finally, the expected value of V_0, which is g^2 times the first term of (9.5) [by (9.2)], is obtained by equating (9.5) and (9.8), namely

$$E_y[V_0] = g^2\left[\frac{N(N-n)}{n}\frac{\sum_{i=1}^{N}(\tau_i - \mu)^2}{N-1} + \left(\frac{N}{n} - 1\right)\frac{\tau(1-g)}{g}\right].$$

Example 9.2   Adaptive Cluster Sampling with Primary and Secondary Units.
We now consider an example, depicted in Figure 9.1, that uses the adaptive sampling scheme with primary and secondary units as described in Section 4.7. In the design an initial random sample of $n_1 = 2$ primary units (elongated clusters or strip plots)

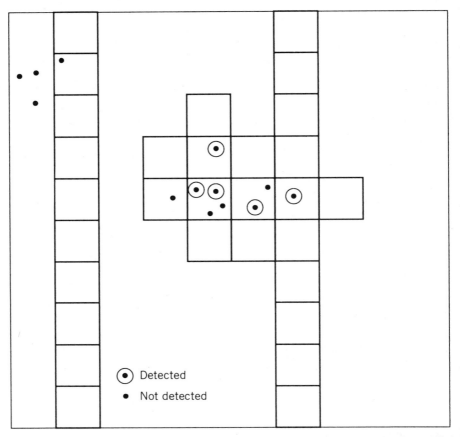

Figure 9.1. An adaptive cluster sample resulting from the initial selection of two strips, when detectability is imperfect. (From S. K. Thompson and G. A. F. Seber, Detectability in conventional and adaptive sampling, 1994, *Biometrics* **50**, 712–724. With permission from the Biometric Society.)

is selected. Each primary unit consists of 10 secondary units (square plots). The total number of primary units in the square study region is $N = 10$. Whenever one or more of the animals is observed, the adjacent secondary units (north, south, east, and west) are added to the sample. Now suppose further that the probability of detecting an animal, given that it is in a unit in the sample, is less than 1.

The study region of Figure 9.1 contains two aggregations of the animals. Each of the two initially selected strips contains one of the animals. In the first (left) sample strip, the animal is not detected by the observers. Thus, no units neighboring that strip are added to the sample and the associated aggregation of animals remains undetected. In the second (right) sample strip, the animal is detected. Then adjacent units are added to the sample, and, when additional animals are detected in some of those units, still more units are added. In all, five of the eight animals in the aggregation are detected.

The complexity of determining the probability that a given unit is included in the sample under an adaptive design with imperfect detectability is illustrated by the figure. Consider the problem of determining the inclusion probability for the unit containing the topmost detected animal of the population (i.e., the unit with one object in the upper center of the figure). That unit was included in the survey of Figure 9.1 because the initial sample intersected the right side of the aggregation *and* the animal on the right side was detected *and* at least one animal was observed in each of the two intervening units. There are other scenarios through which that unit could have been included, for example, by being included in the initial sample. Note also the dependence of detections and samples. For example, given the sample of units shown in the figure, the conditional probability of detection for the object in the second (right) strip is 1.

We now recall the notation and methods of Section 4.7. With imperfect detectability g, networks can be defined only conditionally on the detection vector $\mathbf{y}$. For given $\mathbf{y}$, the networks are distinct and form a partition of the population. Let N be the number of *primary* units in the population and M the number of secondary units in each primary unit. Let K denote the number of (distinct) networks in the population and let y_k^* be the total number of objects *detected* in the kth network. Let x_k be the number of primary units in the population that intersect the kth network. Let κ be the number of distinct networks intersected by the initial sample.

The first adaptive estimator we can use is given by (4.46), namely $\hat{\tau} = \hat{\tau}_0/g$, where

$$\hat{\tau}_0 = MN\hat{\mu}$$

$$= \sum_{k=1}^{\kappa} \frac{y_k^*}{\alpha_k}. \tag{9.9}$$

Here α_k, the probability that the kth network is intersected by the initial sample, is

$$\alpha_k = 1 - \left[\binom{N-x_k}{n_1} \middle/ \binom{N}{n_1} \right],$$

where n_1 is the number of primary units chosen in the initial sample. Then, from (4.47) and (4.48),

$$V_0 = V_0(\mathbf{y}) = \sum_{j=1}^{K} \sum_{k=1}^{K} y_j^* y_k^* \left(\frac{\alpha_{jk} - \alpha_j \alpha_k}{\alpha_j \alpha_k} \right) \tag{9.10}$$

and

$$v_0 = v_0(\mathbf{y}_s) = \sum_{j=1}^{\kappa} \sum_{k=1}^{\kappa} y_j^* y_k^* \left(\frac{\alpha_{jk} - \alpha_j \alpha_k}{\alpha_j \alpha_k \alpha_{jk}} \right), \tag{9.11}$$

where $\mathbf{y}_s$ is the detection vector for the sample. In these expressions, $\alpha_{kk} = \alpha_k$ and the joint intersection probabilities (*conditional* on the detections) are

$$\alpha_{jk} = 1 - \left[\binom{N-x_j}{n_1} + \binom{N-x_k}{n_1} - \binom{N-x_j-x_k+x_{jk}}{n_1}\right] \Bigg/ \binom{N}{n_1},$$

where x_{jk} is the number of primary units that intersect both networks j and k. The variance and estimated variance of $\hat{\tau}$ can then be obtained from (9.2) and (9.3).

The second adaptive estimator we can use is given by (4.51) and (4.52) (but with Mw_i instead of w_i), namely

$$\hat{\tau}_0 = \frac{N}{n_1} \sum_{i=1}^{n_1} w_i = N\bar{w}, \tag{9.12}$$

where

$$w_i = \sum_{k=1}^{\kappa_i} \frac{y_k^*}{x_k}$$

and κ_i is the number of networks that intersect the ith primary unit. Thus, for the ith primary unit selected, a new variable w_i is formed by dividing, for each network intersected, the number of objects detected in that network by the number of primary units that intersect that network and summing over all networks intersected. An unbiased estimate of the population total is given by $\hat{\tau} = \hat{\tau}_0/g$. Although the initial primary units are distinct, a network may be intersected by more than one of the initial primary units, so that repeat values can occur in the estimator.

Noting that with complete detectability $E[\bar{w}] = M\mu = Y/N$, the variance V_0 and variance estimator v_0 are [cf. (4.53) and (4.54) multiplied by N^2]

$$V_0 = \frac{N(N-n_1)}{n_1(N-1)} \sum_{i=1}^{N} \left(w_i - \frac{Y}{N}\right)^2$$

and

$$v_0 = \frac{N(N-n_1)}{n_1(n_1-1)} \sum_{i=1}^{n_1} (w_i - \bar{w})^2. \tag{9.13}$$

By (9.2), $\text{var}[\hat{\tau}] = [E(V_0)/g^2] + \tau(1-g)/g$ and, by (9.3), an unbiased estimate of this variance is $\widehat{\text{var}}[\hat{\tau}] = (v_0/g^2) + \hat{\tau}(1-g)/g$.

In conclusion, we reemphasize that, since the network structure, the numbers x_k of primary units intersecting observed networks, and the intersection probabilities α_k all depend on the pattern of detections, it is advantageous to find means and variances by conditioning first on **y**. The alternative decomposition, which conditions first on the sample s, is more difficult with the adaptive design since the sample selection probabilities depend on the pattern of detections and the detection probabilities given the sample are complex. $\qquad\square$

9.3 ESTIMATED DETECTABILITY

More realistically, suppose the detectability g is not known but a consistent estimator $\hat{g}$ has been obtained from a study independent of the present survey, for example by a method such as double sampling or capture–recapture. Such methods usually also produce an estimate $\widehat{\text{var}}[\hat{g}]$ of the variance $\text{var}[\hat{g}]$. Then a consistent estimator of τ is

$$\tilde{\tau} = \frac{\hat{\tau}_0}{\hat{g}}. \tag{9.14}$$

Using the first term of a Taylor expansion (the so-called delta method) with $E[\hat{\tau}_0] = \tau g$, $\tilde{\tau} = \hat{\tau}_0/g$ and (9.2), an approximate formula for the variance of $\tilde{\tau}$ is (with $\hat{\tau}_0 = \hat{\tau}g$ and $E[\hat{\tau}_0] = \tau g$)

$$\text{var}[\tilde{\tau}] \approx \tau^2 \left(\frac{\text{var}[\hat{\tau}_0]}{(E[\hat{\tau}_0])^2} + \frac{\text{var}[\hat{g}]}{(E[\hat{g}])^2} \right)$$

$$\approx \text{var}[\hat{\tau}] + \frac{\tau^2}{g^2} \text{var}[\hat{g}]$$

$$= \frac{E_y[V_0]}{g^2} + \frac{\tau(1-g)}{g} + \frac{\tau^2}{g^2} \text{var}[\hat{g}]. \tag{9.15}$$

An estimate of this variance is given by

$$\widehat{\text{var}}[\tilde{\tau}] = \frac{v_0}{\hat{g}^2} + \frac{\tilde{\tau}(1-\hat{g})}{\hat{g}} + \frac{\tilde{\tau}^2}{\hat{g}^2} \widehat{\text{var}}[\hat{g}]. \tag{9.16}$$

In some sampling situations, it is more convenient to obtain an estimate (and variance estimate) $\hat{h}$ of $h = 1/g$, the reciprocal of detectability. An exact expression for the variance of a product of two independent random variables $\hat{\tau}_0 \hat{h}$ is available along with an unbiased estimate of the variance (Goodman [1960] and referred to in Seber [1982, p. 9]).

Example 9.2 (***continued***).   We now return to the example described by Figure 9.1. Here each primary unit is a strip consisting of 10 secondary units, and the initial sample consists of $n_1 = 2$ primary units. Suppose that the detectability associated with the survey method has been estimated in an independent experiment (for instance, using double sampling or capture–recapture) to be $\hat{g} = .75$ with $\widehat{\text{var}}[\hat{g}] = .000625$.

Considering the first estimator (9.9), using probabilities of intersection, we have $\alpha_1 = 1 - \binom{10-1}{2} / \binom{10}{2} = 2/10 = .2$ for all the 19 networks of size 1 (with no observed values) that can be intersected by just 1 primary unit. Also, for the network of $y_2^* = 5$ units, which can be intersected by any of $x_k = 3$ primary units, we have $\alpha_2 = 1 - \binom{10-3}{2} / \binom{10}{2} = 1 - 21/45 = .5333$. Thus, $\hat{\tau}_0 = ((0 + \ldots + 0)/.2) + (5/.533) = 9.375$. Taking into account the estimated detectability gives an estimate [by (9.14)] of $\tilde{\tau} = 9.375/.75 = 12.5$, or about 12 or 13 animals in the study region.

For the estimated variance of (9.11) (with only the term y_2^*),

$$v_0 = 5^2(1 - .5333)/.5333^2 = 41.0156,$$

and the estimated variance of $\tilde{\tau}$ is, by (9.16),

$$\widehat{\text{var}}[\tilde{\tau}] = \frac{41.0156}{.75^2} + \frac{12.5(1 - .75)}{.75} + \left(\frac{12.5}{.75}\right)^2 (.000625)$$

$$= 72.92 + 4.17 + 0.17 = 43.23,$$

or a standard error of about 8.8.

Using the second estimator (9.12), we note that for the sample and detected objects shown in Figure 9.1, $w_1 = 0$ for the first (left) primary unit, since no objects were detected from that unit. The second primary unit intersects one observed network ($\kappa_2 = 1$) in which a total of $y_2^* = 5$ objects are detected. As already mentioned, the number of primary units intersecting this observed network is $x_k = 3$. (Note how this depends on the pattern of detections made.) Thus, for the second primary unit, $w_2 = 5/3 = 1.6667$. The estimate using numbers of initial intersections is $\hat{\tau}_0 = (10/2)(0 + 1.6667) = 8.3333$. Taking into account the estimated detectability $\hat{g} = .75$, we have $\tilde{\tau} = 8.3333/.75 = 11.11$, or about 11 animals in the study region. For the estimated variance [cf. (9.13)],

$$v_0 = \frac{10(10 - 2)}{2(2 - 1)}(1.3889) = 55.56,$$

where 1.3889 is the sample variance of the w-values 0 and 1.6667. The estimated variance is, by (9.16),

$$\widehat{\text{var}}[\tilde{\tau}] = \frac{55.56}{.75^2} + \frac{11.11(1 - .75)}{.75} + \left(\frac{11.11}{.75}\right)^2 (.000625)$$

$$= 98.77 + 3.70 + 0.14 = 102.61.$$

The standard error is $\sqrt{102.61} \approx 10.1$.   □

9.4   UNEQUAL DETECTABILITIES AND AN ARBITRARY VARIABLE

In the preceding two sections, the detection probability g was assumed the same for each object in the population. More generally, let g_{ij} be the detection probability for the jth object in the ith unit of the population. As in the preceding sections, the detection probabilities will initially be assumed known. The case in which detectabilities are estimated is considered in the next section.

In Section 9.2, the variable of interest was the number of objects in a unit. More generally, let y_{ij} denote a variable of interest associated with the jth object of the ith unit and let M_i be the number of objects in unit i. We redefine **y** to be

$$\mathbf{y} = (y_{11}, y_{12}, \ldots, y_{1M_1}, \ldots, y_{N1}, \ldots, y_{NM_N})',$$

the population vector of y-values for every object in the population. Let z_{ij} be an indicator random variable equal to 1 if the jth object of the ith unit is detected, and 0 otherwise. Formally, we define the vector

$$\mathbf{z} = (z_{11}, z_{12}, \ldots, z_{1M_1}, \ldots, z_{N1}, \ldots, z_{NM_N})'$$

with an element corresponding to each unit in the population. For object ij, $z_{ij} = 1$ with probability g_{ij} and $z_{ij} = 0$ with probability $1 - g_{ij}$. The z_{ij} are assumed to be independent Bernoulli random variables with $E[z_{ij}] = g_{ij}$ and $\text{var}[z_{ij}] = g_{ij}(1 - g_{ij})$. Let $\tau_i = \sum_{j=1}^{M_i} y_{ij}$ denote the total of the y-values for unit i. The population total is $\tau = \sum_{i=1}^{N} \sum_{j=1}^{M_i} y_{ij} = \sum_{i=1}^{N} \tau_i$.

For estimating the abundance of animals or other objects in the study region (as in the previous sections), the variable of interest y_{ij} equals 1 for each object. Then $y_i = \sum_j y_{ij} z_{ij}$ is the number of the M_i objects detected in unit i. In other surveys, y_{ij} could represent the size of the jth group of animals in the ith plot, the weight of the jth gold nugget in the ith ore unit, or the medical expenses for the jth person in the ith household.

Suppose that for the case in which detectability is perfect we have an estimator $\hat{\tau}_0$ which would be unbiased for τ under the design, with variance V_0, if detectability were perfect. Suppose also that we have an estimator v_0 which is unbiased for V_0. We note that $\hat{\tau}_0$ will be a function of the y_{ij} values in the sample, that is, $\hat{\tau}_0 = \hat{\tau}_0(\mathbf{y}_s)$, where $\mathbf{y}_s = \{y_{ij} : i \in s\}$. Similar comments apply to V_0 and v_0, so that $V_0 = V_0(\mathbf{y})$ and $v_0 = v_0(\mathbf{y}_s)$.

For the case with imperfect detectability of objects, we associate with every object in the population a new variable of interest

$$u_{ij} = \frac{y_{ij} z_{ij}}{g_{ij}}. \tag{9.17}$$

For any object detected ($z_{ij} = 1$), the new variable is formed by dividing the y-value for that object by the detectability of that object. For any object not detected, $u_{ij} = 0$. Conditional on **z**, the u_{ij} are fixed values. Let

$$u_i = \sum_{j=1}^{M_i} u_{ij}$$

denote the total of the u-values for detected objects in unit i. Then, conditional on **z**, the estimator $\hat{\tau}_0$ when applied to the u-variables is design unbiased for the realized population total of the u-variables. Denote the estimator using the sample u-values by $\hat{\tau}$; that is,

$$\hat{\tau} = \hat{\tau}_0(\mathbf{u}_s) \tag{9.18}$$

where $\mathbf{u}_s = \{u_{ij} : i \in s\}$. Then, since $\hat{\tau}_0$ is unbiased for τ,

$$E[\hat{\tau} \mid \mathbf{z}] = \sum_{i=1}^{N} \sum_{j=1}^{M_i} u_{ij} = \sum_{i=1}^{N} \sum_{j=1}^{M_i} \frac{y_{ij} z_{ij}}{g_{ij}}.$$

Unconditionally, the expected value of the estimator $\hat{\tau}$ is

$$E[\hat{\tau}] = \sum_{i=1}^{N} \sum_{j=1}^{M_i} \frac{y_{ij}}{g_{ij}} E[z_{ij}] = \sum_{i=1}^{N} \sum_{j=1}^{M_i} y_{ij} = \tau,$$

so that $\hat{\tau}$ is an unbiased estimator of τ.

For the case with perfect detectability, the variance of $\hat{\tau}_0$ is assumed above to be a known function $V_0(\mathbf{y})$ of the population y-values. Conditional on $\mathbf{z}$, the variance of $\hat{\tau}$ will then be the function V_0 applied to the population u-values; that is, $\mathrm{var}[\hat{\tau} \mid \mathbf{z}] = V$, where $V = V_0(\mathbf{u})$ and

$$\mathbf{u} = (u_{11}, u_{12}, \ldots, u_{1M_1}, \ldots, u_{N1}, \ldots, u_{NM_N})'.$$

The variance of $\hat{\tau}$ can be decomposed by conditioning first on the detection vector $\mathbf{z}$, namely,

$$\mathrm{var}[\hat{\tau}] = E\left[\mathrm{var}(\hat{\tau} \mid \mathbf{z})\right] + \mathrm{var}\left[E(\hat{\tau} \mid \mathbf{z})\right] \tag{9.19}$$

$$= E_{\mathbf{z}}[V] + \mathrm{var}_{\mathbf{z}}\left[\sum_{i=1}^{N} \sum_{j=1}^{M_i} \frac{y_{ij}}{g_{ij}} z_{ij}\right]$$

$$= E_{\mathbf{z}}[V] + \sum_{i=1}^{N} \sum_{j=1}^{M_i} \frac{1 - g_{ij}}{g_{ij}} y_{ij}^2, \tag{9.20}$$

since the z_{ij} are independent Bernoulli variables with variances $g_{ij}(1 - g_{ij})$.

For the perfect detectability case, a design-unbiased estimator $v_0 = v_0(\mathbf{y}_s)$ of $V_0 = V_0(\mathbf{y})$ is assumed to be available. This means that, for any vector of fixed population values, the function v_0 of the sample values is unbiased under the design for the function V_0 of the population values. For the case with imperfect detectability, let v_1 be the function v_0 evaluated with the sample u-values in place of the sample y-values; that is,

$$v_1 = v_0(\mathbf{u}_s).$$

Now conditional on $\mathbf{z}$, $\mathbf{u}$ is a vector of fixed population values. Hence

$$E[v_1 \mid \mathbf{z}] = E[v_0(\mathbf{u}_s) \mid \mathbf{u}] = V_0(\mathbf{u}) = V$$

and

$$E[v_1] = E_z[V].$$

Now for the jth object in the ith unit, define the variable $v_{ij} = [(1 - g_{ij})/g_{ij}]y_{ij}^2$. The sum in the last term on the right in expression (9.20) for $\text{var}[\hat{\tau}]$ is a population total of the v_{ij}. Thus, an unbiased estimator v_2 of this population total is obtained by evaluating the function $\hat{\tau}$ with the sample v_{ij} in place of the sample y_{ij}. This gives

$$v_2 = \hat{\tau}_0(t_s),$$

where

$$t_{ij} = \frac{v_{ij}z_{ij}}{g_{ij}} = \frac{(1 - g_{ij})y_{ij}^2 z_{ij}}{g_{ij}^2}. \tag{9.21}$$

An unbiased estimator of the variance of $\hat{\tau}$ is thus given by

$$\widehat{\text{var}}[\hat{\tau}] = v_1 + v_2. \tag{9.22}$$

Example 9.3 Any Conventional Design with Unequal Detectabilities. With any conventional design in which the inclusion probabilities are known and positive for each unit, and in which there are no nonsampling errors such as detectability, an unbiased estimator of the population total is provided by the Horvitz–Thompson estimator. Let π_i denote the probability that unit i is included in the sample, and let π_{ij} denote the probability that both units i and j are included in the sample. Let ν denote the number of distinct units in the sample. The total of the y-values in unit i is $\tau_i = \sum_{j=1}^{M_i} y_{ij}$. With perfect detectability, an unbiased estimator of τ based on the Horvitz–Thompson estimator would be [cf.(1.18)]

$$\hat{\tau}_0 = \sum_{i=1}^{\nu} \frac{\tau_i}{\pi_i} = \sum_{i=1}^{\nu} \frac{1}{\pi_i} \sum_{j=1}^{M_i} y_{ij},$$

with variance [cf. (1.20)]

$$V_0 = \sum_{i=1}^{N} \sum_{i'=1}^{N} \tau_i \tau_{i'} \left(\frac{\pi_{ii'} - \pi_i \pi_{i'}}{\pi_i \pi_{i'}} \right)$$

and unbiased variance estimator [cf. (1.21)]

$$v_0 = \sum_{i=1}^{\nu} \sum_{i'=1}^{\nu} \tau_i \tau_{i'} \left(\frac{\pi_{ii'} - \pi_i \pi_{i'}}{\pi_{ii'} \pi_i \pi_{i'}} \right),$$

where $\pi_{ii} = \pi_i$.

For notational simplicity in the imperfect detectability case, let $m_i = \sum_{j=1}^{M_i} z_{ij}$ denote the number of objects detected in unit i, so that the sample sum of the u_{ij} for unit i may be written $u_i = \sum_{j=1}^{m_i} y_{ij}/g_{ij}$ instead of the more formal notation with the indicator variable z_{ij}. Then, with probability of detection g_{ij} for the jth object of the ith unit, an unbiased estimator of τ by the results of this section is [cf. (9.18)]

$$\hat{\tau} = \sum_{i=1}^{\nu} \frac{1}{\pi_i} \sum_{j=1}^{M_i} \frac{y_{ij} z_{ij}}{g_{ij}} = \sum_{i=1}^{\nu} \frac{1}{\pi_i} \sum_{j=1}^{m_i} \frac{y_{ij}}{g_{ij}} = \sum_{i=1}^{\nu} \frac{u_i}{\pi_i}, \tag{9.23}$$

with variance given by (9.20), namely

$$\text{var}[\hat{\tau}] = \mathbf{E_z} \left[\sum_{i=1}^{N} \sum_{i'=1}^{N} u_i u_{i'} \left(\frac{\pi_{ii'} - \pi_i \pi_{i'}}{\pi_i \pi_{i'}} \right) \right] + \sum_{i=1}^{N} \sum_{j=1}^{M_i} \frac{1 - g_{ij}}{g_{ij}} y_{ij}^2. \tag{9.24}$$

An unbiased variance estimator of this follows from (9.22), namely

$$\widehat{\text{var}}[\hat{\tau}] = v_1 + v_2, \tag{9.25}$$

where

$$v_1 = v_0(\mathbf{u}_s)$$

$$= \sum_{i=1}^{\nu} \sum_{i'=1}^{\nu} u_i u_{i'} \left(\frac{\pi_{ii'} - \pi_i \pi_{i'}}{\pi_{ii'} \pi_i \pi_{i'}} \right)$$

and

$$v_2 = \hat{\tau}_0(\mathbf{t}_s)$$

$$= \sum_{i=1}^{\nu} \frac{1}{\pi_i} \sum_{j=1}^{m_i} \frac{1 - g_{ij}}{g_{ij}^2} y_{ij}^2.$$

The estimator (9.23) was given by Steinhorst and Samuel [1989, Equation 1], who used the alternative decomposition of the variance (9.7) instead of (9.19) to obtain

$$\text{var}[\hat{\tau}] = V_s + V_d, \tag{9.26}$$

where

$$V_s = \sum_{i=1}^{N} \sum_{i'=1}^{N} \tau_i \tau_{i'} \left(\frac{\pi_{ii'} - \pi_i \pi_{i'}}{\pi_i \pi_{i'}} \right) \tag{9.27}$$

and

$$V_d = \sum_{i=1}^{N} \frac{1}{\pi_i} \sum_{j=1}^{M_i} \frac{1 - g_{ij}}{g_{ij}} y_{ij}^2 = \sum_{i=1}^{N} \frac{1}{\pi_i} \sum_{j=1}^{M_i} v_{ij}. \tag{9.28}$$

Equations (9.24) and (9.26) are equivalent expressions for the variance of $\hat{\tau}$, though the individual terms differ. The alternative decomposition (9.26), in which conditional expectations with respect to the distribution of the detectability vector $\mathbf{z}$ are first taken given a sample s and then unconditional expectations are taken with respect to the sampling design, is feasible here because, with a conventional design, the sampling design probabilities do not depend on the detections (as is the case with an adaptive design). The alternative decomposition (9.26) may be useful for the interpretation of the variance as the sum of a component V_s due to sampling and a component V_d due to detectability.

An unbiased estimate of the detectability component of variance V_d is provided by

$$v_d = \sum_{i=1}^{\nu} \frac{1}{\pi_i^2} \sum_{j=1}^{m_i} \frac{v_{ij}}{g_{ij}} = \sum_{i=1}^{\nu} \frac{1}{\pi_i^2} \sum_{j=1}^{m_i} \frac{1 - g_{ij}}{g_{ij}^2} y_{ij}^2.$$

This follows by using the fact that

$$\hat{\tau} = \sum_{i=1}^{\nu} \frac{1}{\pi_i} \sum_{j=1}^{m_i} \frac{y_{ij}}{g_{ij}}$$

of (9.23) is an unbiased estimate of

$$\tau = \sum_{i=1}^{N} \sum_{j=1}^{M_i} y_{ij}$$

and then by replacing y_{ij} by v_{ij}/π_i. An unbiased estimate of the sampling component of variance V_s is

$$v_s = \sum_{i=1}^{\nu} \sum_{i'=1}^{\nu} u_i u_{i'} \left(\frac{\pi_{ii'} - \pi_i' \pi_{i'}}{\pi_{ii'} \pi_i \pi_{i'}} \right) - \sum_{i=1}^{\nu} \left(\frac{1}{\pi_i^2} - \frac{1}{\pi_i} \right) \sum_{j=1}^{m_i} \frac{1 - g_{ij}}{g_{ij}^2} y_{ij}^2,$$

which is obtained by subtracting v_d from the unbiased variance estimator $\widehat{\mathrm{var}}[\hat{\tau}]$ of (9.25).

Unfortunately, the alternative decomposition (9.26) does not facilitate the search for an estimator of variance as does (9.24). For an estimator of var$[\hat{\tau}]$, Steinhorst and Samuel [1989] gave $v^* = v_1 + v_d$. Although v_d is unbiased for V_d, the estimator v^*, unlike (9.25), is not unbiased for the actual variance [(9.24) or (9.26)] of the estimator $\hat{\tau}$. The bias, obtained by writing $v^* = \widehat{\text{var}}[\hat{\tau}] - v_2 + v_d$, is

$$\mathrm{E}[v^* - \widehat{\text{var}}(\hat{\tau})] = \mathrm{E}[v_d - v_2]$$

$$= \sum_{i=1}^{N} \left(\frac{1}{\pi_i} - 1 \right) \sum_{j=1}^{M_i} (1 - g_{ij}) \frac{y_{ij}^2}{g_{ij}}. \qquad \square$$

9.5 ESTIMATED DETECTABILITIES

Suppose the detectability g_{ij} is estimated by $\hat{g}_{ij}$ from some experiment independent of the present survey, and that estimates of the variances and covariances of the $\hat{g}_{ij}$ are also obtained. Steinhorst and Samuel [1989] give an example of such estimates obtained for aerial surveys of wildlife populations using logistic regression with environmental variables. Let $\tilde{\tau}$ be the estimator formed by replacing the actual detectabilities g_{ij} with the estimated detectabilities $\hat{g}_{ij}$ in (9.18). Let $\hat{u}_{ij}$ denote the value of the new variable using estimated detectability, namely

$$\hat{u}_{ij} = \frac{y_{ij} z_{ij}}{\hat{g}_{ij}}. \qquad (9.29)$$

Then the estimator is

$$\tilde{\tau} = \hat{\tau}_0(\hat{\mathbf{u}}_s), \qquad (9.30)$$

where $\hat{\mathbf{u}}_s = \{\hat{u}_{ij} : i \in s\}$. The estimator $\tilde{\tau}$ is approximately unbiased for τ and, using a first-order Taylor expansion (the "delta method," cf. Rao [1973, pp. 387–388] and Seber [1982, p. 7]),

$$\text{var}[\tilde{\tau}] \approx \text{var}[\hat{\tau}] + \sum_{i=1}^{N} \sum_{j=1}^{M_i} \sum_{i'=1}^{N} \sum_{j'=1}^{M_{i'}} \mathrm{E}\left[\frac{\partial \tilde{\tau}}{\partial \hat{g}_{ij}} \right] \mathrm{E}\left[\frac{\partial \tilde{\tau}}{\partial \hat{g}_{i'j'}} \right] \text{cov}[\hat{g}_{ij}, \hat{g}_{i'j'}], \quad (9.31)$$

where $\text{var}[\hat{\tau}]$ is given by (9.20) and the partial derivatives are evaluated at $\hat{g}_{ij} = g_{ij}$. (The derivation is given in Section 9.7).

For estimating the variance, let

$$\hat{v}_1 = v_0(\hat{\mathbf{u}}_s) \qquad (9.32)$$

and

$$\hat{v}_2 = \hat{\tau}_0(\hat{\mathbf{t}}_s), \qquad (9.33)$$

where

$$\hat{t}_{ij} = \frac{v_{ij} z_{ij}}{\hat{g}_{ij}} = \frac{(1 - \hat{g}_{ij})}{\hat{g}_{ij}^2} z_{ij} y_{ij}^2. \qquad (9.34)$$

Thus, $\hat{v}_1$ and $\hat{v}_2$ are v_1 and v_2 respectively of (9.22) with $\hat{g}_{ij}$ in place of each g_{ij}. An estimate of the variance (9.31) is therefore given by

$$\widehat{\text{var}}[\tilde{\tau}] = \hat{v}_1 + \hat{v}_2 + \hat{v}_3, \tag{9.35}$$

where $\hat{v}_3$ is an estimate of the right-hand term of (9.31). The estimate $\hat{v}_3$ will depend on the design and the form of the estimator $\tilde{\tau}$.

Example 9.4   Linear Homogeneous Estimator.   The expression (9.31) can be written more explicitly if we assume $\hat{\tau}_0$ belongs to the class of linear homogeneous estimators (cf. Faulkenberry and Garoui [1991]), so that

$$\hat{\tau}_0 = \sum_{i=1}^{N} \sum_{j=1}^{M_i} a_{ijs} y_{ij},$$

where the weight a_{ijs} for the jth object of the ith unit may depend on the sample s selected, with $a_{ijs} = 0$ for $i \notin s$ or j not detected. Then

$$\hat{\tau} = \sum_{i=1}^{N} \sum_{j=1}^{M_i} \frac{a_{ijs} y_{ij}}{g_{ij}},$$

$$\tilde{\tau} = \sum_{i=1}^{N} \sum_{j=1}^{M_i} \frac{a_{ijs} y_{ij}}{\hat{g}_{ij}},$$

and

$$\left[\frac{\partial \tilde{\tau}}{\partial \hat{g}_{ij}} \right]_{\hat{g}_{ij}=g_{ij}} = -\frac{a_{ijs} y_{ij}}{g_{ij}^2}.$$

Now $\hat{\tau}$ unbiased for τ implies $E[a_{ijs}/g_{ij}] = 1$ for all i and j, so that $E[a_{ijs}] = g_{ij}$, and hence

$$E\left[\frac{\partial \tilde{\tau}}{\partial \hat{g}_{ij}} \right]_{\hat{g}_{ij}=g_{ij}} = -\frac{y_{ij}}{g_{ij}}. \tag{9.36}$$

Therefore substituting in (9.31), we have

$$\text{var}[\tilde{\tau}] \approx \text{var}[\hat{\tau}] + \sum_{i=1}^{N} \sum_{j=1}^{M_i} \sum_{i'=1}^{N} \sum_{j'=1}^{M_{i'}} \frac{y_{ij} y_{i'j'}}{g_{ij} g_{i'j'}} \text{cov}[\hat{g}_{ij}, \hat{g}_{i'j'}]. \tag{9.37}$$

(A direct proof is given in Section 9.7.)

Example 9.5   Adaptive Cluster Sampling.   Consider an adaptive cluster sampling design with estimator given by (4.6), namely

$$\hat{\tau}_0 = \sum_{k=1}^{\kappa} \frac{y_k^*}{\alpha_k},\tag{9.38}$$

based on the intersection probabilities α_k. This is a special case of Example 9.2 with $M = 1$. Here y_k^* is now defined to be the sum of the y_{ij} values over all the objects in the kth network. Thus,

$$y_k^* = \sum_{i \in B_k} \sum_{j=1}^{M_i} y_{ij},$$

where B_k is the set of units in the kth network and M_i is the number of objects in unit i. We also define u_k^* and t_k^* in a similar fashion, where

$$u_{ij} = \frac{y_{ij}z_{ij}}{g_{ij}} \quad \text{and} \quad t_{ij} = \frac{(1 - g_{ij})y_{ij}^2 z_{ij}}{g_{ij}^2}.$$

If we replace g_{ij} by $\hat{g}_{ij}$, as in (9.29) and (9.34), we get $\hat{u}_k^*$ and $\hat{t}_k^*$, respectively.

The estimator $\tilde{\tau} = \hat{\tau}_0(\hat{\mathbf{u}}_s)$ of (9.30) is obtained from (9.38) by replacing $\mathbf{y}_s$ by $\hat{\mathbf{u}}_s$, that is, by replacing y_k^* by $\hat{u}_k^*$, giving

$$\tilde{\tau} = \sum_{k=1}^{\kappa} \frac{\hat{u}_k^*}{\alpha_k}.$$

Here

$$\hat{u}_k^* = \sum_{i \in B_k} \sum_{j=1}^{M_i} \frac{y_{ij}z_{ij}}{\hat{g}_{ij}}$$

$$= \sum_{i \in B_k} \sum_{j=1}^{m_i} \frac{y_{ij}}{\hat{g}_{ij}},$$

where m_i is the number of objects detected in unit i. For the jth object in the ith unit of the population, the expected partial derivative with respect to $\hat{g}_{ij}$ is then

$$\mathrm{E}\left[\frac{\partial \tilde{\tau}}{\partial \hat{g}_{ij}}\right]_{\hat{g}_{ij}=g_{ij}} = -\frac{y_{ij}}{g_{ij}},$$

which also follows directly from (9.36) as $\hat{\tau}$ is a linear homogeneous estimator.

For the variance of $\hat{\tau} \, [= \hat{\tau}_0(\mathbf{u}_s)]$ we have, first of all, from (9.10) with y_k^* replaced by u_k^*,

$$
\begin{aligned}
V &= \mathrm{var}[\hat{\tau} \,|\, \mathbf{z}] \\
&= V_0(\mathbf{u}) \\
&= \sum_{k=1}^{K} \sum_{k'=1}^{K} u_k^* u_{k'}^* \left(\frac{\alpha_{kk'} - \alpha_k \alpha_{k'}}{\alpha_k \alpha_{k'}} \right).
\end{aligned}
$$

Here, the network structure (i.e., α_k, $\alpha_{kk'}$, and K) now depend on $\mathbf{z}$. Then using (9.20) and substituting in (9.31) we have

$$
\begin{aligned}
\mathrm{var}[\hat{\tau}] &\approx \mathrm{E}_{\mathbf{z}}[V] + \sum_{i=1}^{N} \sum_{j=1}^{M_i} \frac{(1 - g_{ij})}{g_{ij}} y_{ij}^2 \\
&+ \sum_{i=1}^{N} \sum_{j=1}^{M_i} \sum_{i'=1}^{N} \sum_{j'=1}^{M_{i'}} \frac{y_{ij} y_{i'j'}}{g_{ij} g_{i'j'}} \mathrm{cov}[\hat{g}_{ij}, \hat{g}_{i'j'}].
\end{aligned}
\tag{9.39}
$$

For the estimated variance, we need to find the terms in (9.35). Let $M(k)$ denote the total number of objects in network k and let $m(k)$ be the number of these that are detected. Firstly we have $\hat{v}_1 = v_0(\hat{\mathbf{u}}_s)$, where $v_0(\mathbf{y}_s)$ is given by (9.11), so that

$$
\hat{v}_1 = \sum_{k=1}^{K} \sum_{k'=1}^{K} \hat{u}_k^* \hat{u}_{k'}^* \left(\frac{\alpha_{kk'} - \alpha_k \alpha_{k'}}{\alpha_{kk'} \alpha_k \alpha_{k'}} \right).
$$

Secondly, from (9.33), $\hat{v}_2 = \hat{\tau}_0(\hat{\mathbf{t}}_s)$, so that replacing y_k^* by $\hat{t}_k^*$ in (9.38) and letting $y_{(k)j}$ be the y-value for the jth object in the kth cluster with $\hat{g}_{(k)j}$ its corresponding estimated detectability, we obtain

$$
\hat{v}_2 = \sum_{k=1}^{K} \frac{1}{\alpha_k} \sum_{j=1}^{m(k)} \frac{1 - \hat{g}_{(k)j}}{\hat{g}_{(k)j}^2} y_{(k)j}^2.
$$

Defining $z_{(k)j} = 1$ if the jth unit in the kth network is sampled, and 0 otherwise, we finally have

$$
\hat{v}_3 = \sum_{k=1}^{K} \sum_{j=1}^{m(k)} \sum_{k'=1}^{K} \sum_{j'=1}^{m(k')} \frac{y_{(k)j} y_{(k')j'} z_{(k)j} z_{(k')j'}}{\alpha_{kk'} \hat{g}_{(k)j}^2 \hat{g}_{(k')j'}^2} \widehat{\mathrm{cov}}[\hat{g}_{(k)j}, \hat{g}_{(k')j'}],
$$

giving

$$
\widehat{\mathrm{var}}[\hat{\tau}] = \hat{v}_1 + \hat{v}_2 + \hat{v}_3.
$$

The justification for $\hat{v}_3$ is that it would be unbiased for the final term in (9.39) if the actual g_{ij} and covariances are used (call this expression v_3). To see this, define the indicator variable $I_{kk'}$ which, conditional on $\mathbf{z}$, is equal to 1 if the initial sample intersects both networks k and k', and 0 otherwise. Given a realization of $\mathbf{z}$, we can write

$$v_3 = \sum_{k=1}^{K} \sum_{k'=1}^{K} \sum_{j=1}^{M(k)} \sum_{j'=1}^{M(k')} \frac{y_{(k)j} y_{(k')j'}}{\alpha_{kk'} g_{(k)j}^2 g_{(k')j'}^2} I_{kk'} z_{(k)j} z_{(k')j'} \text{cov}[\hat{g}_{(k)j}, \hat{g}_{(k')j'}],$$

where the summation is over every object in the population. Using $\mathrm{E}[I_{kk'} \mid \mathbf{z}] = \alpha_{kk'}$, we have

$$\mathrm{E}[v_3 \mid \mathbf{z}] = \sum_{k=1}^{K} \sum_{k'=1}^{K} \sum_{j=1}^{M(k)} \sum_{j'=1}^{M(k')} \frac{y_{(k)j} y_{(k')j'}}{g_{(k)j}^2 g_{(k')j'}^2} z_{(k)j} z_{(k')j'} \text{cov}[\hat{g}_{(k)j}, \hat{g}_{(k')j'}].$$

Since $z_{(k)j}$ and $z_{(k')j'}$ are independent, $\mathrm{E}[z_{(k)j} z_{(k')j'}] = g_{(k)j} g_{(k')j'}$, giving

$$\mathrm{E}[v_3] = \sum_{k=1}^{K} \sum_{k'=1}^{K} \sum_{j=1}^{M(k)} \sum_{j'=1}^{M(k')} \frac{y_{(k)j} y_{(k')j'}}{g_{(k)j} g_{(k')j'}} \text{cov}[\hat{g}_{(k)j}, \hat{g}_{(k')j'}].$$

which is the last term in (9.39) but with a different way of summing over all the objects in the population. $\qquad\square$

9.6 A NOTE ON WITH-REPLACEMENT DESIGNS

The results in the foregoing sections apply to any sampling design in which the selection of units may be either with replacement or without replacement. When sampling is with replacement, selections (initial selections in the case of adaptive cluster sampling) are independent, and a given unit may be selected more than once. For the preceding results it is assumed that even if a unit is "selected" more than once, it is actually observed only once, producing a unique set of detections for that unit.

If, on the other hand, in sampling with replacement a unit is independently observed each time it is selected, so that independent sets of detections are made, then a very simple estimator and estimator of variance are available (see Rao [1975]). Let $\hat{\tau}_{(i)}$ be an estimator of τ based on the ith of the n independent selections made under the design. With imperfect (but known) detectability, $\hat{\tau}_{(i)}$ will be as given in the earlier sections of this chapter, but based on a sample size of 1. The estimator $\hat{\tau}$ of the population total is the sample mean of the n independent estimators $\hat{\tau}_{(i)}$, that is,

$$\hat{\tau} = \frac{1}{n} \sum_{i=1}^{n} \hat{\tau}_{(i)}.$$

Define the sample variance

$$s^2 = \frac{1}{n-1} \sum_{i=1}^{n} (\hat{\tau}_{(i)} - \hat{\tau})^2.$$

Then, by the standard statistical theory of independent and identically distributed random variables, an unbiased estimate of the variance of $\hat{\tau}$ is $\widehat{\mathrm{var}}[\hat{\tau}] = s^2/n$. With estimated detectability, this simple result would still hold if detectability was independently estimated for each selection. Otherwise, the usual term $\hat{v}_3$ would be added.

9.7 TAYLOR SERIES APPROXIMATIONS

In this section, we show some derivations used in Section 9.5. For simplicity let a single label i index the objects in the population, for $i = 1, \ldots, M$ $(M = \sum_{i=1}^{N} M_i)$. Let the subscripts is denote the ith object in the sample and the subscripts ns the number of objects in the sample. The estimators may be written

$$\hat{\tau} = \hat{\tau}_0 \left(\frac{y_{1s}}{g_{1s}}, \ldots, \frac{y_{ns}}{g_{ns}} \right)$$

and

$$\tilde{\tau} = \hat{\tau}_0 \left(\frac{y_{1s}}{\hat{g}_{1s}}, \ldots, \frac{y_{ns}}{\hat{g}_{ns}} \right).$$

Treating the y-values for the moment as constants, and expanding about the point $(g_{1s}, \ldots, g_{ns})$, the first terms of the Taylor series are

$$\tilde{\tau} \approx \hat{\tau}_0 \left(\frac{y_{1s}}{g_{1s}}, \ldots, \frac{y_{ns}}{g_{ns}} \right) + \sum_{i=1}^{ns} \left[\frac{\partial \tilde{\tau}}{\partial \hat{g}_{is}} \right]_{\hat{g}_{is} = g_{is}} (\hat{g}_{is} - g_{is})$$

$$= \hat{\tau} + \sum_{i=1}^{ns} \hat{f}_{is} (\hat{g}_{is} - g_{is}),$$

where

$$\hat{f}_{is} = \left[\frac{\partial \hat{\tau}_0 \left(\frac{y_{1s}}{\hat{g}_{1s}}, \ldots, \frac{y_{ns}}{\hat{g}_{ns}} \right)}{\partial \hat{g}_{is}} \right]_{\hat{g}_{is} = g_{is}}.$$

The quantity $\hat{f}_{is}$ is a random variable depending on which objects are in the sample.

Next, expand the term $\hat{f}_{is}(\hat{g}_{is} - g_{is})$ about g_{is} and f_{is} $(= E[\hat{f}_{is}])$, giving the approximation

$$\tilde{\tau} \approx \hat{\tau} + \sum_{i=1}^{ns} f_{is}(\hat{g}_{is} - g_{is}).$$

Example 9.6   Homogeneous Linear Estimator.   For a homogeneous linear estimator,

$$\tilde{\tau} = \sum_{i=1}^{N} \sum_{j=1}^{M_i} \frac{a_{ijs} y_{ij}}{\hat{g}_{ij}}.$$

The random variables in this expression are the coefficients a_{ijs}, which depend on the sample selected, and the independent estimates $\hat{g}_{ij}$ of detectabilities. Unbiasedness of $\hat{\tau}$, the estimator with the correct detectabilities inserted, implies $E[a_{ijs}] = g_{ij}$. The partial derivatives are

$$\frac{\partial \tilde{\tau}}{\partial a_{ijs}} = \frac{y_{ij}}{\hat{g}_{ij}}$$

and

$$\frac{\partial \tilde{\tau}}{\partial \hat{g}_{ij}} = -\frac{a_{ijs} y_{ij}}{\hat{g}_{ij}^2}.$$

Since $E[a_{ijs}] = g_{ij}$, the first-order Taylor series approximation of $\tilde{\tau}$ with both the a_{ijs} and the $\hat{g}_{ij}$ expanded about g_{ij} gives

$$\begin{aligned}
\tilde{\tau} &\approx \sum_{i=1}^{N} \sum_{j=1}^{M_i} y_{ij} + \sum_{i=1}^{N} \sum_{j=1}^{M_i} \frac{y_{ij}}{g_{ij}}(a_{ijs} - g_{ij}) - \sum_{i=1}^{N} \sum_{j=1}^{M_i} \frac{y_{ij}}{g_{ij}}(\hat{g}_{ij} - g_{ij}) \\
&= \tau + \hat{\tau} - \tau - \sum_{i=1}^{N} \sum_{j=1}^{M_i} \frac{y_{ij}}{g_{ij}}(\hat{g}_{ij} - g_{ij}) \\
&= \tau + \hat{\tau} - \sum_{i=1}^{N} \sum_{j=1}^{M_i} \frac{y_{ij} \hat{g}_{ij}}{g_{ij}},
\end{aligned}$$

since $\tau = \sum \sum y_{ij}$. Thus, $E[\tilde{\tau}] \approx E[\hat{\tau}] = \tau$ if $E[\hat{g}_{ij}] \approx g_{ij}$. Now

$$\tilde{\tau} - \tau = \hat{\tau} - \tau - \sum_{i=1}^{N} \sum_{j=1}^{M_i} \frac{y_{ij}(\hat{g}_{ij} - g_{ij})}{g_{ij}}.$$

If we square and take expected values, the covariance of the two terms on the right-hand side is zero, as the $\hat{g}_{ij}$ are independent of the sampling design and therefore of $\hat{\tau}$. Hence

$$\text{var}[\tilde{\tau}] \approx \text{var}[\hat{\tau}] + \text{var}\left[\sum_{i=1}^{N}\sum_{j=1}^{M_i}\frac{y_{ij}}{g_{ij}}(\hat{g}_{ij} - g_{ij})\right]$$

$$= \text{var}[\hat{\tau}] + \sum_{i=1}^{N}\sum_{j=1}^{M_i}\sum_{i'=1}^{N}\sum_{j'=1}^{M_{i'}}\frac{y_{ij}y_{i'j'}}{g_{ij}g_{i'j'}}\text{cov}[\hat{g}_{ij}, \hat{g}_{i'j'}]. \qquad \square$$

CHAPTER 10

Optimal Sampling Strategies

10.1 INTRODUCTION

A sampling *strategy* for estimating or predicting a population quantity $z(\mathbf{y})$ consists of a design $p(s \mid \mathbf{y})$ together with an estimator or predictor $\hat{z}(d)$. We recall that $d_O = (s_O, \mathbf{y}_O)$ denotes the data in the order selected, s consists of the reduced sample s_R but with the labels listed from the smallest to the largest label, and $d = (s, \mathbf{y}_s)$ is the corresponding reduced data vector. Since d is a sufficient statistic, it will be unnecessary to deal separately with the original sample s_O in this chapter.

An optimal unbiased strategy, for the purposes of this chapter, will consist of a design and an unbiased estimator or predictor giving the smallest mean-square error among the class of unbiased estimators or predictors and among a specified class of designs. In this chapter, we shall look at results on optimal strategies in sampling in the fixed population setting and with varying degrees of population model assumptions.

10.2 FIXED POPULATION SETTING

In the fixed population setting, the population vector $\mathbf{y}$ is considered a vector of fixed but unknown values and probability enters the situation only through the design. The optimal strategy $(p_{opt}, \hat{z}_{opt})$ would satisfy

$$\mathrm{E}\left[\hat{z}_{opt}(d) \mid \mathbf{y}\right] = z(\mathbf{y})$$

for all $\mathbf{y} \in \mathcal{Y}$ and

$$\mathrm{var}\left[\hat{z}_{opt}(d) \mid \mathbf{y}\right] = \min_{(p,\hat{z}) \in B} \mathrm{var}\left[\hat{z}(d) \mid \mathbf{y}\right]$$

for all $\mathbf{y} \in \mathcal{Y}$, where B is the class of all design-unbiased strategies. We assume that each y_i can potentially take on any real value, so that $\mathcal{Y} = \mathbb{R}^N$.

235

It can be shown that (cf. Theorem 2.7 in Section 2.6), for any given design, there does not exist any minimum-variance unbiased estimator, so that in the fixed population setting no optimal strategy exists (Godambe [1955] and Godambe and Joshi [1965]; see also Basu [1971], Chaudhuri and Stenger [1992], and Hedayat and Sinha [1991]). Godambe [1955] showed that with any conventional design there does not exist a uniformly minimum-variance *linear* unbiased estimator of the population mean. For the wider class of sequential (adaptive) designs, Stenger [1977] showed that the universality of Godambe's nonexistence result did not extend to this class by producing an example of a sequential design for which a best linear unbiased estimator does exist. The example involves a sequential design in which some of the possible samples contain only one unit. However, the best linear unbiased estimator obtained depends on order of selection, and so is not a function of the minimal sufficient statistic. Further, when improved by the Rao–Blackwell method, the resulting estimator is no longer linear and is not uniformly minimum variance among the wider class of unbiased estimators (Sinha [1991]).

In spite of the lack of sweeping optimality results for conventional designs, many practical results are available regarding conditions under which one strategy will be better than another among designs, such as simple random sampling, stratified random sampling, cluster sampling, systematic sampling, and two-stage sampling (cf. Cochran [1977], Levy and Lemeshow [1991], and other standard sampling texts). Some similar types of results for adaptive designs, describing conditions under which a specific adaptive strategy is more efficient than a comparable nonadaptive strategy, have been summarized in this monograph.

10.3 KNOWN POPULATION MODEL

We assume that $\mathbf{y}$ is a realization of a random vector $\mathbf{Y}$ having a density function $f(\mathbf{y}; \boldsymbol{\phi})$ which is assumed to be known exactly; that is, the form of f is known and the value of the parameter $\boldsymbol{\phi}$ is known, or $\boldsymbol{\phi}$ has a known prior distribution. Then inference about a function $z(\mathbf{Y})$ such as the population total involves only pure prediction. By the usual prediction results applied to survey designs, the best unbiased predictor of Z is $\hat{Z} = \mathrm{E}[Z \mid d]$, that is, $\mathrm{E}[\hat{Z} \mid s] = \mathrm{E}[Z \mid s]$ and $\mathrm{E}[(\hat{Z} - Z)^2 \mid s]$ is minimized.

Among conventional designs, the best design would be to purposively select the sample s giving the lowest value of $\mathrm{E}[(\hat{Z} - Z)^2 \mid s]$. However, it is possible to do better still with an adaptive design. Basu [1969] argued that optimal designs would be of the form $p(s \mid \mathbf{y}_s)$, that is, adaptive. The theoretically optimal design, based on prior knowledge of the population distribution, with fixed sample size n was described by Zacks [1969]. The optimal design is in general adaptive; sufficient conditions under which it would be nonadaptive were also given by Zacks [1969]. Necessary and sufficient conditions for the optimal design to be a conventional one in the closely related two-phase situation were given in Thompson [1988]. In addition, for a given expected sample size, it is in general better to adaptively choose the sample size than to fix it in advance (Thompson [1988]). Reviews of the search for optimal sampling strategies that are not restricted to conventional designs are found in Solomon and Zacks [1970] and Chaudhuri [1988].

In the optimal fixed sample size design described by Zacks, selection of the first sample unit depends on the known distribution of $\mathbf{Y}$. Selection of the second unit depends on the conditional distribution of unobserved values given the initially selected unit and the associated observation. At each step, selection is based on the conditional distribution of unobserved values given the units already selected and the y-values already observed. The actual optimization is in general computationally complicated.

Thus, with a known population model, the optimal strategy is adaptive. Only in special cases will a nonadaptive strategy do as well. In one sense this result is trivially true, since the class of adaptive designs can be viewed as containing the conventional designs as special cases. The more valuable results in the above-mentioned papers are the descriptions of optimal selection of units for fixed sample size designs, of optimal adaptive choice of sample size, and of the exact conditions under which the optimal design is nonadaptive. For example, if the Y_i are independent and identically distributed random variables, the distribution of the unobserved Y-values is independent of the ones already observed, and one might as well use a nonadaptive design.

10.3.1 Phases in Adaptive Designs

A "phase" of a survey is defined to be a point at which a selection of units may be made. Conventional designs including multistage designs are single stage in that the entire sample of units to observe may be selected at one time, prior to the survey. An example of a two-phase design is one in which an initial simple random sample of n_1 units are selected and the y-values for these units observed; then an additional sample of n_2 units is selected, with the second sample size depending on the values of the initial observations. A fixed size adaptive design such as Zacks [1969] described is n-phase: each of the n units is selected taking into account the y-values of the previously selected units. Adaptive cluster sampling is a class of L-phase designs where the number L of phases is random. At the first phase, a conventional probability sample of units is selected. Based on the y-values of these units, additional units may be added (phase 2) in the neighborhood of any unit for which the y-value satisfies a specified condition. At the third phase, the recently added units are observed and additional units are added from the neighborhood of any of those that satisfy the condition. The process continues until the sample contains all units in the neighborhood of any unit satisfying the condition.

10.3.2 Optimal Two-Phase Strategy

The essence of the optimal design question is contained in a two-phase design. Consider a design with a fixed sample size n, and suppose n_1 of those units have already been selected and observed. The initial sample of n_1 units will be denoted by s_1 and the vector of associated y-values by $\mathbf{y}_{s_1}$. Here s_1 is the set of distinct labels but uniquely ordered from the smallest to the largest label. The remaining sample to be selected is s_2, with associated values $\mathbf{y}_{s_2}$. With a slight redefinition of s, the complete sample will therefore be $s = (s_1, s_2)$, having associated values $\mathbf{y}_s = (\mathbf{y}_{s_1}, \mathbf{y}_{s_2})$.

The object is to predict a population quantity $Z = z(\mathbf{Y})$, such as the population total or mean, with a function $T(d)$ of the final data $d = (s, \mathbf{y}_s)$. We require T to be model-unbiased, that is, $\mathrm{E}[T \mid s] = \mathrm{E}[Z \mid s]$. The function of the data minimizing the mean-square prediction error $\mathrm{E}[(T - Z)^2 \mid s]$ is the conditional expectation

$$T(d) = \mathrm{E}[Z \mid d],$$

which can be computed since the population density is known. We recall from Section 3.5 (with s instead of s_O) that, for conventional or adaptive designs, the conditional density given the data does not depend on what design is used. Any other type of predictor $T(d)$ could be used in place of the optimal one and the following results would apply equally well, so long as T is a function that can be computed uniquely from any final set of data without regard to what design was used to obtain the data. For example, T could be a maximum likelihood predictor of Z.

The question now is, should the selection of the remaining sample units s_2 depend on the values $\mathbf{y}_{s_1}$ in the initial sample? The ideal adaptive two-phase design would select the second phase units s_2 to minimize

$$g_{s_2}(s_1, \mathbf{y}_{s_1}) = \mathrm{E}\{[t(s_1, \mathbf{y}_{s_1}, s_2, \mathbf{Y}_{s_2}) - z(\mathbf{Y})]^2 \mid s_1, \mathbf{y}_{s_1}\}$$

$$= \int (t - z)^2 f(\mathbf{y}_{\bar{s}_1} \mid s_1, \mathbf{y}_{s_1}; \boldsymbol{\phi}) \, d \, \mathbf{y}_{\bar{s}_1} \qquad (10.1)$$

where $\bar{s}_1$ is the vector of units not in the first-phase sample (i.e, those units in s_2 and those not in s) and $f(\mathbf{y}_{\bar{s}_1} \mid s_1, \mathbf{y}_{s_1}; \boldsymbol{\phi})$ is the conditional density, given the initial set of data, of the components of $\mathbf{y}$ not in the initial sample. By the results in Section 3.5 applied to s_1, the conditional density $f(\mathbf{y}_{\bar{s}_1} \mid s_1, \mathbf{y}_{s_1}; \boldsymbol{\phi})$ does not depend on how the initial sample was selected, provided it was obtained with a conventional or adaptive design, not a nonstandard design. The integration in (10.1) is over all the components of $\mathbf{y}$ in $\bar{s}_1$, that is, $d \, \mathbf{y}_{\bar{s}_1} = \prod_{j \notin s_1} dy_j$. Note that the dependence on s_2 is not in the conditioning but in the computation of the predictor t. For purposes of comparison, the initial sample s_1 is assumed to be conventionally selected and is taken as given.

In dealing with the function $g_{s_2}(s_1, \mathbf{y}_{s_1})$, it is convenient to assume that the set $\mathcal{S}_2$ of all possible second-phase samples is countable so that we can order them arbitrarily and identify each possible second-phase sample with an integer i, for $i = 1, 2, \ldots, \#\{\mathcal{S}_2\}$, where $\#\{\mathcal{S}_2\}$ is the number of samples in $\mathcal{S}_2$. The conditional mean-square error for the optimal choice is then

$$\min_{s_2} g_{s_2}(s_1, \mathbf{y}_{s_1}) = \min_i \mathrm{E}\{[t(s_1, \mathbf{y}_{s_1}, s_2, \mathbf{Y}_{s_2}) - Z]^2 \mid s_1, \mathbf{y}_{s_1}, s_2 = i\}.$$

Thus, starting with initial sample s_1 and taking the optimal choice for s_2 (say, $s_2 = j$), the overall mean-square error is

$$E[(T - Z)^2 \,|\, s_1, s_2 = j] = \int \min_{s_2} g_{s_2}(s_1, \mathbf{y}_{s_1}) f(\mathbf{y}; \boldsymbol{\phi}) \, d\, \mathbf{y}.$$

The best nonadaptive design, on the other hand, would select s_2 to minimize the mean-square error without taking the first-phase observations $\mathbf{y}_{s_1}$ into account. If this occurs at $s_2 = k$, the mean-square error for the optimal conventional design is

$$E[(T - Z)^2 \,|\, s_1, s_2 = k] = \min_{s_2} \int g_{s_2}(s_1, \mathbf{y}_{s_1}) f(\mathbf{y}; \boldsymbol{\phi}) \, d\, \mathbf{y}.$$

We can compare this with the adaptive design above by noting that, by a basic property of integration,

$$\int \min_{s_2} g_{s_2}(s_1, \mathbf{y}_{s_1}) f(\mathbf{y}; \boldsymbol{\phi}) \, d\, \mathbf{y} \le \min_{s_2} \int g_{s_2}(s_1, \mathbf{y}_{s_1}) f(\mathbf{y}; \boldsymbol{\phi}) d\, \mathbf{y}.$$

Hence the optimal adaptive procedure will always be as good as or better than the best nonadaptive procedure. The following theorem gives necessary and sufficient conditions for equality to hold in the above expression, that is, for the optimal design to be conventional.

Theorem 10.1. Let $\{g_n\}$ be a sequence of functions. Define a sequence $\{h_n\}$ by $h_n = g_i \in \{g_n\}$ such that $\int g_i \, dF = \min_{k \le n} \int g_k \, dF$. Suppose there exists an integrable function q such that $|h_n| \le q$ almost everywhere (a.e.) and suppose $\lim_{n \to \infty} h_n = h$ a.e. Then $\int \inf_n g_n \, dF = \inf_n \int g_n \, dF$ if and only if $h = \inf_n g_n$ a.e.

Proof: The proof writes the infimum as a limit and uses Lesbegue's Dominated Convergence Theorem (D.C.T.) on the integration of a limit. We shall denote $\lim_{n \to \infty}$ by lim.

First, suppose $h = \inf_n g_n$ a.e. Then

$$\int \inf_n g_n \, dF = \int h \, dF = \int \lim h_n \, dF = \lim \int h_n \, dF$$

by the D.C.T. Also, by the definition of h_n,

$$\inf_n \int g_n \, dF = \lim \min_{k \le n} \int g_k \, dF = \lim \int h_n \, dF,$$

establishing the sufficiency of the condition.

Second, suppose $\int \inf_n g_n \, dF = \inf_n \int g_n \, dF$. By the definition of h_n, we can write

$$\inf \int g_n \, dF = \lim \min_{k \leq n} \int g_k \, dF = \lim \int h_n \, dF.$$

By the D.C.T., $\lim \int h_n \, dF = \int \lim h_n \, dF = \int h \, dF$, so that $\int (h - \inf g_n) \, dF = 0$. But $h_n \in \{g_n\}$ so that $h_n \geq \inf_n g_n$. Since the right-hand side of this last equation is constant, we can take limits and get $(h - \inf g_n) \geq 0$. Hence $h = \inf_n g_n$ almost everywhere, thus establishing necessity and completing the proof. $\qquad \square$

For our application of the theorem, the function g_i is given by (10.1), with $s_2 = i$. The initial sample s_1 is taken as given and fixed. Suppose the number of possible second samples is finite so that there are a finite number of g_i. Then h in the theorem is the g_i having the smallest integral over all possible values $\mathbf{y}_{s_1}$ of the variable of interest in the initial sample. But at each value of $\mathbf{y}_{s_1}$, $h(\mathbf{y}_{s_1})$ is not necessarily the minimum of the $g_i(s_1, \mathbf{y}_{s_1})$. Only if $h(\mathbf{y}_{s_1})$ equals the minimum over i of the $g_i(s_1, \mathbf{y}_{s_1})$ for all values of $\mathbf{y}_{s_1}$ will the optimal design be conventional. This will occur only if there is a function g_i in the collection satisfying $g_i(s_1, \mathbf{y}_{s_1}) \leq g_k(s_1, \mathbf{y}_{s_1})$ for all $k \neq i$ and for all values of $\mathbf{y}_{s_1}$, that is, g_i is uniformly minimum.

The theorem says that a nonadaptive design, in which the entire sample may be selected ahead of time, will be optimal if and only if there is some possible selection of second phase units which is best for every possible outcome of the first-phase observations. One would expect this condition to be satisfied for example if the y-values in the population are independent, so that the sample outcome of the first-phase units says nothing about the remaining units.

In the above comparison we are taking the selection s_1 of first-phase units as the same for the adaptive and conventional procedure. The procedure for selecting the first-phase units could be purposive selection, simple random sampling, or some other probability design. Alternatively, the first-phase sample could be chosen to minimize the unconditional mean-square error taking into account the design to be used at the second phase. In any case, the result says that the best adaptive procedure will be as good as or better than the best nonadaptive procedure.

Example 10.1 Optimal Two-Phase Strategy.   Consider the population of Examples 3.1 through 3.11 with $N = 4$ units and with six possible values of the population vector $\mathbf{y}$, each with equal probability under the assumed model. We will now determine the optimal sampling strategy with fixed sample size $n = 2$ for that population under the assumed model. Let Z be the population total and $\hat{Z} = \mathrm{E}[Z|d]$.

To find expectations conditional on the first unit selected, it will be convenient to take the ordered outcomes of Table 3.1 and rearrange the rows in order of first unit selected. The values of $\hat{Z}$ from Table 3.9 in Section 3.10 are then entered for each outcome (Table 10.1).

Consider the outcomes in which the first unit selected is unit number 1, which are given in the first three rows of Table 10.1. Suppose that when unit 1 is selected and observed, the associated value is $y_1 = 0$. For those first three rows, the first four

Table 10.1. Values of $\hat{Z}$ for Each Possible Outcome Under the Specified Model $f(y) = 1/6$ for Each of the Six Possible Values of y.

y:	(0,0,0,1)	(0,0,1,1)	(0,0,1,0)	(0,1,0,0)	(1,1,0,0)	(1,0,0,0)
z(y):	1	2	1	1	2	1
$s_O \setminus f(\mathbf{y})$:	1/6	1/6	1/6	1/6	1/6	1/6
(1,2)	1.33	1.33	1.33	1	2	1
(1,3)	1	1.5	1.5	1	1.5	1.5
(1,4)	1.5	1.5	1	1	1.5	1.5
(2,1)	1.33	1.33	1.33	1	2	1
(2,3)	1	1.5	1.5	1.5	1.5	1
(2,4)	1.5	1.5	1	1.5	1.5	1
(3,1)	1	1.5	1.5	1	1.5	1.5
(3,2)	1	1.5	1.5	1.5	1.5	1
(3,4)	1	2	1	1.33	1.33	1.33
(4,1)	1.5	1.5	1	1	1.5	1.5
(4,2)	1.5	1.5	1	1.5	1.5	1
(4,3)	1	2	1	1.33	1.33	1.33

columns have a y-value of 0 for the first unit, so each outcome in the 3 by 4 block has initial data value $d_1 = (s_1, y_{s_1}) = (1, 0)$. The conditional probability of each of the four possible values of **y** given the initial data $d_1 = (1, 0)$, is $(1/6)/(4/6) = 1/4$. Given that unit 1 was the first selection, the choices for the second selection are units 2, 3, and 4. Using Equation (10.1),

$$g_2(1, 0) = (1.33 - 1)^2(1/4) + (1.33 - 2)^2(1/4) + (1.33 - 1)^2(1/4) + (1 - 1)^2(1/4)$$
$$= .167$$

$$g_3(1, 0) = (1 - 1)^2(1/4) + (1.5 - 2)^2(1/4) + (1.5 - 1)^2(1/4) + (1 - 1)^2(1/4)$$
$$= .125$$

$$g_4(1, 0) = (1 - 1)^2(1/4) + (1.5 - 2)^2(1/4) + (1.5 - 1)^2(1/4) + (1 - 1)^2(1/4)$$
$$= .125.$$

Suppose on the other hand that with unit 1 as the first selection, the associated value is $y_1 = 1$. The initial data are then $d_1 = (1, 1)$, corresponding to the final two columns of the first three rows. The two possible values of **y** given $d_1 = (1, 1)$ each have probability 1/2. Using Equation (10.1),

$$g_2(1, 1) = (2 - 2)^2(1/2) + (1 - 1)^2(1/2) = 0$$
$$g_3(1, 1) = (1.5 - 2)^2(1/2) + (1.5 - 1)^2(1/2) = .25$$
$$g_4(1, 1) = (1.5 - 2)^2(1/2) + (1.5 - 1)^2(1/2) = .25.$$

Thus, if the initial data $d_1 = (s_1, y_{s_1})$ are $(1,0)$, the optimal second selection s_2 is either of units 3 or 4, since they minimize $g_{s_2}(1, 0)$. It does not matter which of units 3 or 4 is chosen. If the initial data are $(1,1)$, then the optimal second selection s_2 is unit 2, because it minimizes $g_{s_2}(1, 1)$.

By symmetry, the optimal design starting with any of the other units follows the same pattern: If the initial y-value is 0, select the second sample unit from one of the two units on the other side of the population from the initial unit. If the initial y-value is 1, select the unit on the same side of the population as the initial unit.

We have just shown that the adaptive strategy described in Example 3.1 is an optimal two-phase strategy under the assumed model. In Examples 3.1 and 3.11 we had already noted that the adaptive strategy did better than the optimal conventional strategy for this example.

Although the example is small and oversimplified, the optimal strategy derived is suggestive of a class of practical adaptive procedures for real populations such as rare animal species with clustering tendencies. If low abundance is observed in initially selected sites, observe subsequent sites in a distant part of the study region; for initial sites at which high abundance is observed, add subsequent sites nearby. The optimal procedure may be some combination of additional sites near initial high-abundance observations and additional sites far away from initial low-abundance observations or selected at random.

Example 10.2 Multivariate Normal Models With Known Covariance Structure. Suppose the population vector $\mathbf{Y}$ has an N-dimensional multivariate normal distribution with known mean vector $\boldsymbol{\mu}$ and known variance–covariance matrix $\boldsymbol{\Sigma}$. For the coordinates given by s with n elements, we can reorder and partition the random vector $\mathbf{Y}$ into $\mathbf{Y} = (\mathbf{Y}_s, \mathbf{Y}_{\bar{s}})'$ and correspondingly partition the mean vector $\boldsymbol{\mu}$ into $\boldsymbol{\mu} = (\boldsymbol{\mu}_s, \boldsymbol{\mu}_{\bar{s}})'$ and the covariance matrix into $\boldsymbol{\Sigma}_{ss}, \boldsymbol{\Sigma}_{s\bar{s}}, \boldsymbol{\Sigma}_{\bar{s}s}$, and $\boldsymbol{\Sigma}_{\bar{s}\bar{s}}$. The conditional density of $\mathbf{Y}_{\bar{s}}$ given $\mathbf{Y}_s = \mathbf{y}_s$ is $(N - n)$-dimensional multivariate normal with mean vector $\boldsymbol{\mu}_{\bar{s}} + \boldsymbol{\Sigma}_{\bar{s}s}\boldsymbol{\Sigma}_{ss}^{-1}(\mathbf{y}_s - \boldsymbol{\mu}_s)$ and variance–covariance matrix $\boldsymbol{\Sigma}_{\bar{s}\bar{s}} - \boldsymbol{\Sigma}_{\bar{s}s}\boldsymbol{\Sigma}_{ss}^{-1}\boldsymbol{\Sigma}_{s\bar{s}}$ (cf. Arnold [1990, p. 214] or Seber [1977, p. 36]).

To predict any function $Z(\mathbf{Y})$ of the population values from the data $d = (s, \mathbf{y}_s)$, the predictor minimizing the mean-square prediction error is $\hat{Z} = \mathrm{E}[Z \mid d]$. The mean-square prediction error of $T = \hat{Z}$ is

$$\mathrm{E}\big[(\hat{Z} - Z)^2 \mid s\big] = \mathrm{E}\big[(Z - \hat{Z})^2 \mid s\big]$$

$$= \mathrm{E}\big\{\mathrm{E}[(Z - \mathrm{E}[Z \mid d])^2 \mid d] \mid s\big\}$$

$$= \mathrm{E}\big[\mathrm{var}(Z \mid d) \mid s\big].$$

If $Z = \sum_{i=1}^{N} Y_i$, the population total, the conditional variance is

$$\mathrm{var}[Z \mid d] = \mathrm{var}\left[\sum_{i=1}^{N} Y_i \mid d\right]$$

$$= \sum_{i=1}^{N} \sum_{j=1}^{N} \mathrm{cov}[Y_i, Y_j \mid d].$$

Further, since Z can be written as $Z = \sum_s y_i + \sum_{\bar{s}} Y_i$ when $\mathbf{Y}_s$ is fixed at $\mathbf{y}_s$, the conditional variance is

$$\text{var}[Z \mid d] = \sum_{i \in \bar{s}} \sum_{j \in \bar{s}} \text{cov}[Y_i, Y_j \mid d]. \tag{10.2}$$

Now let $d = (d_1, d_2) = (s_1, \mathbf{y}_{s_1}, s_2, \mathbf{y}_{s_2})$ and suppose that the initial data d_1 have been obtained and the question remains how to optimally choose the second-phase sample s_2. With $T = \hat{Z}$, we have from (10.1)

$$\begin{aligned}
g_{s_2}(s_1, \mathbf{y}_{s_1}) &= \mathrm{E}_{\mathbf{Y}_{\bar{s}_1}} \left\{ [Z - \mathrm{E}(Z \mid s_1, \mathbf{y}_{s_1}, s_2, \mathbf{Y}_{s_2})]^2 \mid s_1, \mathbf{y}_{s_1} \right\} \\
&= \mathrm{E}_{\mathbf{Y}_{s_2}} \left(\mathrm{E}_{\mathbf{Y}_{\bar{s}}} \{ [Z - \mathrm{E}(Z \mid s_1, \mathbf{y}_{s_1}, s_2, \mathbf{y}_{s_2})]^2 \mid s_1, \mathbf{y}_{s_1}\, s_2, \mathbf{y}_{s_2} \} \mid s_1, \mathbf{y}_{s_1} \right) \\
&= \mathrm{E}(\mathrm{E}\{ [Z - \mathrm{E}(Z \mid d)]^2 \mid d \} \mid d_1) \\
&= \mathrm{E}(\text{var}[Z \mid d] \mid d_1) \\
&= \mathrm{E}\left(\sum_{\bar{s}} \sum_{\bar{s}} \text{cov}[Y_i, Y_j \mid d] \mid d_1 \right),
\end{aligned}$$

by (10.2). But, as noted above from the conditional distribution of $\mathbf{Y}_{\bar{s}}$, $\text{cov}[Y_i, Y_j \mid d]$ is the ijth element of the matrix $\mathbf{\Sigma}_{\bar{s}\bar{s}} - \mathbf{\Sigma}_{\bar{s}s} \mathbf{\Sigma}_{ss}^{-1} \mathbf{\Sigma}_{s\bar{s}}$, and this does not depend on Y-values. Therefore $\text{cov}[Y_i, Y_j \mid d]$ is independent of $\mathbf{y}_{s_1}$ and $g_{s_2}(s_1, \mathbf{y}_{s_1})$ is constant over $\mathbf{y}_{s_1}$, so that the choice of s_2 that minimizes g will do so for all values of $\mathbf{y}_{s_1}$. Thus, by Theorem 10.1, the optimal design for a population with known normal distribution is a conventional design.

The problem of optimally locating n sample sites in a study region or of optimally locating n_2 additional sites to an existing pattern of sample sites can be computationally challenging (Cressie [1993, pp. 319–322], Guttorp et al. [1993], and Sacks and Schiller [1988]). The remarkable property of the normal models with known covariance structure is that in the conditional distributions given sample data, the conditional means depend on the observed sample y-values, but the conditional variances and covariances do not. Suppose for example that air pollution has been monitored at n_1 sites and we wish to optimally add n_2 more sites. With an assumed normal model for air pollution distribution, the optimization needs to take account of the locations of the existing sites but not the values recorded at them, even if some of the sites have recorded zero pollution levels while others have recorded extremely high levels.

According to the multivariate normal model, the variable of interest can take on both positive and negative values. Under this model, if a value near zero has been observed near one site and a value of 10,000 has been observed at a second site, the conditional variability of a site near the first is the same as that of a site near the second.

In addition to the assumption of normality, the assumption of known variance structure is also crucial to the optimality of the conventional strategy with the normal models. For example, with the model for stratified sampling in which the Y-values are independent with means and variances depending on strata, if the variances are known for each stratum, then the optimal design uses fixed sample sizes in each stratum determined by Neyman allocation. The use of adaptive allocation plans with real populations can be motivated by the variances not being known in advance, as well as by models in which the variances depend conditionally on initial sample values.

In the next example, we consider a multivariate model in which values of the variable of interest are nonnegative.

Example 10.3   Multivariate Lognormal Model.   Suppose that the distribution of the N-dimensional population vector $\mathbf{Y}$ is multivariate lognormal. We suppose that the random vector $\mathbf{W}$ defined by $W_i = \log Y_i$, for $i = 1, \ldots, N$, has an N-dimensional multivariate normal distribution with known mean vector $\boldsymbol{\mu}$ and known variance–covariance matrix $\boldsymbol{\Sigma}$. Then the mean vector of $\mathbf{Y}$ has components

$$E[Y_i] = e^{\mu_i + \sigma_{ii}}$$

and the variance–covariance matrix of $\mathbf{Y}$ has components

$$\mathrm{var}[Y_i] = \big[E(Y_i)\big]^2 (e^{\sigma_{ii}-1})$$

and

$$\mathrm{cov}[Y_i, Y_j] = \big[E(Y_i)\big]\big[E(Y_j)\big](e^{\sigma_{ij}-1})$$

(cf. Aitchison and Brown [1957] and Cressie [1993, p. 135]).

For the coordinates given by s with n elements, we can reorder and partition the random vectors $\mathbf{Y}$ and $\mathbf{W}$ as in the previous example into $\mathbf{Y} = (\mathbf{Y}_s, \mathbf{Y}_{\bar{s}})'$, $\mathbf{W} = (\mathbf{W}_s, \mathbf{W}_{\bar{s}})'$, and correspondingly partition the mean vector $\boldsymbol{\mu}$ into $\boldsymbol{\mu} = (\boldsymbol{\mu}_s, \boldsymbol{\mu}_{\bar{s}})'$ and the covariance matrix into $\boldsymbol{\Sigma}_{ss}$, $\boldsymbol{\Sigma}_{s\bar{s}}$, $\boldsymbol{\Sigma}_{\bar{s}s}$, and $\boldsymbol{\Sigma}_{\bar{s}\bar{s}}$. The conditional density of $\mathbf{W}_{\bar{s}}$ given $\mathbf{W}_s = \mathbf{w}_s$ is an $(N - n)$-dimensional multivariate normal with mean vector $\boldsymbol{\nu} = \boldsymbol{\mu}_s + \boldsymbol{\Sigma}_{\bar{s}s}\boldsymbol{\Sigma}_{ss}^{-1}(\mathbf{w}_s - \boldsymbol{\mu}_{\bar{s}})$ and variance–covariance matrix $\boldsymbol{\Gamma} = \boldsymbol{\Sigma}_{\bar{s}\bar{s}} - \boldsymbol{\Sigma}_{\bar{s}s}\boldsymbol{\Sigma}_{ss}^{-1}\boldsymbol{\Sigma}_{s\bar{s}}$. Let the components of $\boldsymbol{\nu}$ and $\boldsymbol{\Gamma}$ be denoted by ν_i and γ_{ij} respectively. Since the transformation from $\mathbf{W}_s$ to $\mathbf{Y}_s$ is diagonal (and therefore invertible), the conditional density of $\mathbf{W}_{\bar{s}}$ given $\mathbf{Y}_s$ is the same as that given $\mathbf{W}_s$. The conditional density of $\mathbf{Y}_{\bar{s}}$ given $\mathbf{Y}_s$ (i.e., given d) is thus lognormal with mean and covariance components given by

$$E[Y_i \mid d] = e^{\nu_i + \gamma_{ii}},$$
$$\mathrm{var}[Y_i \mid d] = \big[E(Y_i) \mid d\big]^2 (e^{\gamma_{ii}-1})$$

and

$$\mathrm{cov}[Y_i, Y_j \mid d] = \big[E(Y_i) \mid d\big]\big[E(Y_j) \mid d\big](e^{\gamma_{ij}-1}).$$

For the prediction of the population total $Z = \sum_{i=1}^{N} Y_i$, we have, as in the previous example,

$$g_{s_2}(\mathbf{y}_{s_1}, s_1) = E\left(E\{[Z - E(Z|s_1, \mathbf{y}_{s_1}, s_2, \mathbf{y}_{s_2})]^2|s_1, \mathbf{y}_{s_1} s_2, \mathbf{y}_{s_2}\}|s_1, \mathbf{y}_{s_1}\right)$$

$$= E\left(\sum_{\bar{s}}\sum_{\bar{s}} \text{cov}[Y_i, Y_j \,|\, d] \,|\, d_1\right).$$

The conditional variances and covariances $\text{cov}[Y_i, Y_j | d]$ are functions of the conditional means that depend on the values $\mathbf{y}_{s_1}$ in the initial sample through the ν_i. Therefore $\text{cov}[Y_i, Y_j | d]$ and hence $g_{s_2}(\mathbf{y}_{s_1}, s_1)$ are functions of the initial values $\mathbf{y}_{s_1}$. Thus, the optimal design for a population with a lognormal distribution will in general be adaptive.

With a lognormal model with positive covariances, the conditional variances and covariances would tend to be large where the conditional means were large. With such a model in the air pollution monitoring study, the optimal selection of sites to add to an existing sample would have to take into account the values observed at the existing sites as well as their locations.   □

10.3.3 Optimal Multiphase Strategy

Consider a design with fixed sample size n in which after each unit is selected we are free to examine the data up to that point before deciding which unit to select next. The optimal n-phase design was described by Zacks [1969] with a Bayes prior providing a known distribution for $\mathbf{Y}$, together with a general loss function, and using the Bayes estimator. In this section, we describe the optimal procedure when the population density $f(\mathbf{y})$ is assumed known, using for concreteness squared error for the loss function.

We now have a change of notation. Let s_i denote the ith unit selected for the sample so that the ordered sample is $s_O = (s_1, s_2, \ldots, s_n)$. Let y_{s_i} denote the y-value associated with the ith selected unit so that the ordered data are $d_O = ((s_i, y_{s_i}) : i \in s_O)$. Suppose $T = t(D_O)$ is the predictor and $Z = z(\mathbf{Y})$ is a population quantity to be predicted. As we are going to be looking at subsets of d_O, it is convenient for expository purposes to write $d_O = d_n$. The following results will apply not only to the optimal predictor $T = E[Z | d_n]$ but to other predictors, such as those based on maximum likelihood, that are well-defined functions of the final data d_n.

Define $m_0(d_n)$, the conditional mean-square error of the predictor T given the data, that is,

$$m_0(d_n) = E\{[t(d_n) - z(\mathbf{Y})]^2 \,|\, d_n\}$$

$$= \int \{t[(s_1, y_{s_1}), \ldots, (s_n, y_{s_n})] - z(\mathbf{y})\}^2 f(\mathbf{y}_{\bar{s}(d_n)} \,|\, d_n) d\,\mathbf{y}_{\bar{s}(d_n)},$$

where $\mathbf{y}_{\bar{s}(d_n)}$ represents the vector of components of $\mathbf{y}$ not in the sample and

$$d\,\mathbf{y}_{\bar{s}(d_n)} = \prod_{j \notin \{s_1, \ldots, s_n\}} dy_j.$$

Suppose that $(n - 1)$ units have been selected so far. Let d_{n-1} represent the data collected up to that point, that is,

$$d_{n-1} = \left[(s_1, y_{s_1}), (s_2, y_{s_2}), \ldots, (s_{n-1}, y_{s_{n-1}})\right].$$

For the selection of s_n, the nth unit of the sample, compute the function

$$g_i^{(1)}(d_{n-1}) = \mathrm{E}\left[m_0(d_{n-1}, (i, Y_i)) \mid d_{n-1}\right]$$
$$= \mathrm{E}\{[t(d_{n-1}, (i, Y_i)) - Z]^2 \mid d_{n-1}\},$$

for all i not already selected, that is, for all $i \notin \{s_1, s_2, \ldots, s_{n-1}\}$. Explicitly, the conditional expectation $g_i^{(1)}(d_{n-1})$ is [as in Equation (10.1) for the two-phase strategy]

$$g_i^{(1)}(d_{n-1}) = \int \{t[(s_1, y_{s_1}), \ldots, (s_{n-1}, y_{s_{n-1}}), (i, y_i)] - z(\mathbf{y})\}^2$$
$$\times f(\mathbf{y}_{\bar{s}(d_{n-1})} \mid d_{n-1}) \prod_{j \notin \{s_1, \ldots, s_{n-1}\}} dy_j$$
$$= \int m_0\left[(s_1, y_{s_1}), \ldots, (s_{n-1}, y_{s_{n-1}}), (i, y_i)\right] f_i(y_i \mid d_{n-1}) dy_i,$$

where $\mathbf{y}_{\bar{s}(d_{n-1})}$ represents the vector of components of $\mathbf{y}$ not in the initial sample $(s_1, s_2, \ldots, s_{n-1})$ and f_i is the conditional marginal density of the single component of $\mathbf{y}$ corresponding to the unit i. Note [as in (10.1)] that the dependence of g_i on the choice $s_n = i$ is in the computation of the predictor function t, not in the relevant conditional density.

The optimal choice for the nth unit is the unit $s_n = i$ giving the smallest value of $g_i^{(1)}(d_{n-1})$. The conditional mean-square error with this optimal choice will be denoted

$$m_1(d_{n-1}) = \min_i g_i^{(1)}(d_{n-1}).$$

The conditionally minimized mean-square error $m_1(d_{n-1})$ is a function of the initial data vector d_{n-1}.

Now suppose that $(n - 2)$ units have been selected, and the associated data are $d_{n-2} = [(s_1, y_{s_1}), \ldots, (s_{n-2}, y_{s_{n-2}})]$. Compute the function

$$g_i^{(2)}(d_{n-2}) = \mathrm{E}\left[m_1(d_{n-2}, (i, Y_i)) \mid d_{n-2}\right]$$
$$= \int m_1\left[(s_1, y_{s_1}), \ldots, (s_{n-2}, y_{s_{n-2}}), (i, y_i)\right] f_i(y_i \mid d_{n-2}) dy_i$$

for all units i not already selected. The optimal choice of the $(n - 1)$st unit in the sample is the unit $s_{n-1} = i$, which minimizes $g_i^{(2)}(d_{n-2})$, giving conditional mean-square error

$$m_2(d_{n-2}) = \min_i g_i^{(2)}(d_{n-2}).$$

More generally, suppose that $(n - k - 1)$ units have been selected so far, producing data $d_{n-k-1} = [(s_1, y_{s_1}), \ldots, (s_{n-k-1}, y_{s_{n-k-1}})]$, for $k = 1, \ldots, (n - 1)$. Consider the function

$$g_i^{(k+1)}(d_{n-k-1}) = \mathrm{E}\big[m_k(d_{n-k-1}, (i, Y_i)) \,|\, d_{n-k-1}\big]$$

$$= \int m_k\big[(s_1, y_{s_1}), \ldots, (s_{n-k-1}, y_{s_{n-k-1}}), (i, y_i)\big] \, f_i(y_i \,|\, d_{n-k-1}) \, dy_i.$$

The optimal $(n - k)$th selection is the unit $s_{n-k} = i$, which minimizes $g_i^{(k+1)}(d_{n-k-1})$, giving the minimized conditional mean-square error

$$m_{k+1}(d_{n-k-1}) = \min_i g_i^{(k+1)}(d_{n-k-1}).$$

The optimal first selection is the unit $s_1 = i$, which minimizes

$$g_i^{(n)} = \mathrm{E}[m_n(i, Y_i)]$$

$$= \int m_n(i, y_i) f_i(y_i) \, dy_i,$$

where f_i is the unconditional marginal density for the ith component of $\mathbf{y}$.

Note that the optimal selections described above are not necessarily easy to compute (see Solomon and Zacks [1970]). The procedure is much more computationally intensive than for example simply adding at each phase the unit whose value has the largest conditional variance given the data obtained so far. The optimal selection procedure will be adaptive unless a condition of the type in Theorem 10.1 holds at every step.

10.3.4 Optimal Choice of Sample Size

Now consider a two-phase design as in Section 10.2.2 with a well-defined procedure $p(s_2 \,|\, s_1, \mathbf{y}_{s_1})$ for selecting the second-phase units s_2 given the data from the n_1 units s_1 in first phase. The procedure could for example be simple random sampling from the remaining units $\bar{s}_1$ or the optimal selection described in the previous section. Suppose the first-phase sample has been selected giving data $(s_1, \mathbf{y}_{s_1})$. We now allow the sample size of the second phase to depend on the observed y-values associated with the first-phase sample.

Let the function $k(\mathbf{y}_{s_1})$ give the rule for choosing the total sample size as a function of the initial data. With the given initial sample and the prescribed design for selecting the second-phase sample, the conditional mean-square error is a function, which we call G, of the initial data and of the rule k. That is,

$$G\big[\mathbf{y}_{s_1}, k(\mathbf{y}_{s_1})\big] = \mathrm{E}\big[(T - Z)^2 \,|\, s_1, \mathbf{y}_{s_1}\big],$$

where T is a function of s_2. The initial sample s_1 is taken as fixed for purposes of comparison, so the dependence of G on s_1 is left implicit. Then, given $s = (s_1, s_2)$, the model-based unconditional expectation is

$$E[(T - Z)^2 \mid s_1, s_2] = E\{E[(T - Z)^2 \mid s_1, \mathbf{y}_{s_1}, \mid s\}$$

$$= \int G[\mathbf{y}_{s_1}, k(\mathbf{y}_{s_1})] f(\mathbf{y}) \, d\, \mathbf{y}.$$

For a given expected sample size $E[k(\mathbf{y}_{s_1})] = \nu$, the optimal choice of sample size would be given by the function $k(\mathbf{y}_{s_1})$ that minimizes the mean-square error $E[(T - Z)^2 \mid s]$ subject to having expected sample size ν. Writing $dF(\mathbf{y})$ for $f(\mathbf{y}) \, d\, \mathbf{y}$, the problem can be stated as follows: Find a function $k(\mathbf{y}_{s_1})$ that minimizes $\int G[\mathbf{y}_{s_1}, k(\mathbf{y}_{s_1})] dF(\mathbf{y})$ subject to the constraint $\int k(\mathbf{y}_{s_1}) dF(\mathbf{y}) = \nu$.

Finding a function that minimizes an integral subject to a constraint in the form of another integral is the "isoperimetric" problem of the calculus of variations. The following theorem states the necessary condition for a function $k(\mathbf{y}_{s_1})$ to be the optimal second-phase sample size choice.

Theorem 10.2.   The function $k(\mathbf{y}_{s_1})$ which minimizes the integral

$$\int G[\mathbf{y}_{s_1}, k(\mathbf{y}_{s_1})] dF(\mathbf{y})$$

subject to the constraint $\int k(\mathbf{y}_{s_1}) dF(\mathbf{y}) = \nu$ must satisfy $\partial G / \partial k = \lambda$ for some constant λ.

Proof: By Euler's equation from the calculus of variations, a necessary condition for the function $k(\mathbf{y}_{s_1})$ to minimize $\int G \, dF$ subject to fixed $\int k \, dF$ is that k satisfy $\partial / \partial k \{ G(\mathbf{y}_{s_1}, k) - \lambda k \} = 0$, where λ is a Lagrange multiplier. Hence the necessary condition is $\partial G / \partial k = \lambda$ for all outcomes $\mathbf{y}_{s_1}$, completing the proof. $\square$

An important special case is the one in which the conditional variance has an inverse relationship with sample size; that is, $E[(T - Z)^2 \mid s_1, \mathbf{y}_{s_1}] = c^2(\mathbf{y}_{s_1})/(k + b)$, where c is a function of $\mathbf{y}_{s_1}$ only and b is a constant. This is the case for example when second-phase units are selected by simple random sampling without replacement, so that the conditional mean-square prediction error is divided by sample size and the constant b arises from a finite population correction factor. For this special case, the optimal sample size based on the first-phase observations is given by $c^2/k^2 = $ constant, that is,

$$k(\mathbf{y}_{s_1}) = \frac{\nu c(\mathbf{y}_{s_1})}{E[c(\mathbf{y}_{s_1})]}.$$

The optimal sample size result is superficially similar to the well-known optimal allocation formula in stratified sampling, with the difference that here the allocation is not over strata but over the set of all possible outcomes of the first-phase observations.

The results of this and the preceding section say that the optimal sampling strategy among fixed sample size designs is in general adaptive and, further, that one can do better still by letting the sample size be adaptive as well. The application of the results extend readily from two-phase designs to multiple phase designs as in the optimal strategy of Zacks [1969].

The optimality results depend however on the model $f(\mathbf{y})$ for the population, and the implementation of the optimal strategies would be computationally intensive (cf. Solomon and Zacks [1970]. Even so, the fact that, in principle, with any known population model, the optimal adaptive strategy is as good or better than the optimal conventional strategy suggests that in many real situations conventional strategies may be improved upon by considering the larger class of adaptive strategies.

10.4 PARTIALLY KNOWN POPULATION MODEL

When population models are used in sampling, often no single distribution or known prior is assumed. Rather a family of distributions is assumed. The distributions may be assumed to be of known parametric form, such as multivariate normal, with unknown parameter values. In other cases, no specific form is assumed for the densities but more general characteristics, such as independence or a covariance matrix proportional to a known matrix are assumed. For such models, results are available describing optimal strategies among the class of conventional designs. More general optimality results considering adaptive designs along with conventional designs are not at present available.

For instance, results on best unbiased strategies and best linear unbiased strategies at present rely on the assumption of a conventional design. The aim in best prediction is to find a predictor $\hat{Z}$ that minimizes the mean-square prediction error $E[(\hat{Z} - Z)^2 \mid s]$ subject to model unbiasedness $E[\hat{Z} \mid s] = E[Z \mid s]$. For best linear unbiased prediction, the object is to find a linear function of the sample Y-values $\hat{Z} = \sum_{i \in s} a_i Y_i$ that minimizes the mean-square prediction error subject to model unbiasedness.

With a conventional design, the model expectation (given the sample) does not depend on the design (i.e., $E[Z \mid s] = E[Z]$), so that best (model) unbiased prediction and best (model) unbiased linear prediction with a given sample is based solely on the population model. With an assumed covariance and linear model structure for the population, the best linear predictor (BLUP) of a population quantity $z(\mathbf{Y})$, given the sample s, consists of the regression predictors described by Royall [1970, 1971, 1988] and others and the kriging predictors of spatial statistics (cf. Cressie [1993]). With an assumed parametric form to the model, one can obtain (with a complete family of distributions) best unbiased predictors; for the normal model, these coincide with the best linear unbiased predictors (Tam [1987] and Bolfarine and Zacks [1992]).

The optimal design to use with the best (linear) predictor is to purposively select the specific sample s that gives the lowest value of the mean-square prediction error $E[(\hat{Z} - Z)^2 \mid s]$. The optimal model-based conventional strategy is thus to use that particular sample with the best (linear) unbiased estimator.

With the same model, perhaps a better strategy is possible among the larger class including adaptive as well as conventional designs. The problem is complicated however by the fact that with adaptive designs the usual predictors are not in general unbiased (see Section 3.8). Unbiased estimation with such designs needs to take into account the design as well as the population model.

Example 10.4   Environmental Restoration.   Spatial linear prediction methods are used by Englund and Heravi [1995] in an investigation of two-phase adaptive strategies for environmental restoration sampling. Restoration of soils contaminated with hazardous waste is very expensive. Restoration decisions are based on sampling and prediction methods. Contaminant concentration is measured for a sample of soil units within the study region. Linear prediction methods are then used to produce a map of predicted concentrations for the entire study region. For each 100 square meter "restoration unit" in the study region, the decision whether to restore is based on the predicted contamination concentration. Restoration is carried out if the predicted concentration exceeds the "action level." With the kriging model assumed, the variance associated with a predicted value depends only on the locations of the sampled sites and not on the contamination levels observed at those sites. With the variance structure independent of observed values, one would not expect any gain through adaptive procedures for predicting total or average contamination for the whole study region. The purpose of the study however was not estimation but cost-effective environmental restoration.

Englund and Heravi propose a cost function giving the cost of each wrong decision—not restoring a unit which actually exceeds the action level and restoring a unit that does not exceed the action level. To reduce total cost, the following two-phase adaptive design was proposed. Select an initial sample of n_1 sites with stratified random sampling, with one sample unit per stratum so that the sample is fairly evenly spread out. From observations of contamination at these sites, initial predicted values are computed for the entire study region. These predicted values are used to estimate the probabilities of wrong decisions and resulting expected cost at each site. The second-phase sample consists of sites in the n_2 units having the highest expected loss. Because wrong decisions are most likely to be made at units with concentrations very near the action level, many of the second-phase sites are located by the adaptive procedure near the action level contour in the study region. Thus, the adaptive procedure approximates an optimal stratification that might have been made with prior knowledge about the distribution of pollutant in the study region. In a refinement of the adaptive procedure, the estimated prediction variances are recomputed after each second-phase sample site is located. The procedure is still two-phase because contaminant levels for the second-phase sites are not available until the end of the survey. A more nearly optimal, but computationally daunting, two-phase procedure would select the set of n_2 second-phase sites giving the smallest estimated expected cost.

In simulation studies using environmental spatial data sets Englund and Heravi found that the two-phase strategies gave consistently lower cost than the nonadaptive strategy with the same total sample size. The best proportion of the total sample size to put in the first phase was found to be about 70–80%. However, the proportional cost savings were only a small percentage of total cost. □

10.4.1 Combined and Modified Approaches

From a frequentist model-based viewpoint, the strategy of using the best linear unbiased predictor and purposively selecting the sample that gives the lowest mean-square error with that predictor is optimal among strategies using conventional designs. The optimality within that class depends on the correctness of the assumed model, however. A number of approaches have been developed to provide robustness against departures from the assumed model.

One approach is to select, rather than the optimal sample, a "balanced" sample (Royal and Herson [1973a, b], Royal and Pfeffermann [1982], Royal [1988], and Cumberland and Royal [1988]). For instance, a sample balanced in an auxiliary variable could be selected to have the sample mean of the auxiliary variable equal to its known population mean. Balancing on higher moments is also possible. In addition, estimators known to be robust can be used in place of the estimator that is optimal under the assumed model.

For conventional designs in which units have unequal selection probabilities, generalized regression estimators incorporating the inclusion probabilities can be used in place of the best linear unbiased predictor. The best linear unbiased predictor, which ignores the design, is optimal if the assumed model is correct but may not have good properties if the assumed model is not the true one. The generalized estimators, combining both model and design based features, may not be optimal under any model, nor are they design unbiased, but they have desirable consistency properties under the design regardless of the model (Brewer [1979], Brewer et al. [1988], Cassel et al. [1976, 1977, 1979], Hájek [1981], Isaki and Fuller [1982], Little [1983], Särndal [1980a,b], Särndal and Wright [1984], Tam [1988], and Wright [1983]; also see Särndal et al. [1992], Thompson [1992].)

An approach combining design and model considerations seeks a strategy of lowest possible unconditional mean-square error subject to design unbiasedness, namely a strategy $(p, \hat{z})$ that minimizes $E\{[\hat{z}(d) - z(\mathbf{y})]^2\}$ subject to $E[\hat{z}(d) \mid \mathbf{y}] = z(\mathbf{y})$ for all $\mathbf{y} \in \mathcal{Y}$. Optimal conventional strategies with this approach have been derived for a variety of models and specific estimators by Godambe [1955, 1982], Cassell et al. [1976], Tam [1984, 1986, 1987], Hájek [1981], and Mukerjee and Sengupta [1990]; see also Chaudhuri and Stenger [1992]. It is possible that similar results for the larger class of design-unbiased strategies including adaptive as well as conventional designs are possible, but no such results are available at present. In any case, theoretically optimal sampling strategies tend to be demanding in terms of prior knowledge required about the population and complex in implementation and computation. Such results can be enlightening and useful however in suggesting practical strategies and improvements in currently used designs and inference methods.

References

Aitchison, J., and Brown, J. A. C. (1957). *The Lognormal Distribution, with Special Reference to its Uses in Economics*. Cambridge: Cambridge University Press.

Arnold, S. F. (1990). *Mathematical Statistics*. Englewood Cliffs, New Jersey: Prentice-Hall.

Barry, D. A., and Fristedt, B. (1985). *Bandit Problems: Sequential Allocation of Experiments*. London: Chapman and Hall.

Basu, D. (1969). Role of the sufficiency and likelihood principles in sample survey theory. *Sankhyā A* **31**, 441–454.

Basu, D. (1971). An essay on the logical foundations of survey sampling, Part I. *In* V. P. Godambe and D. A. Sprott (Eds.), *Foundations of Statistical Inference*, pp. 203–242. Toronto: Holt, Rinehart, Winston.

Berger, J. O., and Wolpert, R. L. (1988). *The Likelihood Principle*. Hayward, California: Institute of Mathematical Statistics.

Bethel, J. (1989). Sample allocation in multivariate surveys. *Survey Methodology* **15**, 47–57.

Biemer, P. P., Groves, R. M., Lyberg, L. E., Mathiowetz, N. A., and Sudman, S. (1991). *Measurement Errors in Surveys*. New York: Wiley.

Birnbaum, Z. W., and Sirken, M. G. (1965). Design of Sample Surveys to Estimate the Prevalence of Rare Diseases: Three Unbiased Estimates. *Vital and Health Statistics,* Ser. 2, No. 11. Washington, D.C.: Government Printing Office.

Bjørnstad, J. F. (1990). Predictive likelihood: a review. *Statistical Science* **5**, 242–265.

Blackwell, D. (1947). Conditional expectation and unbiased sequential estimation. *Annals of Mathematical Statistics* **18**, 105–110.

Bolfarine, H., and Zacks, S. (1992). *Prediction Theory for Finite Populations*. New York: Springer Verlag.

Brewer, K. R. W. (1979). A class of robust sampling designs for large-scale surveys. *Journal of the American Statistical Association* **74**, 911–915.

Brewer, K. R. W., and Hanif, M. (1983). *Sampling With Unequal Probabilities*. New York: Springer Verlag.

Brewer, K. R. W., Hanif, M., and Tam, S. M. (1988). How nearly can model-based prediction and design-based estimation be reconciled? *Journal of the American Statistical Association* **83**, 128–132.

Brown, J. A. (1994). The application of adaptive cluster sampling to ecological studies. *In* D. J. Fletcher and B. F. J. Manly (Eds.), *Statistics in Ecology and Environmental Monitoring*, pp. 86–97. Otago Conference Series No. 2. Dunedin, New Zealand: University of Otago Press.

Butler, R. W. (1986). Predictive likelihood inference with applications (with discussion). *Journal of the Royal Statistical Society B* **48**, 1–38.

Butler, R. W. (1990). Comment on J. F. Bjørnstad, Predictive likelihood: A review. *Statistical Science* **5**, 255–259.

Cassel, C. M., Särndal, C. E., and Wretman, J. H. (1976). Some results on generalized difference estimation and generalized regression estimation for finite populations. *Biometrika* **63**, 615–620.

Cassel, C. M., Särndal, C. E., and Wretman, J. H. (1977). *Foundations of Inference in Survey Sampling*. New York: Wiley.

Cassel, C. M., Särndal, C. E., and Wretman, J. H. (1979). Prediction theory for finite populations when model-based and design-based principles are combined. *Scandinavian Journal of Statistics* **6**, 97–106.

Chaudhuri, A. (1988). Optimality of sampling strategies. *In* P. R. Krishnaiah and C. R. Rao (Eds.), *Handbook of Statistics, Vol. 6*, pp. 47–96. Elsevier Publishers B.V.

Chaudhuri, A., and Stenger, H. (1992). *Survey Sampling: Theory and Methods*. New York: Marcel Dekker.

Chaudhuri, A., and Vos, J. W. E. (1988). *Unified Theory and Strategies of Survey Sampling*. Amsterdam: North Holland.

Cochran, W.G. (1977). *Sampling Techniques*, 3rd edition. New York: Wiley.

Cook, R. D., and Martin, F. B. (1974). A model for quadrat sampling with "visibility bias." *Journal of the American Statistical Association* **69**, 345–349.

Cormack, R. M. (1988). Statistical challenges in the environmental sciences: A personal view. *Journal of the Royal Statistical Society A* **151**, 201–210.

Cressie, N. (1993). *Statistics for Spatial Data*, 2nd edition. New York: Wiley.

Cumberland, W. G., and Royall, R. M. (1988). Does simple random sampling provide adequate balance? *Journal of the Royal Statistical Society B* **50**, 118–124.

Czaja, R. F., Snowdon, C. B., and Casady, R. J. (1986). Reporting bias and sampling errors in a survey of a rare population using multiplicity counting rules. *Journal of the American Statistical Association* **81**, 411–419.

Danaher, P. J., and King, M. (1994). Estimating rare household characteristics using adaptive sampling. *The New Zealand Statistician* **29**, 14–23.

Diggle, P. J. (1983). *Statistical Analysis of Spatial Point Patterns*. New York: Academic Press.

Draper, N. R., and Guttman, I. (1968). Some Bayesian stratified two-phase sampling results. *Biometrika* **55**, 131–138.

Drummer, T. D., and McDonald, L. L. (1987). Size bias in line transect sampling. *Biometrics* **43**, 13–21.

Englund, E. J., and Heravi, N. (1995). Phased sampling for soil remediation. *Environmental and Ecological Statistics*, to appear.

Faulkenberry, G. D., and Garoui, A. (1991). Estimating a population total using an area frame. *Journal of the American Statistical Association* **86**, 445–449.

Francis, R. I. C. C. (1984). An adaptive strategy for stratified random trawl surveys. *New Zealand Journal of Marine and Freshwater Research* **18**, 59–71.

Francis, R. I. C. C. (1991). Statistical properties of two-phase surveys: comment. *Canadian Journal of Fisheries and Aquatic Sciences* **48**, 1228.

Gasaway, W. C., DuBois, S. D., Reed, D. J., and Harbo, S. J. (1986). *Estimating Moose Population Parameters from Aerial Surveys*. Biological Papers of the University of Alaska (Institute of Arctic Biology) Number 22. Fairbanks: University of Alaska.

Geiger, H. J. (1994). A Bayesian approach for estimating hatchery contribution in a series of salmon fisheries. *Alaska Fishery Research Bulletin* **1**, 66–75.

Godambe, V. P. (1955). A unified theory of sampling from finite populations. *Journal of the Royal Statistical Society B* **17**, 269–278.

Godambe, V. P. (1966). A new approach to sampling from finite populations, I, II. *Journal of the Royal Statistical Society B* **28**, 310–328.

Godambe, V. P. (1982). Estimation in survey sampling: Robustness and optimality. *Journal of the American Statistical Association* **77**, 393–403.

Godambe, V. P., and Joshi, V. M. (1965). Admissibility and Bayes estimation in sampling finite populations, I. *Annals of Mathematical Statistics* **36**, 1701–1722.

Godambe, V. P., and Rajarshi, M. B. (1989). Optimal estimation for weighted distributions: semiparametric models. *In* Y. Dodge (Ed.), *Statistical Data Analysis and Inference*, pp. 199–208. Amsterdam: Elsevier Science/North Holland.

Goodman, L. A. (1960). On the exact variance of products. *Journal of the American Statistical Association* **55**, 708–713.

Grosh, D. L. (1969). Bayes one- and two-stage stratified sampling schemes for finite populations with beta-binomial prior distributions. Ph.D. thesis, Department of Statistics, Kansas State University.

Guttorp, P., Le, N.D., Sampson, P.D., and Zidek, J.V. (1993). Using entropy in the redesign of an environmental network. *In* G. P. Patil and C. R. Rao (Eds.), *Multivariate Environmental Statistics*, pp. 175–202. New York: North Holland/Elsevier Science Publishers.

Hájek, J. (1971). Discussion of An essay on the logical foundations of survey sampling, Part one, by D. Basu. *In* V. P. Godambe and D. A. Sprott (Eds.), *Foundations of Statistical Inference*, p. 236. Toronto: Holt, Rinehart, Winston.

Hájek, J. (1981). *Sampling From a Finite Population.* New York: Marcel Dekker.

Hansen, M. M., and Hurwitz, W. N. (1943). On the theory of sampling from finite populations. *Annals of Mathematical Statistics* **14**, 333–362.

Hedayat, A. S., and Sinha, B. K. (1991). *Design and Inference in Finite Population Sampling.* New York: Wiley.

Hiby, A. R., and Hammond, P. S. (1989). Survey techniques for estimating abundance of cetaceans. *In* G. P. Donovan (Ed.), *The Comprehensive Assessment of Whale Stocks: The Early Years,* Special Issue 11, Reports of the International Whaling Commission, pp. 47–80. Cambridge: International Whaling Commission.

Hinkley, D. V. (1979). Predictive likelihood. *Annals of Statistics* **7**, 718–729. Corrigendum **8**, 694.

Horvitz, D. G., and Thompson, D. J. (1952). A generalization of sampling without replacement from a finite universe. *Journal of the American Statistical Association* **47**, 663–685.

Husch, B., Miller, C. I., and Beers, T. W. (1982). *Forest Mensuration.* New York: Wiley.

Isaki, C. T., and Fuller, W. A. (1982). Survey design under the regression superpopulation model. *Journal of the American Statistical Association* **77**, 89–96.

Johnson, N. L., and Kotz, S. (1977). *Urn Models and Their Application: An Approach to Modern Discrete Probability Theory.* New York: Wiley.

Jolly, G. M. (1993). Bias in two-phase stratified random sampling for optimum allocation. Unpublished manuscript.

Jolly, G. M., and Hampton, I. (1990). A stratified random transect design for acoustic surveys of fish stocks. *Canadian Journal of Fisheries and Aquatic Science* **47**, 1282–1291.

Jolly, G. M., and Hampton, I. (1991). Reply to comment by R.I.C.C. Francis. *Canadian Journal of Fisheries and Aquatic Science* **48**, 1228–1229.

Kalton, G., and Anderson, D. W. (1986). Sampling rare populations, *Journal of the Royal Statistical Society A* **149**, 65–82.

Kishino, H., and Kasamatsu, F. (1987). Comparison of the closing and passing mode procedures used in sighting surveys. *Report of the International Whaling Commission* **37**, 253–258.

Kokan, A. R., and Khan, S. (1967). Optimum allocation in multivariate surveys: an analytical solution. *Journal of the Royal Statistical Society B* **29**, 115–125.

Kremers, W. K. (1987). Adaptive sampling to account for unknown variability among strata. Preprint No. 128. Institut für Mathematik, Universität Augsburg, Germany.

Lauritzen, S. L. (1974). Sufficiency, prediction and extreme models. *Scandinavian Journal of Statistics* **1**, 128–134.

Lehmann, E. L. (1983). *Theory of Point Estimation*. New York: Wiley.

Lehmann, E. L. (1986). *Testing Statistical Hypotheses*, 2nd edition. New York: Wiley.

Lejeune, M., and Faulkenberry, G. D. (1982). A simple predictive density function. *Journal of the American Statistical Association* **77**, 654–657.

Lessler, J. T., and Kalsbeek, W. D. (1992). *Nonsampling Error in Surveys*. New York: Wiley.

Levy, P. S. (1977). Optimum allocation in stratified random network sampling for estimating the prevalence of attributes in rare populations. *Journal of the American Statistical Association* **72**, 758–763.

Levy, P. S., and Lemeshow, S. (1991). *Sampling of Populations: Methods and Applications*. New York: Wiley.

Little, R. J. A. (1983). Estimating a finite population mean from unequal probability samples. *Journal of the American Statistical Association* **78**, 596–604.

Matérn, B. (1986). *Spatial Variation*, 2nd edition. Berlin: Springer Verlag.

Mukerjee, R., and Sengupta, S. (1989). Optimal estimation of finite population total under a general correlated model. *Biometrika* **76**, 789–794.

Munholland, P. L., and Borkowski, J. J. (1993a). Latin square sampling +1 designs. Technical Report No. 1-11-93, Department of Mathematical Sciences, Montana State University, Bozeman.

Munholland, P. L., and Borkowski, J. J. (1993b). Adaptive Latin square sampling +1 designs. Technical Report No. 3-23-93, Department of Mathematical Sciences, Montana State University, Bozeman.

Nathan, G., and Holt, D. (1980). The effect of survey design on regression analysis. *Journal of the Royal Statistical Society B* **42**, 377–386.

Neyman, J. (1934). On the two different aspects of the representative method: the method of stratified sampling and the method of purposive selection. *Journal of the Royal Statistical Society A* **97**, 558–606.

Otto, M. C., and Pollock, K. H. (1990). Size bias in line transect sampling: a field test. *Biometrics* **46**, 239–245.

Quang, P. X. (1991). A nonparametric approach to size-biased line transect sampling. *Biometrics* **47**, 269–279.

Quinn, T. J. II (1981). The effect of group size on line transect estimators of abundance. *In* C. J. Ralph and J. M. Scott (Eds.), *Estimating the Numbers of Terrestrial Birds, Studies in Avian Biology* **6**, 502–508. Oxford: Pergamon.

Ramsey, F. L., and Sjamsoe'oed, R. (1995). Habitat association in conjuction with adaptive cluster samples. *Environmental and Ecological Statistics*, to appear.

Rao, C. R. (1945). Information and accuracy attainable in estimation of statistical parameters. *Bulletin of the Calcutta Mathematical Society* **37**, 81–91.

Rao, C. R. (1973). *Linear Statistical Inference and its Applications*, 2nd edition. New York: Wiley.

Rao, J. N. K. (1975). Unbiased variance estimation for multistage designs. *Sankhyā C* **37**, 133–139.

Robbins, H. (1952). Some aspects of the sequential design of experiments. *Bulletin of the American Mathematical Society* **58**, 527–535.

Roesch, F. A., Jr. (1993). Adaptive cluster sampling for forest inventories. *Forest Science* **39**, 655–669.

Royall, R. M. (1968). An old approach to finite population sampling theory. *Journal of the American Statistical Association* **63**, 1269–1279.

Royall, R. M. (1970). On finite population sampling theory under certain linear regression models. *Biometrika* **57**, 377–387.

Royall, R. M. (1971). Linear regression models in finite population sampling theory. *In* V. P. Godambe and D. A. Sprott (Eds.), *Foundations of Statistical Inference*, pp. 259–279. Toronto: Holt, Rinehart, Winston.

Royall, R. M. (1976). The linear least-squares prediction approach to two-stage sampling. *Journal of the American Statistical Association* **71**, 657–664.

Royall, R. M. (1988). The prediction approach to sampling theory. *In* P. R. Krishnaiah and C. R. Rao (Eds.), *Handbook of Statistics, Vol. 6 (Sampling)*, pp. 399–413. Amsterdam: Elsevier Science Publishers.

Royall, R. M., and Herson, J. (1973a). Robust estimation in finite populations I. *Journal of the American Statistical Association* **68**, 880–889.

Royall, R. M., and Herson, J. (1973b). Robust estimation in finite populations II. *Journal of the American Statistical Association* **68**, 890–893.

Royall, R. M., and Pfeffermann, D. (1982). Balanced samples and robust Bayesian inference in finite population sampling. *Biometrika* **69**, 401–409.

Rubin, D. B. (1987). *Multiple Imputation for Nonresponse in Surveys*. New York: Wiley.

Sacks, J., and Schiller, S. (1988). Spatial designs. *In* S. S. Gupta and J. O. Berger (Eds.), *Statistical Decision Theory and Related Topics IV, Vo. 2*, pp. 385–389. New York: Springer Verlag.

Salehi, M., and Seber, G. A. F. (1995). Two-stage adaptive cluster sampling. Unpublished manuscript.

Särndal, C. E. (1980a). A two-way classification of regression estimation strategies in probability sampling. *Canadian Journal of Statistics* **8**, 165–177.

Särndal, C. E. (1980b). On π-inverse weighting versus best linear unbiased weighting in probability sampling. *Biometrika* **67**, 639–50.

Särndal, C. E., Swensson, B., and Wretman, J. (1992). *Model Assisted Survey Sampling*. New York: Springer Verlag.

Särndal, C. E., and Wright, R. L. (1984). Cosmetic form of estimators in survey sampling. *Scandinavian Journal of Statistics* **11**, 146–156.

Schweder, T., Øien, N., and Høst, G. (1993). Estimates of abundance of northeastern Atlantic minke whales in 1989. *Report of the International Whaling Commission* **43**, 323–331.

Scott, A. J. (1977). On the problem of randomization in survey sampling. *Sankhyā C* **39**, 1–9.

Seber, G. A. F. (1977). *Linear Regression Analysis*. New York: Wiley.

Seber, G. A. F. (1982). *The Estimation of Animal Abundance*, 2nd edition. London: Griffin.

Seber, G. A. F. (1986). A review of estimating animal abundance. *Biometrics* **42**, 267–292.

Seber, G. A. F. (1992). A review of estimating animal abundance II. *International Statistical Review* **60**, 129–166.

Seber, G. A. F., and Thompson, S. K. (1994). Environmental adaptive sampling. *In* G. P. Patil and C. R. Rao (Eds.), *Handbook of Statistics, Vol. 12 (Environmental Sampling)*, pp. 201–220. New York: North Holland/Elsevier Science Publishers.

Sen, A. R. (1953). On the estimate of the variance in sampling with varying probabilities. *Journal of the Indian Society of Agricultural Statistics* **5**, 119–127.

Siegmund, D. (1985). *Sequential Analysis; Tests and Confidence Intervals*. New York: Springer Verlag.

Sinha, B. K. (1991). Sequential Methods for Finite Populations. *In* B. K. Ghosh and P. K. Sen (Eds.), *Handbook of Sequential Analysis*, pp. 313–329. New York: Marcel Dekker.

Sirken, M. G. (1970). Household surveys with multiplicity. *Journal of the American Statistical Association* **63**, 257–266.

Sirken, M. G. (1972a). Stratified sample surveys with multiplicity. *Journal of the American Statistical Association* **67**, 224–227.

Sirken, M. G. (1972b). Variance components of multiplicity estimators. *Biometrics* **28**, 869–873.

Sirken, M. G., and Levy, P. S. (1974). Multiplicity estimation of proportions based on ratios of random variables. *Journal of the American Statistical Association* **69**, 68–73.

Smith, D. R., Conroy, M. J., and Brakhage, D. H. (1995). Efficiency of adaptive cluster sampling for estimating density of wintering waterfowl. *Biometrics* **51**, 777–788.

Solomon H., and Zacks, S. (1970). Optimal design of sampling from finite populations: a critical review and indication of new research areas. *Journal of the American Statistical Association* **65**, 653–677.

Steinhorst, R. K., and Samuel, M. D. (1989). Sightability adjustment methods for aerial surveys of wildlife populations. *Biometrics* **45**, 415–425.

Stenger, H. (1977). Sequential sampling from finite populations. *Sankhyā C* **39**, 10–20.

Sudman, S., Sirken, M. G., and Cowan, C. D. (1988). Sampling rare and elusive populations. *Science* **240**, 991–996.

Tam, S. M. (1984). Optimal estimation in survey sampling under a regression super-population model. *Biometrika* **71**, 645–647.

Tam, S. M. (1987). Optimality of predictor under a Gaussian superpopulation model. *Biometrika* **74**, 659–660.

Tam, S. M. (1988). Some Results on robust estimation in finite population sampling. *Journal of the American Statistical Association* **83**, 242–248.

Thompson, S. K. (1988). Adaptive sampling. *Proceedings of the Section on Survey Research Methods of the American Statistical Association*, 784–786.

Thompson, S. K. (1990). Adaptive cluster sampling. *Journal of the American Statistical Association* **85**, 1050–1059.

Thompson, S. K. (1991a). Adaptive cluster sampling: Designs with primary and secondary units. *Biometrics* **47**, 1103–1115.

Thompson, S. K. (1991b). Stratified adaptive cluster sampling. *Biometrika* **78**, 389–397.

Thompson, S. K. (1992). *Sampling*. New York: Wiley.

Thompson, S. K. (1993). Multivariate aspects of adaptive cluster sampling. *In* G. P. Patil and C. R. Rao (Eds.), *Multivariate Environmental Statistics*, pp.561–572. New York: North Holland/Elsevier Science Publishers.

Thompson, S. K. (1994). Factors influencing the efficiency of adaptive cluster sampling. Technical Report 94-0301, Center for Statistical Ecology and Environmental Statistics, Department of Statistics, Pennsylvania State University, University Park.

Thompson, S. K. (1996). Adaptive cluster sampling based on order statistics. *Environmetrics* **7**, to appear.

Thompson, S. K., and Ramsey, F. L. (1983). *Adaptive Sampling of Animal Populations*. Technical Report 82, Dept. of Statistics, Corvallis: Oregon State University,

Thompson, S. K., and Ramsey, F. L. (1987). Detectability functions in observing spatial point processes. *Biometrics* **43**, 355–362.

Thompson, S. K., Ramsey, F. L., and Seber, G. A. F. (1992). An adaptive procedure for sampling animal populations. *Biometrics* **48**, 1195–1199.

Thompson, S. K., and Seber, G. A. F. (1994). Detectability in conventional and adaptive sampling. *Biometrics* **50**, 712–724.

Wald, A. (1947). *Sequential Analysis*. New York: Wiley.

Wright, R. L. (1983). Finite Population sampling with multivariate auxiliary information. *Journal of the American Statistical Association* **78**, 879–884.

Yates, F., and Grundy, P. M. (1953). Selection without replacement from within strata with probability proportional to size. *Journal of the Royal Statistical Society B* **15**, 253–261.

Zacks, S. (1969). Bayes sequential designs of fixed size samples from finite populations. *Journal of the American Statistical Association* **64**, 1342–1349.

Zacks, S. (1970). Bayesian design for single and double stratified sampling for estimating proportion in finite population. *Technometrics* **12**, 119–129.

Author Index

Subject Index

WILEY SERIES IN PROBABILITY AND STATISTICS

ESTABLISHED BY WALTER A. SHEWHART AND SAMUEL S. WILKS

Editors
*Vic Barnett, Ralph A. Bradley, Nicholas I. Fisher, J. Stuart Hunter,
J. B. Kadane, David G. Kendall, David W. Scott, Adrian F. M. Smith,
Jozef L. Teugels, Geoffrey S. Watson*

Probability and Statistics

ANDERSON · An Introduction to Multivariate Statistical Analysis, *Second Edition*
*ANDERSON · The Statistical Analysis of Time Series
ARNOLD, BALAKRISHNAN, and NAGARAJA · A First Course in Order Statistics
BACCELLI, COHEN, OLSDER, and QUADRAT · Synchronization and Linearity:
An Algebra for Discrete Event Systems
BARTOSZYNSKI and NIEWIADOMSKA-BUGAJ · Probability and Statistical Inference
BERNARDO and SMITH · Bayesian Statistical Concepts and Theory
BHATTACHARYYA and JOHNSON · Statistical Concepts and Methods
BILLINGSLEY · Convergence of Probability Measures
BILLINGSLEY · Probability and Measure, *Second Edition*
BOROVKOV · Asymptotic Methods in Queuing Theory
BRANDT, FRANKEN, and LISEK · Stationary Stochastic Models
CAINES · Linear Stochastic Systems
CAIROLI and DALANG · Sequential Stochastic Optimization
CHEN · Recursive Estimation and Control for Stochastic Systems
CONSTANTINE · Combinatorial Theory and Statistical Design
COOK and WEISBERG · An Introduction to Regression Graphics
COVER and THOMAS · Elements of Information Theory
CSÖRGÖ and HORVÁTH · Weighted Approximations in Probability Statistics
*DOOB · Stochastic Processes
DUDEWICZ and MISHRA · Modern Mathematical Statistics
ETHIER and KURTZ · Markov Processes: Characterization and Convergence
FELLER · An Introduction to Probability Theory and Its Applications, Volume 1,
Third Edition, Revised; Volume II, *Second Edition*
FREEMAN and SMITH · Aspects of Uncertainty: A Tribute to D. V. Lindley
FULLER · Introduction to Statistical Time Series, *Second Edition*
FULLER · Measurement Error Models
GIFI · Nonlinear Multivariate Analysis
GUTTORP · Statistical Inference for Branching Processes
HALD · A History of Probability and Statistics and Their Applications before 1750
HALL · Introduction to the Theory of Coverage Processes
HANNAN and DEISTLER · The Statistical Theory of Linear Systems
HEDAYAT and SINHA · Design and Inference in Finite Population Sampling
HOEL · Introduction to Mathematical Statistics, *Fifth Edition*
HUBER · Robust Statistics
IMAN and CONOVER · A Modern Approach to Statistics
JUREK and MASON · Operator-Limit Distributions in Probability Theory
KAUFMAN and ROUSSEEUW · Finding Groups in Data: An Introduction to Cluster
Analysis
LAMPERTI · Probability: A Survey of the Mathematical Theory
LARSON · Introduction to Probability Theory and Statistical Inference, *Third Edition*
LESSLER and KALSBEEK · Nonsampling Error in Surveys
LINDVALL · Lectures on the Coupling Method
MANTON, WOODBURY, and TOLLEY · Statistical Applications Using Fuzzy Sets

*Now available in a lower priced paperback edition in the Wiley Classics Library.

*Now available in a lower priced paperback edition in the Wiley Classics Library.

*Now available in a lower priced paperback edition in the Wiley Classics Library.

*Now available in a lower priced paperback edition in the Wiley Classics Library.